DIABETIC RENAL-RETINAL SYNDROME

Diabetic Renal-Retinal Syndrome

Pathogenesis and Management Update 2002

Edited by

Eli A. Friedman, M.D.

State University of New York,
Downstate Medical Center, New York, U.S.A.

and

Francis A. L'Esperance, Jr. M.D.

Colombia University,
New York, U.S.A.

SPRINGER-SCIENCE+BUSINESS MEDIA, B.V.

A C.I.P. Catalogue record for this book is available from the Library of Congress.

ISBN 978-94-010-3930-7 ISBN 978-94-010-0614-9 (eBook)
DOI 10.1007/978-94-010-0614-9

Printed on acid-free paper

Table of Contents

Dedication

From their inception in 1976, these Symposia devoted to the Diabetic Renal-Retinal Syndrome derived their energy from clinicians delivering care to diabetic individuals in the seemingly unyielding grasp of progressive co-morbid complications. A quarter-century ago, efforts to attain *reasonable* rehabilitation after onset of uremia and severe vision loss attributed to diabetes uniformly resulted in marginal short-term benefit with minimal life extension. Gradually and inexorably the prognosis has improved with an increasing cohort of patients returning to employment, education, or home responsibilities.

We salute our brave, intrepid, and trusting patients who have followed the difficult course we prescribe to preserve sight while substituting for vision loss. In so many instances, it has been the life force of our patients that has prompted the medical, surgical, and pharmacologic advances detailed in this volume. By permitting us to address their threatened existence and need for remedy, our patients gave us the stimulus and motivation to join the struggle against a disease about to yield to the wonders of molecular medicine. With exceptional optimism for a near term conquest of its subject, we dedicate these proceedings to our patients.

Eli A. Friedman
Francis A. L'Esperance, Jr.

Foreword

Diabetic Renal-Retinal Syndrome Proactive not Reactive Intervention

Conference held March 23, 2001

In the thirty years since this series of symposia on the Diabetic Renal-Retinal syndrome were begun, there has been an epidemiologic explosion in the incidence and prevalence of end-stage renal disease (ESRD) attributed to diabetes. Furthermore, according to the most current estimates by the world health organization, the growth rate of diabetes world-wide is accelerating resulting in a pandemic of ESRD [1]. Whereas in 1995 there were an estimated 133,286,000 diabetic persons throughout the world, by 2,025 there will be a minimum of 299,974,000 with diagnosed diabetes, an increase of 210% [2].

James V. Neel in 1962 proposed that the basic defect in type 2 diabetes was a quick insulin trigger meaning that the pancreas secretes insulin too quickly in response to elevated blood sugar assisting our tribal, hunter-gatherer ancestors through intermittent, sometimes feast-or-famine food availability, minimizing renal loss of glucose [3]. Analogous to the advantage afforded by the sickle hemoglobin mutation against malarial parasites in the wild, the so called thrifty gene that is beneficial during food deprivation converts to a disease promoter when food is abundent. Under genetic control, a too quick insulin trigger might be disadvantageous and might take the form of a balanced polymorphism. By the end of the last century, it was evident that at least 95% of newly diagnosed people with diabetes had type 2 diabetes reflecting emergence and dominance of the thrifty gene [4].

The basic premise underlying the thrifty gene hypothesis is that genes that determine increased fat storage, which in times of famine represent a survival advantage, in a plentiful food environment result in obesity and type 2 diabetes. The peroxisome proliferator-activated receptor gamma (PPARgamma) has over the last decade been shown to have a key role in adipogenesis and may be a master controller of the 'thrifty gene response' leading to efficient energy storage. Its synthetic ligands, the thiazolidinediones, are insulin sensitizing drugs, currently utilized in treating type 2 diabetes mellitus [5]. High rates of type 2 diabetes are repeatedly detected in historically undernourished, recently urbanized peoples characterized by Nelson as 'transitional and disadvantaged populations [6].'

Extrarenal comorbidity, especially heart disease and blindness has accompanied the kidney failure in pandemic type 2 diabetes. For example, among the aboriginal Oji-Cree of northern Ontario coronary heart disease has tripled over the past 20 years and is inextricably linked to the approximately 40% who manifest typical obesity-related type 2 diabetes [7]. Unacculturated Yanomama and Mambo of the Brazilian Amazon Basin did not express the glucose intolerance of the highly acculturated and very obese adult North American Pima. Validating Neel's findings, Valencia et al. compared the rate of type 2 diabetes in Pima Indians of southern Arizona with the Pimas in the Sierra Madre Mountains of northern Mexico. Both groups of Pima descended from the Hohokam group that settled in the valleys of the Gila and Salt rivers in Arizona 2000 years ago and separated about 700 years ago. Government assistance programs and a cash economy replaced traditional farming activities for the Arizona Pima who have the highest reported prevalence worldwide of type 2 diabetes, and are uniformly obese. By contrast, the Mexican Pima who practice traditional, non-mechanized agriculture, lacking electricity and running water in their homes use no modern household devices have a 5.6% rate of type 2 diabetes [8].

Based on the Pima diabetes story, there has been a quest for a single major locus for type 2 diabetes. Mitchell reported that a (dominant) allele controls insulin levels during insulin challenge tests in Mexican Americans and this allele accounts for 35% of changes in 2-hour insulin levels. Recently a susceptibility locus on chromosome 2 was thought to account for 30% of type 2 diabetes in Mexican Americans [9].

Bearing on the explosion of incident type 2 diabetic people is the interest in Syndrome X, the combination of type 2 diabetes, hypertension, and a truncal/abdominal (android) obesity also called an insulin resistance syndrome. But in surveys conducted by Neel and associates, obesity and hypertension were more closely correlated than obesity and type 2 diabetes suggesting that Syndrome X may not be a true syndrome governed by a common genetic pathogenesis [10]. Whatever the interrelationships between Syndrome X and type 2 diabetes [11], it seems clear that the life style changes that have made food abundant and exercise a sometime endeavor translate into an era of increasing obesity coupled with hypertension and type 2 diabetes. Projecting the expanded base of those with type 2 diabetes along side the pandemic of obesity and hypertension defines a coming era of dominant metabolic disease leading to the threat of multiple organ failure centering on the eyes, the, heart, kidneys, and peripheral vasculature.

The American Diabetes Association's most recent guidelines (2001) for blood pressure regulation aim at a target of 135/85 mmHg and for glucose regulation a hemoglobin A1c of $< 7\%$. Additional components of contemporary care, not yet attaining an 'affirmed by evidence' label include reduction of hyperlipidemia and aggressive treatment of microalbuminuria even when blood pressure measurements are normal. Indeed, the most aggressive strategists advocate treatment with an angiotensin converting enzyme inhibitor for normalbuminuric normotensive people with diabetes [12]. What is now clear is that strategic planning for 'long-term' care in diabetes should be initiated as a component of the first contact with the health care team. By means of a check list (Table 1), periodic and repetitive evaluations of the diabetic patient's most vulnerable organ systems can be initiated along with a continuing education program. Comprehensive management for the diabetic patient is a daunting undertaking demanding large amounts of professional time, especially when enervating complications such as impotence, vision loss, stroke residual, limb amputation, and the need for uremia therapy threaten to interdict any forward motion in life planning. The American Diabetes

Table 1. Nephrologist's comprehensive diabetes care check list.

System	Specialist	Procedure
Eyes	Ophthalmologist	Fluorescein angiogram Eye pressure
Cardiovascular	Cardiologist	Dobutamine stress test Coronary angiography
Peripheral vascular	Vascular surgeon	Doppler vessel flow MRI angiogram
Podiatry	Podiatrist	Prescription shoes, routine care
Neurologic	Neurologist	EEG, Carotid flow stroke rehabilitation
Transplant	Transplant Surgeon	Consider kidney or kidney/ pancreas transplant
Gastroenterologic	Gastroenterologist	Endoscopy Sonography for cholelithiasis
Endocrinologic	Endocrinologist	Pituitary-thyroid integrity Bone density and maintenance
Genitourinary	Urologist	Urodynamics, sonography, CT-scan, Penile prosthesis
Gynecologic	Gynecologist	Cytology, irregular bleeding
Overall health	Nutritionist, Nurse Educator	Weight adjustment diet prescription Learn medications, instruction in self-injection of insulin, erythropoietin
Social service	Social worker	Obtain benefits, restructure home
Total patient care	Local and national organization	Social interaction, facilitate proximity to vital information

Association prepares and updates a clinical practice guide that is an important resource when structuring a life plan for individual diabetic patients [13, 14].

While the specter of a potentially overwhelming 21st century burden of type 2 diabetes is frightening, the progress of the past decade in interdicting diabetic complications provides reason to remain hopeful that medical progress will prove equal to the challenge. Gradually, almost unperceptively, the formerly stark prognosis for combined diabetic nephropathy and retinopathy has been transformed from a tragedy signaling near-term death to a clarion call for effective interventive measures. This transition reflects the cumulative effect of multiple small increments in understanding and therapy. Central to all current regimens for managing diabetes is the normalization of hypertensive blood pressure together with the best possible approximation of normoglycemia.

If the 'diabetes wars' of 2002–2004 earn the label of detente or even truce because the rate of complications in properly treated patients can be slowed to that of aging

Table 2. Interdicting diabetes: Research initiatives.

Approach	**Strategy**	**Authors**
Pancreas transplants	Improved results in diabetes types 1 and 2	Friedman A [16]
Human islets repeated	Sirolimus, tacrolimus, and daclizumab	Shapiro et al. [17]
Intrahepatic islets	Anti-CD154 (hu5c8)	Kenyon et al. [18]
Microencapsulated islets	Purified alginate encapsulation	Rayat et al. [19]
Bypass insulin receptor	Demethylasterriquinone B-1 (DMAQ-B1) from endophytic fungus, Pseudomassaria sp	Salituro et al. [20]
Inhale insulin	Preprandial insulin given by inhalation	Skyler et al. [21]
Beta cells	Immortalize insulin secreting cells	Andrikopoulos et al. [22]
Bionic pancreas	Implantable insulin secreting device	Stotcum et al. [23]
Gut stem cells	Nurture regrowth of pancreas	Silverstein and Rosenbloom [24]
Genetic cell engineering	Pancreatic and duodenal homebox genes-1 to liver by adenovirus	Ferber S [25]

without the cloud of diabetes, then the years that follow should register unrestricted triumph. While molecular biology will provide blocking and diverting techniques to mute the panoply of tissue and organ complications the elusive goal of total prevention of diabetes as a disease is within reach. The interdiction of diabetes has followed a multi-front attack as shown in Table 2. Other promising approaches include: (1) insulin sensitizers including protein tyrosine phosphatase-1B and glycogen synthase kinase 3, (2) inhibitors of gluconeogenesis like pyruvate dehydrogenase kinase inhibitors, (3) lipolysis inhibitors, (4) fat oxidation including carnitine palmitoyltransferase I and II inhibitors, and (5) energy expenditure by means of beta 3-adrenoceptor agonists [15]. The pages of this brief volume recount a tale of adventure, high hopes, and pursuit of enormous treasure – the main prize being elimination of a human scourge. What is clear is that the hunt for a cure is the intoxicant that makes repetitive failures along the way bearable. From the essays that follow, the reader perceives the message that 'it can be done, and soon.'

Eli A. Friedman, M.D.
Francis A. L'Esperance, Jr. M.D.

REFERENCES

1. Ritz E, Rychlik I, Locatelli F, Halimi S. End-stage renal failure in type 2 diabetes: a medical catastrophe of worldwide dimensions. Am J Kidney Dis 1999; 34: 795–808.
2. Http//www.who.int/ Revised 14 March 2001.
3. Diabetes mellitus: A 'thrifty' genotype rendered detrimental by 'progress'? Am J Hum Genet 1962; 14: 353–352.

4. Neel JV, Weder A, Julius S. Type II diabetes, essential hypertension, and obesity as 'syndromes of impaired genetic homeostasis:' the 'thrifty genotype' hypothesis enters the 21 st century. Perspect Biol Med 1998; 42: 44–74.
5. Auwerx J. PPARgamma, the ultimate thrifty gene. Diabetologia 1999; 42: 1033–1049.
6. Nelson RG. Diabetic renal disease in transitional and disadvantaged populations. Nephrology 2001; 6: 9–17.
7. Hegele RA. Genes and environment in type 2 diabetes and atherosclerosis in aboriginal Canadians. Curr Atheroscler Rep 2001; 3: 216–221.
8. Ravussin E, Valencia ME, Esparza J, Bennett PH, Schulz LO. Effects of a traditional lifestyle on obesity in Pima Indians. Diabetes Care 1994; 17: 1067–1074.
9. Mitchell BD, Kammerer LM, Hixon JE, Atwood LD, Hackleman S, Blangero J, Haffner SM, Stern MP, MacCluer JW. Evidence for a major gene affecting post-challenge insulin levels in Mexican Americans. Diabetes 1995; 44: 284–289.
10. Neel JV, Weder A, Julius S. Type II diabetes, essential hypertension, and obesity as 'syndromes of impaired genetic homeostasis:' the 'thrifty genotype' hypothesis enters the 21 st century. Perspect Biol Med 1998; 42: 44–74.
11. Ezenwaka CE, Davis G and Offiah NV. Evaluation of features of syndrome X in offspring of Caribbean patients with Type 2 diabetes. Scand J Clin Lab Invest 2001; 61: 19–26.
12. O'Hare P, Bilbous R, Mitchell T, O' Callaghan CJ, Viberti GC. Low-dose ramipril reduces microalbuminuria in type 1 diabetic patients without hypertension: results of a randomized controlled trial. Diabetes Care 2000; 23: 1823–1829.
13. American Diabetes Association: Clinical practice recommendations Supplement 1. Diabetes Care 2000; 23 (Suppl 1): S1–S116.
14. Clark CM, Fradkin JE, Hiss RG, Lorenz RA, Vinicor F, Warren-Boulton E. Promoting early diagnosis and treatment of type 2 diabetes: the National Diabetes Education Program. JAMA 2000; 284: 363–365.
15. Wagman AS, Nuss JM. Current therapies and emerging targets for the treatment of diabetes. Curr Pharm Des 2001; 7: 417–450.
16. Friedman AL. Appropriateness and timing of kidney and/or pancreas transplants in type 1 and type 2 diabetes. Adv Ren Replace Ther 2001; 8: 70–82.
17. Ryan EA, Lakey JR, Rajotte RV, Korbutt GS, Kin T, Imes S, Rabinovitch A, Elliott JF, Bigam D, Kneteman NM, Warnock GL, Larsen I, Shapiro AM. Clinical outcomes and insulin secretion after islet transplantation with the Edmonton protocol. Diabetes 2001; 50: 710–719.
18. Kenyon NS, Chatzipetrou M, Masetti M, Ranuncoli A, Oliveira M, Wagner JL, Kirk AD, Harlan DM, Burkly LC, Ricordi C. Long-term survival and function of intrahepatic islet allografts in rhesus monkeys treated with humanized anti-CD154. Proc Natl Acad Sci USA 1999; 96: 8132–8137.
19. Rayat GR, Rajotte RV, Ao Z, Korbutt GS. Microencapsulation of neonatal porcine islets: protection from human antibody/complement-mediated cytolysis in vitro and long-term reversal of diabetes in nude mice. Transplantation 2000; 69: 1084–1090.
20. Salituro GM, Pelaez F, Zhang BB. Discovery of a small molecule insulin receptor activator. Recent Prog Horm Res 2001; 56: 107–126.
21. Skyler JS, Cefalu WT, Kourides IA, Landschulz WH, Balagtas CC, Cheng SL, Gelfand RA. Efficacy of inhaled human insulin in type 1 diabetes mellitus: a randomised proof-of-concept study. Lancet 2001; 357: 331–335.
22. Andrikopoulos S, Verchere CB, Teague JC, Howell WM, Fujimoto WY, Wight TN, Kahn SE. Two novel immortal pancreatic beta-cell lines expressing and secreting human islet amyloid polypeptide do not spontaneously develop islet amyloid. Diabetes 1999; 48: 1962–1970.
23. Stocum DL. Regenerative biology and engineering: strategies for tissue restoration. Wound Repair Regen 1998; 6: 276–290.
24. Silverstein JH, Rosenbloom AL. New developments in type 1 (insulin-dependent) diabetes. Clin Pediatr 2000; 39: 257–266.
25. Ferber S. Can we create new organs from our own tissues? Isr Med Assoc J 2000; 2 (Suppl): 32–36.

Conference Participants

Brooklyn, New York, USA, March 23, 2001

David A. Antonetti, Ph.D.
Assistant Professor of Cellular and Molecular Physiology
Penn State University
College of Medicine
Room C 4800
Hershey, PA 17033
New Insights into the Molecular Mechanisms of Vascular Permeability in Diabetes

Morrell Michael Avram, M.D.
Professor of Medicine
Chief of Nephrology
The Long Island College Hospital
340 Henry Street
Brooklyn, New York 11201
Impact of Nutrition in Uremic Diabetics

Alistair J. Barber, Ph.D.
Assistant Professor of Ophthalmology
Penn State University College of Medicine
Room C4800, H166
Hershey Medical Center
Hershey, PA 17033
Apoptosis and Neurodegeneration in Diabetes: Lessons From the Retina

David H. Berman, M.D.
Clinical Associate Professor of Ophthalmology
Director Retinal Service
The Brooklyn Hospital Center
240 Willoughby Street #6D
Brooklyn, NY 11201
Ocular Findings at Onset of Uremia

Clinton D. Brown, M.D.
Assistant Professor of Medicine
Department of Medicine
450 Clarkson Avenue
Brooklyn, New York 11203
Erythrocyte Glycation in Uremic Diabetic Patients

Anthony Cerami, Ph.D.
Adjunct Professor of Medicine
The Kenneth S. Warren Laboratories
765 Old Saw Mill River Road
Tarrytown, New York 10591
Novel Applications for Recombinant Human Erythropoietin

Steven T. Charles, M.D.
Clinical Professor of Ophthalmology
University of Tennessee
College of Medicine
6401 Popular Avenue, Suite 190
Memphis, TN 38119
Diabetic Vitrectomy Update

Luther T. Clark, M.D.
Chief, Division of Cardiology
Professor of Medicine
SUNY, Downstate Medical Center
Department of Medicine
450 Clarkson Avenue
Brooklyn, New York 11203
Prevention, Diagnosis, and Management of Heart Disease in Renal Failure

Frederick L. Ferris, M.D.
Director, Division of Biometery and Epidemilogy
National Eye Institute
31 Center Drive, MSC 2510 Bldg.31
Room 6A52, Bethesda, MD 20892-2510
Diabetic Macular Edema

Amy L. Friedman, M.D.
Assistant Professor of Surgery
Department of Surgery
Yale University, College of Medicine
FMB 112, 233 Cedar Street
New Haven, CT 06625
Islet Transplantation in ESRD Due to Type 2 Diabetes

Eli A. Friedman, M.D.
Chief, Division of Renal Disease
Distinguished Teaching Professor
Department of Medicine
SUNY, Downstate Medical Center
450 Clarkson Avenue
Brooklyn, New York 11203
Preempting Diabetic Complications: A Realistic Prediction

Ronald I. Klein, M.D.
Professor, Department of Ophthalmology and Visual Sciences
University of Wisconsin
610 North Walnut Street, WARF 460
Madison, Wisconsin 53705
Medical Approaches to Preventing Visual Loss from Diabetic Retinopathy

Francis A. L'Esperance, Jr., M.D.
Clinical Professor of Ophthalmology
College of Physicians and Surgeons
1 East 71st Street
New York, NY 10021
Promising Initiatives in Diabetic Eye Disease

Mariana S. Markell, M.D.
Associate Professor of Medicine
Department of Medicine
450 Clarkson Avenue
Brooklyn, New York 11203
Consequences of Post-Transplant Diabetes

S. Michael Mauer, M.D.
Professor of Pediatrics
University of Minnesota
Box 491, 420 Deleware Street SE
Minneapolis, MN 55455
Which Comes First: Renal Injury or Microalbuminuria?

Carl Erik Mogensen, M.D.
Head Physician
Professor of Medicine
Aarhus Kommunehospital
DK-8000 Aarhus C
Denmark
Microalbuminuria in Perspective

John S. Najarian, M.D.
Professor of Surgery
Regent's Professor Emeritus
University of Minnesota Hospital Center
Box 195
Minneapolis, MN 55455
Pancreas Transplantation: Does it have a future?

Dimitrios G. Oreopoulos, M.D.
Professor of Medicine
University of Toronto
Room A44 Research Wing
Toronto Western Hospital
Toronto, Ontario M5T 2S8
Canada
Long-Term Perspective on Utility of Peritoneal Dialysis in Diabetes

Jonathan A. Sheindlin, M.D.
Assistant Professor of Ophthalmology
Long Island College Hospital Center
339 Hicks Street
4th Floor - Othmer Building
Brooklyn, New York 11201
Progressive Proliferative Diabetic Retinopathy-Current Treatment Strategies

Yalemzewd Worerdekal, M.D.
Assistant Professor of Medicine
Department of Medicine
SUNY, Downstate Medical Center
450 Clarkson Avenue
Brooklyn, New York 11203
Reducing Mortality in Hemodialysis for Diabetes

CLINTON D. BROWN, ZHONG ZHAO, LORAINE THOMAS AND ELI A. FRIEDMAN

1. Effect of Aminoguanidine on Induced-diabetic Rabbits and Diabetic Patients with ESRD

Editors' Comment: Pursuing the lead established by Brownlee and Cerami in 1986 (Brownlee M, Vlassara H, Kooney A, Ulrich P, Cerami A. Aminoguanidine prevents diabetes-induced arterial wall protein cross-linking. Science. 1986;232:1629–32), Brown et al. demonstrate that the adverse effect of diabetes on erythrocyte deformability is reversed by treatment with aminoguanidine. Whether treatment with aminoguanidine holds potential in human diabetes is now unclear due to the blurred outcome of initial clinical trials indicating drug hepatotoxicity. Evidence that advanced glycosylated end-products (AGEs) are somehow involved in the pathogenesis of diabetic micro- and macrovascular complications, however, continues to mount. It is probable that derivative studies of aminoguanidine or related compounds that block formation or break established AGEs will answer the question of the value of pharmacologic intervention to prevent AGE effect.

INTRODUCTION

Perturbation in blood rheology are thought to contribute to the genesis of progressive micro- and macrovascular complications in diabetes mellitus [1]. Specifically, hyperviscosity of whole blood and plasma, and reduced RBC deformability (RBC-df), have been linked to the pathogenesis and progression of overt nephropathy in diabetic persons [2, 3]. In a previous study [4] we reported a highly significant correlation between RBC-df impairment and the loss of renal function in diabetic subjects.

RBC-df is defined as the passive change in the shape of erythrocytes in response to shear forces, is crucial for normal flow and function in the microcirculation [5–7].

Although the association of impaired RBC-df with diabetes has not been observed universally [8, 9], nevertheless, there is considerable evidence which suggests that diabetes-induced impairment of RBC-df may be the consequence of a direct effect of advanced glycation endproducts (AGE) or an AGE-mediated modification of the RBC [10]. If abnormal RBC-df is truly an important contributor to diabetic vascular problems, there are unfortunately few therapeutic agents available that have proven efficacy [11, 12] and no agent has been subjected to a long-term trial. Herein, we review two experiments which tests the efficacy of Aminoquanidine (Ag), a nucleophilic hydrazine and AGE blocker, on RBC-df in an animal model and in human subjects with diabetes.

RBC-DF INDEX

The measurement of RBC-df is defined by the rate of filtration of a dilute (4% hematocrit) suspensions of washed RBCs (Phosphate buffered saline (PBS) solution) through

E.A. Friedman and F.A. L'Esperance, Jr. (eds.), Diabetic Renal-Retinal Syndrome, 1–5.

a polycarbonate membrane with straight channels of 3 μm diameter pores (Nucleopore, corning, Acton, Ma) under a constant negative pressure (–20 cm H_2O).

The RBC-df index (DI) is defined as the filtration rate of suspended RBCs divided by the filtration rate of an equal volume of buffer. In the diabetic patients study, DI was normalized for a reference mean RBC-df, and inversely related to RBC-df.

INDUCED DIABETIC RABBIT MODEL

Male NZW rabbits were made hyperglycemic following intravenous injection of alloxan (65–120 mg/Kg). Six healthy rabbits with sustained hyperglycemia (293 ± 25 mg/dl) for 12 months were used to study the effect of Ag on RBC-df. Prior to the administration of Ag (oral gavage), the induced-diabetic rabbit group, demonstrated mean RBC

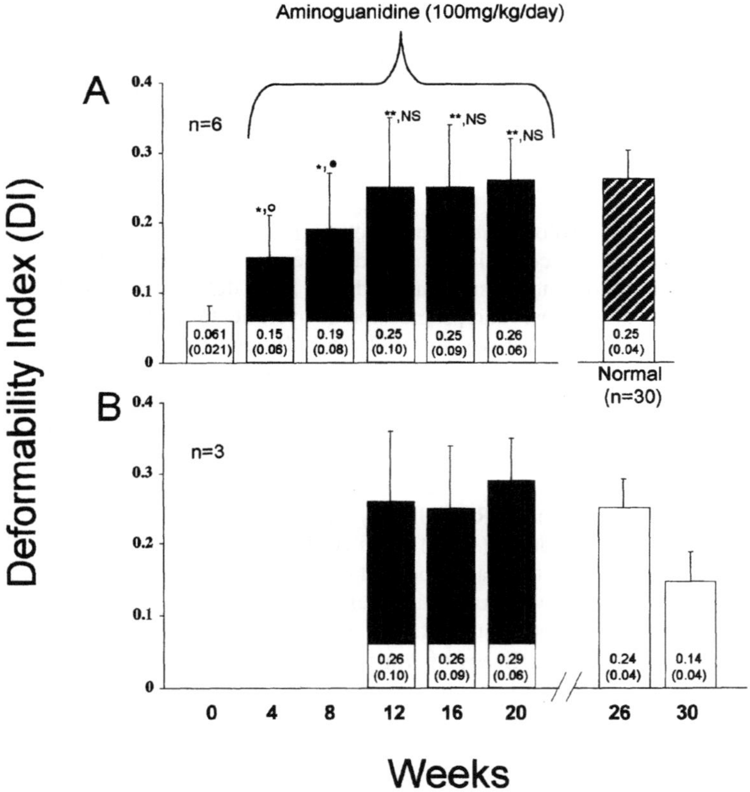

Figure 1. A: Improvement in RBC deformability in ALX-induced diabetic rabbits after treatment with aminoguanidine. • $P < 0.05$; ** $P < 0.01$, compared with pretreatment value. ○ $P < 0.0001$, ● $P = 0.02$, NS, compared with mean normal deformability index. *B:* Deterioration in RBC deformability of –50% 10 week after discontinuing aminoguanidine.

stiffness that was five fold greater than nondiabetic control rabbits ($P = 0.0005$). There was steady and significant improvement in RBC-df at 4 and 8 weeks of treatment ($P = 0.001$), and RBC-df was restored to normal by week 12. Conversely, RBC-df deteriorated and returned to pre-treatment impairment by week 10 following Ag withdrawal (Figure 1, A, B).

TRIAL OF AG IN DIABETIC PATIENTS

In the diabetic human trial of AG, 20 patients were enrolled in a randomized (2 × 1) double-blind, placebo-controlled paralled group trial for 12 months. All patients had ESRD and had begun hemodialysis 4 months prior to study enrollment. Eight patients received a low dose of Ag (100 mg) by mouth ever other day, Six patients received a higher dose of Ag (200 mg), and Six patients received placebo. RBC-df was measured at baseline, 6 and 12 months while receiving one of two dosages of Ag or placebo, and at 2 and 4 months after Ag was withdrawn.

Diabetic patients given high dose Ag had a marked and highly significant ($P = 0.001$) improvement in RBC that approached normally at 12 months. Diabetic patients given the lower dose of Ag showed a mild but not significant ($P = 0.13$) improvement in RBC-df at 12 months. There was no change in RBC-df throughout the trial period for patients randomized to receive placebo (Figure 2). Four months after Ag was withdrawn RBC-df returned to pre-treatment impairment (Figure 3).

DISCUSSION/CONCLUSION

The difference in efficacy between the two treatment groups (low vs high dose) is noted in Figure 2. An incomplete response observe in diabetic patients receiving Low dose

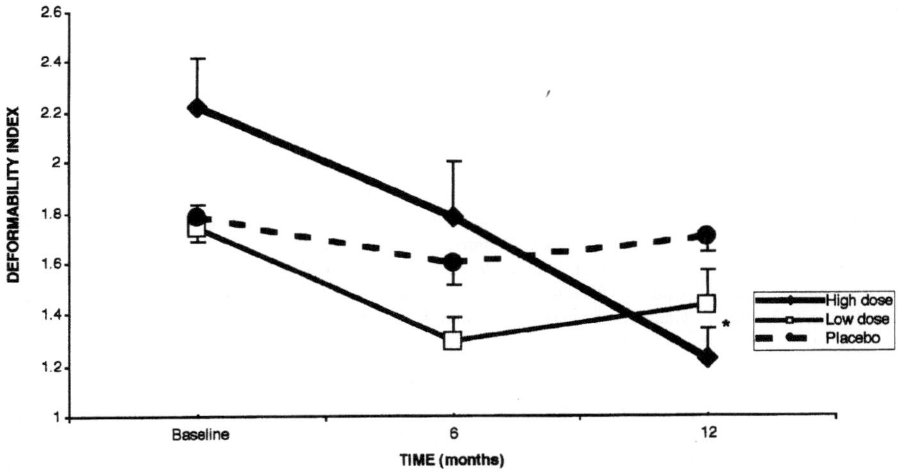

Figure 2. Effect of aminoguanidine (Ag) on RBC deformability in diabetic patients with ESRD.

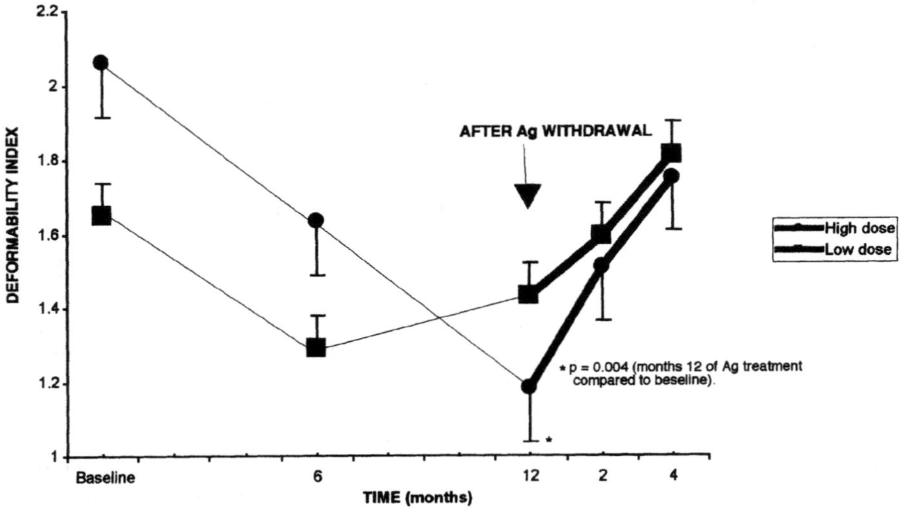

Figure 3. Effect of aminoguanidine (Ag) on RBC deformability in diabetic patients with ESRD.

(100 mg) Ag. We believe that low dose Ag was not sufficient to effectively block the formation of AGEs and therefore had a minor and insignificant effect on RBC-df. The response to high dose Ag was that of continuous improvement in RBC-df throughout the study course. We speculate that Ag when given at this dosage, protects nascent RBCs against alteration in RBC-df that might occur from RBCs and AGEs interaction.

We conclude that Ag significantly corrects impaired RBC-df in induced-diabetic rabbits and in diabetic patients with ESRD. The mechanism(s) underlying diabetes linked impaired RBC-df has not been confirmed nor have the mechanism(s) for the apparent efficacy of Ag on diabetes-induced abnormal RBC-df.

REFERENCES

1. Barnes AJ, Loeke P, Scudder PR, Dormandy TL, Dormandy JA, Slack J. Is hyperviscosity a treatable component of diabetic microcirculatory disease? Lancet 1977; ii: 789–791.
2. Simpson LO. Intrinsic stiffening of red blood cells as the fundamental cause of diabetic nephropathy and microangiopathy: a new hypothesis. Nephron 1985; 39: 344–351.
3. Vermes I, Steimetz ET, Zeyen LJ, van der Veen EA. Rheological properties of white blood cells are changed in diabetic patients with microvascular complications. Diabetologia 1987; 30: 434–436.
4. Brown CD, Zhao Z, Ghali H, Thomas L, Friedman EA. Association of reduced RBC deformability and diabetic nephropathy. ASIAO 1999; 45: 209.
5. Shiga T, Maeda N Influence of membrane fluidity on erythrocyte functions. Biorhology 1980; 17: 485–499.
6. La celle PL, Smith BD. Biochemical factors influencing erythrocyte deformability and capillary entrance phenomena. Scand J Clin Lab Invest 1981; 41 (Suppl 156): 145–149.
7. Mazzanti L, Rabini RA, Test I, Bertoli E. Modifications induced by diabetes on the physiochemical and functional properties of erythrocyte plasma membrane. Eur J Clin Invest 1989; 19: 84–89.
8. Williamson JR, Gardner RA, Boylan CW, Carroll GL, Chang K, Marvel JS, Gonen B, Kilo C, Tay

T-S, Sutera SP. Microrheologic investigation of erythrocyte deformability in diabetes mellitus. Blood 1985; 65: 283–288.
9. Rand PW, Norton JM, Richards AL, Lacombe EH, Pirone LA. Effects of diabetes mellitus on red cell properties. Clin Hemorheology 1981: 1, 373–385.
10. Meada N, Kon K, Imaizumi K, Sekiya M, Shiga T. Alteration of rheological properties of human erythrocytes by crosslinking of membrane proteins. Biochem et Biophysica Acta 1983; 735: 104–112.
11. Brown CD, Zhao ZH, de Alvaro F, Chan S, Friedman EA. Correction of Eryrhrocyte Deformability Defect in ALX-induced diabetic rabbits after treatment with Aminoguanidine. Diabetes 1993; 16: 590–593.
12. Vague P, Juhan I. Red cell deformability, platelet aggregation, and insulin action. Diabetes 1983; 32 (Suppl 2): 88–91.

MORRELL MICHAEL AVRAM

2. Impact of Nutrition in Uremic Diabetics

Editor's Comment: Long concerned with the adverse impact of uremia on nutrition, Avram underscores the utility of measuring serum levels of albumin, creatinine and prealbumin as not only predictors of present nutritional status but as markers for survival in diabetic patients undergoing maintenance hemodialysis and continuous ambulatory peritoneal dialysis. Noting the extensive prevalence of malnutrition in his own patients, Avram advises reduction of atherogenic lipid and lipoproteins levels as well as attention to hyperhomocysteinemia in order to prevent nutritional wasting with associated excess comorbidity and mortality.

INTRODUCTION

Throughout the world, the incidence of diabetes mellitus is increasing at an alarming rate [1]. During the period between 1990 and 1998, the prevalence of diabetes in the United States increased by 33% [2]. In 1997, health care costs associated with diabetes were an estimated $98 billion [3]. At present, some 16 million Americans have diabetes and the number is expected to rise to 22 million by 2025. Diabetes is the leading cause of end-stage renal disease (ESRD) worldwide. Because of increased incidence of diabetes mellitus, diabetic nephropathy has become a silent epidemic worldwide. According to the recent reports of the United States Renal Data System (USRDS), there is a continuous increase in the incidence of ESRD with primary etiology as diabetes mellitus between 1980 and 1998 even when adjusted for age, gender and race [4]. In 1998, diabetic nephropathy accounted for 45.4% and 43.5% of all new cases in hemodialysis (HD) and peritoneal dialysis (PD) patients respectively. Medicare expenditures for the ESRD program have been increasing steadily, from $5 billion in 1991 to $12 billion in 1998. Diabetes mellitus contributes to the rising cost of ESRD therapy more than any other disease. Increase in physician/supplier costs was highest for diabetic patients at 34% [4].

With advances in dialysis technology and therapeutic management, mortality rate of ESRD patients has slightly declined. In the United States, there has been a reduction in mortality rate of ESRD patients by 13% (adjusted) and 7.5% (unadjusted) during the period from 1989 to 1996. Nevertheless, annual unadjusted mortality rate of ESRD patients still remains high at 23.5% in 1996 [5]. Co-morbid factors such as advanced age, malnutrition, infection, cardiovascular disease, malignancy, and technical factors like adequacy of dialysis, vascular access complications, and peritoneal membrane characteristics may all contribute to this high mortality rate during HD and PD. Protein energy malnutrition is highly prevalent in HD patients and a major factor contributing to the high mortality rate in these patients [6]. This has led to the search for various nutritional markers predictive of long-term and short-term survival of dialysis patients

E.A. Friedman and F.A. L'Esperance, Jr. (eds.), Diabetic Renal-Retinal Syndrome, 7–21.

[7]. Over the past decade, various nutritional parameters have emerged as powerful predictors of mortality in dialysis patients [6, 8]. We and others have reported that, in addition to advanced age and diabetes, serum measures of nutritional markers such as albumin, cholesterol, creatinine, and prealbumin are important predictors of survival in HD and PD patients [9–14]. We have shown that biochemical indices reflecting nutritional status can predict very long-term survival in HD (up to 30 years) and PD (up to 10 years) patients [15, 16]. In most of these studies, prognostic importance of single enrollment levels of serum nutritional markers has been evaluated in dialysis patients. We have also reported that serial measurement of nutritional parameters provide important prognostic information on survival in HD patients [17].

Over the past 10 years, mortality rates of diabetic dialysis patients have gradually declined. In the United States, there is a reduction in first year mortality rate from 40.4 per 100 patient years in 1986 to 23.2 per 100 patient years in 1996 [5]. Patients with diabetes undergoing maintenance dialysis therapy continue to do worse with respect to survival compared to non-diabetic patients. In the United States, 5-year actuarial survival rate of non-diabetic and diabetic dialysis patients is 35% and 21%, respectively [18]. In Australia, 5-year actuarial survival rate of non-diabetic and diabetic dialysis patients was 60% and 27%, respectively [19]. According to Catalonia (Spain) registry, 5-year survival in non-diabetic patients was 60% on HD and 29.3% on PD compared with 30% on HD and 25% on PD in diabetic patients [20]. The causes and factors contributing to high mortality of diabetic dialysis patients need to be investigated. Cardiovascular disease has been reported to be the important factor contributing to the high mortality in diabetic dialysis patients. But many diabetic patients suffer from malnutrition during renal replacement therapy, which may also contribute to high mortality in these patients. We at the Long Island College Hospital for the first time successfully treated a diabetic renal failure patient as reported in the literature in 1966 [21]. For the past two decades, we have been searching for markers of morbidity and mortality in our increasing number of diabetic ESRD patients. We have added protein energy malnutrition as a risk factor for mortality in diabetic ESRD patients. Previously, we reported various demographic and enrollment biochemical and nutritional markers as predictors of mortality in diabetic dialysis patients [22]. In the present study, we have extended the follow-up period with a larger number of patients and investigated, in detail, the role of demographics and nutrition as the risk factors for long term survival in our single center HD and PD patients. In addition to enrollment values, we have also examined the prognostic importance of monthly serial data in these patients.

METHODS

Patients

Five hundred eighty three (583) HD patients and four hundred sixty-eight (468) PD patients treated at the Avram Center for Kidney Diseases, The Long Island College Hospital were enrolled in this study from January 1987 and December 1983, respectively. For the prealbumin study, 131 HD and 128 PD patients were enrolled beginning June 1991. All the patients were followed until December 2000. On enrollment, clinical and laboratory data were collected. Follow-up was censored on transplantation, transfer to different dialysis modality, or transfer to another dialysis center.

Clinical and Biochemical Data

Clinical data included age, race, gender, etiology of ESRD, total months on dialysis prior to enrollment, hypertension and diabetic status. On enrollment, nonfasting baseline predialysis blood samples were drawn at the routine monthly visit and multiphasic biochemistry screening was obtained for each patient. In addition, serial monthly measures of serum nutritional markers such as albumin, creatinine, and cholesterol were also collected and entered into the database. 'Enrollment' value was the first value of a given nutritional marker for the patient obtained at or immediately after enrollment. 'Mean' value was the mean of all subsequent values of the given nutritional marker for the patient. Prealbumin was measured by rate nephelometry on a Beckman Array Protein System (Beckman Instruments Inc, Brea, CA). Serum intact parathyroid hormone (PTH) was measured by radioimmunoassay (Nichols Institute Diagnostics, San Juan Capistrano, CA). Six month moving average values for albumin, creatinine, and cholesterol were calculated separately for each month on the study as the average of the preceding six monthly values in each patient.

Dialysis Prescription

HD patients were treated on Cobe Centry system 3 machines with bicarbonate-based dialysate and volumetric ultrafiltration control. Cellulose-based membranes (predominantly cellulose acetate and cellulose triacetate) were used for more than 98% of all treatments. Dialyzers were not reprocessed. The hemodialysis prescription was guided (from 1987) by the goal of achieving urea reduction ratio of > 60% on the first treatment of the week (i.e. 68 hours after the preceding dialysis). This was later increased to > 65%. Before 1992, continuous ambulatory peritoneal dialysis (PD) prescription was 2-L Dianeal solution, 4 exchanges/day in most patients before dialysis adequacy measurements were performed. After 1992, PD prescription was guided to achieve a target Kt/V of 1.7, later increased to 2.1. Dextrose concentration in the dialysate varied with the need for ultrafiltration.

Study Design and Data Analysis

Continuous variables are reported as the mean ± standard deviation (SD). In cross-sectional analyses, comparisons of continuous variables and proportions were made using ANOVA and the chi-squared test, respectively, unless otherwise stated. For selected comparisons between two group means, parametric (*t*-test) or non-parametric (Mann-Whitney) tests were used wherever applicable. Observed patient survival between groups of patients was analyzed by Kaplan-Meier method [23]. Log-rank test was used to compare two survival curves. Survival was evaluated by first univariate and then by multivariate Cox's proportional hazards model using age, race, gender, diabetes, prior months on dialysis and nutritional markers as independent variables [24, 25]. Non-diabetic and diabetic patients were analyzed separately using Cox's proportional hazards model. Calculations were performed using SPSS for Windows 10.0.1 (SPSS Inc., Chicago, IL).

RESULTS

HD patients

Table 1 shows the demographic characteristics of the HD population studied. The mean age was 60 years. Forty eight percent of the patients were diabetic and 47% were male. Diabetic patients were older than non-diabetics ($P < 0.0001$). The etiology of ESRD was diabetes in 40%, hypertension in 31%, glomerulonephritis in 10%, polycystic kidney disease in 4%, obstruction in 3%, HIV in 2% and other/unknown in 12%. Mean follow-up period was 3.35 ± 2.96 years with range from 36.5 days to maximum 13.55 years. Levels of enrollment biochemical markers such as albumin ($P < 0.0001$), creatinine ($P < 0.0001$), prealbumin ($P = 0.002$), predialysis BUN ($P = 0.004$) and PTH ($P = 0.034$) were significantly lower in diabetic HD patients compared to non-diabetic patients (Table 2). Mean levels of biochemical markers showed similar results.

Observed cumulative survival of non-diabetic patients were significantly ($P < 0.0001$) higher than that of diabetic patients (Figure 1). Survival adjusted for enrollment age, race, gender, and months on HD also showed similar results ($P = 0.001$). Even the difference in survival of patients 50 years or less was highly significant between non-diabetic and diabetic patients ($P = 0.0002$). Results from Cox's proportional univariate analysis indicate enrollment age (RR = 1.034, $P < 0.0001$) and glucose (RR = 1.002, $P < 0.0001$) were significant positive predictors of mortality in all HD patients. Significant negative predictors of mortality in all HD patients included albumin (RR = 0.47, $P < 0.0001$), creatinine (RR = 0.94, $P < 0.0001$), total cholesterol (RR = 0.997, $P = 0.019$), prealbumin (RR = 0.92, $P < 0.0001$) and PTH (RR = 0.63, $P < 0.0001$). Diabetic patients had 1.63 times the relative risk of mortality compared to non-

Table 1. Demographics of HD patients.

	All patients (n = 583)	Non-diabetic (n = 309)	Diabetic (n = 274)	P
Age at enrollment (Year)				
(Mean ± SD)	60.4 ± 15	58.2 ± 18	63.0 ± 12	< 0.0001
Gender (%)				
Male	47	51	44	0.04
Race (%)				
White	24	28	21	
Black	59	61	58	0.001
Hispanic	17	11	21	
Diabetes (%)	48			

Table 2. Enrollment biochemical markers in HD patients.

	Non-diabetic	Diabetic	*P*
Albumin (g/dL)	3.75 ± 0.49	3.55 ± 0.53	< 0.0001
Creatinine (mg/dL)	12.4 ± 5.0	9.2 ± 3.8	< 0.0001
Total cholesterol (mg/dL)	178 ± 50	181 ± 57	0.46
Prealbumin (mg/dL)	28.1 ± 5.7	24.5 ± 7.0	0.002
Glucose (mg/dL)	103 ± 26	170 ± 92	< 0.0001
Predialysis blood urea nitrogen (mg/dL)	80.4 ± 25	74.5 ± 23	0.004
Parathyroid hormone (pg/mL)	409 ± 512	305 ± 345	0.034

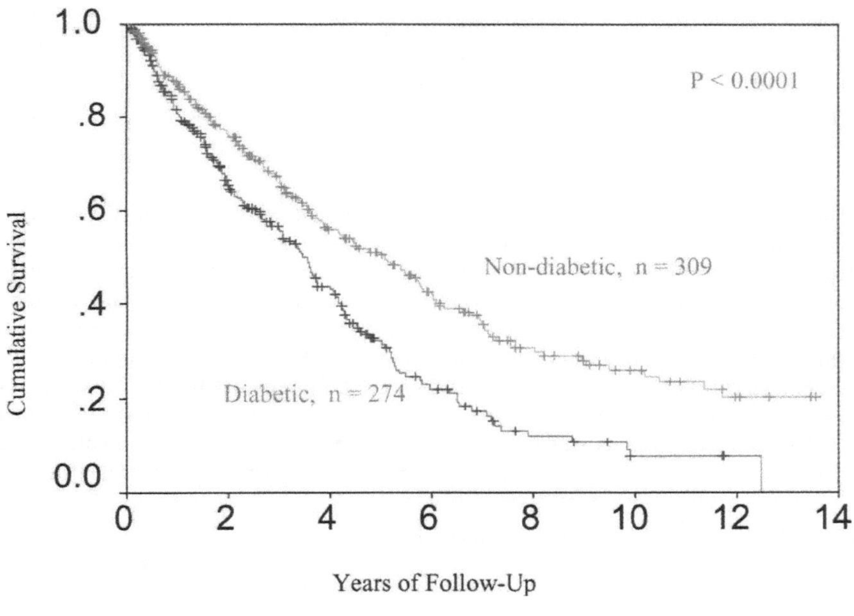

Figure 1. Survival in HD patients by diabetic status.

diabetic patients. Analyses of non-diabetic and diabetic patients indicate that age (RR = 1.04, $P < 0.0001$ versus RR = 1.025, $P = 0.0001$), albumin (RR = 0.42, $P < 0.0001$ versus RR = 0.64, $P < 0.0001$), creatinine (RR = 0.95, $P = 0.001$ versus RR = 0.95, $P = 0.011$), and prealbumin (RR = 0.92, $P = 0.003$ versus RR = 0.94, $P = 0.013$) were significant predictors of mortality. One unit increase in serum albumin, creatinine, and prealbumin values decreased the relative risk of death by 58%, 5%, and 8% for non-diabetic patients and 36%, 5%, and 6% for diabetic patients, respectively. Among the diabetics, risk of mortality in black patients was 51% lower compared to white patients ($P = 0.001$) (Table 3). Observed cumulative survival of white HD diabetics was significantly lower compared to Hispanic or black diabetic HD patients (Figure not shown).

Table 3. Univariate analysis of predicators of mortality in non-diabetic and diabetic HD.

	Non-diabetic		Diabetic	
	Relative risk	*P*	Relative risk	*P*
Age (year)	1.04	< 0.0001	1.025	0.001
Race				
Black vs. White	0.88	0.44	0.49	< 0.0001
Albumin (g/dL)	0.42	< 0.0001	0.64	< 0.0001
Creatinine (mg/dL)	0.95	0.001	0.95	0.011
Total cholesterol (mg/dL)	0.99	0.15	0.99	0.15
Prealbumin (mg/dL)	0.92	0.003	0.94	0.013
Glucose (mg/dL)	1.009	0.005	1	0.79
Predialysis BUN (mg/dL)	0.997	0.39	0.99	0.34
Log PTH (pg/mL)	0.65	0.012	0.69	0.06

The multivariate Cox's proportional hazards model for HD patients is shown in Table 4. In this model, factors such as age, diabetes, albumin, and total cholesterol were independent predictors of mortality. Adjusting for other confounding variables such as age, race, gender, months on dialysis, serum albumin, creatinine, and total cholesterol, diabetic patients had a 1.4-fold increase in mortality risk compared to non-diabetic patients ($P = 0.009$). For non-diabetic and diabetic patients, age, serum albumin, and total cholesterol were significant independent predictors of mortality (Table 5). Race was an independent predictor of mortality in diabetic HD patients. Risk of death for black HD diabetic patients were 48% lower compared to white HD diabetic patients ($P = 0.004$) (Table 5). Results (Table 6) show six-month moving average values of

Table 4. Predictors of mortality in HD patients.

	Relative risk	*P*
Age (years)	1.03	< 0.0001
Gender		
Female vs. male	1.03	0.8
Race		
Black vs. White	0.79	0.1
Diabetes (yes or no)	1.41	0.009
Albumin (g/dL)	0.55	< 0.0001
Creatinine (mg/dL)	1	0.83
Total Cholesterol (mg/dL)	0.99	0.005
Months on dialysis at enrollment	1.003	0.07

Multivariae Cox's proportional hazards analysis.

Table 5. Predicators of mortality in non-diabetic and diabetic HD patients.

	Non-diabetic		Diabetic	
	Relative risk	*P*	Relative risk	*P*
Age (year)	1.04	< 0.0001	1.02	0.039
Gender				
Female vs. male	0.84	0.34	1.29	0.16
Race				
Black vs. White	0.97	0.86	0.52	0.004
Albumin (g/dL)	0.48	< 0.0001	0.64	0.014
Creatinine (mg/dL)	1.003	0.89	1.001	0.98
Total cholesterol (mg/dL)	0.99	0.05	0.99	0.048
Months on dialysis at enrollment	1.003	0.13	1.003	0.21

Multivariae Cox's proportional hazards analysis.

Table 6. Six month moving average values of nutritional markers in HD patients.

	Non-diabetic (n = 124)	Diabetic (n = 140)	*P*
Albumin (g/dL)	3.77 ± 0.39	3.59 ± 0.41	0.001
Creatinine (mg/dL)	10.4 ± 3.7	8.7 ± 3.0	< 0.0001
Total cholesterol (mg/dL)	184 ± 47	183 ± 51	0.92

nutritional markers for a subset of patients whose monthly serial data were analyzed. Six month moving average serum albumin ($P = 0.001$) and serum creatinine ($P < 0.0001$) values are significantly higher for non-diabetic compared to diabetic patients.

PD patients

Table 7 describes the demographics of PD patients. The etiology of renal failure in PD patients was similar to that of HD patients except for a higher proportion of human immunodeficiency virus (HIV) infection in PD patients (7.0% for PD vs. 2.0% for HD). Mean follow-up was 2.18 ± 2.1 years with range from 1 month to maximum 13.43 years. Diabetic PD patients had significantly lower enrollment levels of biochemical markers such as albumin ($P < 0.0001$), creatinine ($P < 0.0001$), prealbumin ($P < 0.0001$), glucose ($P < 0.0001$) and PTH ($P = 0.001$) compared to non-diabetic patients (Table 8). Cumulative observed survival was significantly higher in non-diabetics ($P < 0.0001$) (Figure 2). After adjustment for age, race, gender, and months on PD at enrollment, non-diabetic patients had a better rate of survival ($P = 0.002$). In a subset of younger

Table 7. Demographics of PD patients.

	All patients (n = 468)	Non-diabetic (n = 271)	Diabetic (n = 197)	P
Age at enrollment (year)				
(Mean ± SD)	54.6 ± 16	51.4 ± 17	58.6 ± 12	< 0.0001
Gender (%)				
Male	48	49	44	0.14
Race (%)				
White	23	23	23	
Black	60	64	52	0.002
Hispanic	17	13	25	
Diabetes (%)	42			

Table 8. Enrollment biochemical markers in PD patients.

	Non-diabetic	Diabetic	P
Albumin (g/dL)	3.50 ± 0.77	3.41 ± 0.56	< 0.0001
Creatinine (mg/dL)	12.0 ± 4.9	9.1 ± 3.3	< 0.0001
Total cholesterol (mg/dL)	200 ± 61	207 ± 56	0.23
Prealbumin (mg/dL)	37.7 ± 11.7	30.3 ± 8.7	< 0.0001
Glucose (mg/dL)	99 ± 31	192 ± 118	< 0.0001
Predialysis blood urea nitrogen (mg/dL)	66.8 ± 26.7	69.2 ± 25.0	0.37
Parathyroid hormone (pg/mL)	425 ± 382	280 ± 268	0.001

PD patients with age 50 years or less, survival of diabetic patients was still lower than non-diabetics (P = 0.002).

Using Cox's univariate analysis, enrollment age (RR = 1.027, P < 0.0001) and glucose (RR = 1.002, P < 0.014) were significant positive predictors of mortality. Albumin (RR = 0.46, P < 0.0001), creatinine (RR = 0.89, P < 0.0001), prealbumin (RR = 0.96, P = 0.003) and PTH (RR = 0.44, P < 0.0001) were significant negative predictors of mortality in all PD patients. Risk of mortality was 1.77 times higher for diabetics compared to non-diabetics. Black patients had 47% lower mortality risk than white patients (P = 0.003). Analyses for non-diabetic and diabetic PD patients indicate that age (RR = 1.026, P < 0.0001 versus RR = 1.027, P = 0.002), albumin (RR = 0.39, P < 0.0001 versus RR = 0.52, P < 0.0001), creatinine (RR = 0.89, P < 0.0001 versus RR = 0.89, P < 0.0001), and PTH (RR = 0.45, P < 0.0001 versus RR = 0.51, P = 0.002) were significant predictors of mortality. Risks of mortality in black non-diabetic and diabetic PD patients were 37% and 34% lower than in white patients, respectively. For every unit increase in the albumin value, risk of death decreased by 61% and 48% for non-diabetics and diabetics respectively (Table 9).

In multivariate Cox's Proportional Hazards model, age, diabetes, AIDS, albumin, and total cholesterol were independent predictors of mortality. Adjusting for other confounding variables, such as age, race, gender, months on dialysis, AIDS, serum

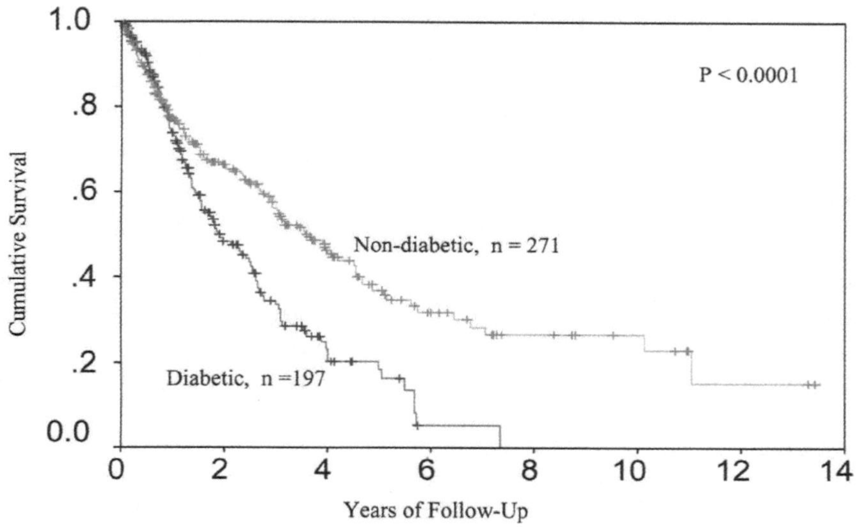

Figure 2. Survival in PD patients by diabetic status.

Table 9. Univariate analysis of predicators of mortality in non-diabetic and diabetic PD patients.

	Non-diabetic		Diabetic	
	Relative risk	*P*	Relative risk	*P*
Age (year)	1.026	< 0.0001	1.027	0.002
Race				
Black vs. White	0.63	0.03	0.66	0.06
Albumin (g/dL)	0.39	< 0.0001	0.52	< 0.0001
Creatinine (mg/dL)	0.89	< 0.0001	0.89	< 0.0001
Total cholesterol (mg/dL)	0.99	0.44	1	0.82
Prealbumin (mg/dL)	0.95	0.02	0.99	0.73
Glucose (mg/dL)	1.006	0.09	1	0.89
Predialysis BUN (mg/dL)	0.992	0.04	0.99	0.84
Log PTH (pg/mL)	0.45	< 0.0001	0.51	0.002

albumin, creatinine, and total cholesterol, diabetes was associated with 1.44-fold increase in mortality risk compared to non-diabetic patients. In non-diabetic and diabetic patients, age (RR = 1.03, $P < 0.0001$ versus RR = 1.02, $P = 0.029$) and serum albumin (RR = 0.44, $P < 0.0001$ versus RR = 0.56, $P = 0.001$) were significant independent predictors of mortality. In diabetic PD patients, serum creatinine was a significant predictor of mortality. Each 1 mg/dL increase in serum creatinine was associated with

Table 10. Predicators of mortality in non-diabetic and diabetic PD patients.

	Non-diabetic		Diabetic	
	Relative risk	*P*	Relative risk	*P*
Age (year)	1.033	< 0.0001	1.021	0.029
Gender				
Female vs. male	0.96	0.85	0.87	0.55
Race				
Black vs. White	0.73	0.19	0.77	0.29
Albumin (g/dL)	0.44	< 0.0001	0.56	0.001
Creatinine (mg/dL)	0.97	0.18	0.94	0.05
Total cholesterol (mg/dL)	1.002	0.15	1.001	0.65
Months on dialysis at enrollment	1.003	0.9	1.003	0.81
AIDS (yes vs. no)	1.92	0.05	0.62	0.47

Multivariae Cox's proportional hazards analysis.

a 6% decrease in mortality risk (Tables 10). In non-diabetic PD patients, patients with AIDS had almost 2-fold mortality risk compared to those without AIDS. There were only 5 HIV positive patients in the diabetic patient population.

DISCUSSION

Almost 40 years ago, our physicians at Long Island College Hospital were the first to dialyze diabetic patients with ESRD [21]. Since then, there has been significant improvement in the treatment and medical management of diabetic dialysis patients. The mortality rate of diabetic dialysis patients, although declining during the past decade, still remains unacceptably high. Our present data show that long term observed and adjusted survival of non-diabetic dialysis patients is significantly higher compared to diabetic dialysis patients (Figures 1 and 2). Even in the younger HD and PD patients, there was highly significant difference in survival between diabetics and non-diabetics. This is in contrary to the previously published report that the survival rate for PD patients 55 years or less was not significantly different between diabetic and non-diabetic patients [26]. Five year actuarial survival rates of diabetic PD patients have been reported to be lower than that of age matched non-diabetic PD patients [27]. Diabetes is a strong independent negative predictor of long term survival in both HD and PD patients.

The results of this study clearly demonstrate a significant difference in nutritional status between diabetic and non-diabetic patients on HD and PD (Tables 2 and 8). Serum levels of albumin, creatinine and prealbumin were significantly lower in diabetic HD and PD patients compared to non-diabetic HD and PD patients. This indicates a poorer nutritional status in the former group, this may contribute to higher mortality in diabetic dialysis patients. The results of this study confirm our previously published papers [22]. Lowrie et al. suggested that higher mortality in diabetic HD patients could be explained by the poorer nutritional status in diabetic HD patients [28]. By logistic regression analysis, diabetes and female sex have been reported to be the strongest predictors of moderate and severe malnutrition in PD patients [29]. Much of the difference in survival rates between diabetics and non-diabetics can be accounted for by the poorer nutritional status in the former group [30]. Serum albumin, a marker for visceral protein status, has been reported to be the strongest predictor of mortality in dialysis patients [10, 31]. Diabetes is one of the strongest predictor of low serum albumin level in PD patients [32, 33]. It has been recently reported that serum albumin levels of diabetic HD patients were significantly lower than that of non-diabetics [34]. In one study, after 24 months on PD, diabetic patients showed significant decrease in albumin [35]. Serum creatinine, a marker of somatic protein status has been shown to be a negative predictor of mortality in HD and PD patients [9, 10, 36, 37]. Diabetes as a cause of ESRD is a significant correlate of incident creatinine level in ESRD patients [37]. Serum creatinine level is affected by the deterioration of muscle mass and nutritional status. Lower serum creatinine levels in diabetic dialysis patients indicate poorer somatic protein status. It is now well established that, protein malnutrition is highly prevalent in HD and PD patients and is worse among diabetic patients. In addition to cardiovascular and other causes, malnutrition may be one of the most important factors contributing to higher mortality in diabetic dialysis patients. In one study on PD patients, 91% of diabetic and 76% of non-diabetic patients showed some degree of malnutrition with serum albumin, creatinine were significantly lower in diabetic patients compared to non-diabetic patients [29]. Higher incidence of mild to moderate malnutrition in diabetics than in non-diabetics have been reported by Young et al. [38]. Among predialysis chronic renal failure patients, diabetics had worse nutrition than non-diabetics as indicated by lower serum albumin, transferrin, prealbumin, IGF-1. Diabetic patients had more severe proteinuria and less dietary intake than non-diabetic patients [39]. Zimmerman et al reported that albumin, predialysis BUN and predialysis serum cholesterol are significant risk factors for diabetic patient survival [26].

Prealbumin, a serum protein with low turnover rate and small pool size, is a highly sensitive nutritional marker in general population [40]. We have reported for the first time the prognostic importance of prealbumin in HD and PD patients [11, 41–43]. Recently, Chertow et al reported that in HD patients, serum prealbumin provides prognostic value independent of serum albumin and other established predictors of mortality in this population [44]. Prealbumin concentration was significantly lower among patients with diabetes compared to those without diabetes [44]. Lower levels of this highly sensitive nutritional marker in diabetic HD and PD patients further confirms that diabetic patients are more malnourished than non-diabetic patients.

It is interesting to note that race is an important risk factor for mortality in diabetic HD patients which is in agreement with our previously published report [22]. Higher mortality rate of white diabetic ESRD patients compared to patients from other racial backgrounds has been previously reported [45]. Owen also reported the improved survival enjoyed by blacks with ESR [46]. The reason for greater risk of death in white diabetic HD patients than blacks or Hispanics is unknown. Ethnicity may be a predictor of survival in type 1 but not in type 2 diabetes [47]. In the present study, we cannot draw any conclusion due to lack of information regarding the type of diabetes our patients possess. When we analyzed demographics and biochemical markers by race, however, white patients were significantly older than Hispanic and black patients

(65.9 ± 16.0 for white, 58.6 ± 14.8 for black and 56.9 ± 14.3 years ± SD for Hispanics, $P < 0.0001$). Findings also reveal that enrollment serum creatinine (9.3 ± 4.0 for whites, 11.7 ± 4.9 for blacks and 10.2 ± 4.6 for Hispanics, $P < 0.0001$) and PTH (245 ± 350 pg/mL for whites, 433 ± 494 pg/mL for blacks and 249 ± 281 pg/mL for Hispanics, $P = 0.003$) levels were significantly higher for blacks than whites.

Recently, several investigators have reported that new dialysis patients experience a progressive rise in serum albumin and creatinine [48, 49]. Serum levels of nutritional markers may increase or decrease during the entire study period depending on nutritional status of the patients and possibly other factors. We have compared six-month moving average values of serial measures of nutritional markers such as albumin, creatinine, and cholesterol between non-diabetic and diabetic HD patients. Six-month moving average serum albumin and serum creatinine values are significantly higher for non-diabetic patients compared to diabetic patients which confirms that diabetic patients are more malnourished than non-diabetic patients (Table 6).

Our earlier observations conclude that enrollment serum PTH was significantly lower in diabetic patients than non-diabetics. Hyperparathyroidism and elevated levels of serum PTH are common complications among dialysis patients [50, 51]. Elevated levels of PTH in dialysis patients have been associated with various abnormalities. We reported that lower enrollment PTH in HD and PD patients are associated with higher mortality [52]. Also, PTH is directly correlated with nutritional markers such as albumin, creatinine and prealbumin [52, 53]. Recently other investigators have also reported that lower PTH levels reflect poor nutritional status and become an indicator for the risk of mortality in dialysis patients [54–56]. In contrast, Chertow et al recently reported that, with adequate adjustment for case mix and nutritional indicators, there is a direct relationship between serum PTH and mortality [53].

The factors which contribute to higher degree of malnutrition in diabetic patients similar to those of non-diabetic dialysis patients include inadequate dietary intake, impaired taste acuity, metabolic acidosis, comorbid diseases, and increased nutrient requirement due to the loss of proteins and amino acids during dialysis procedure. Diabetic dialysis patients suffer from more comorbid diseases such as heart disease than non-diabetic dialysis patients. One study reported that comorbid disease is the most important determinant of nutritional status. Protein synthesis may be depressed in poorly controlled diabetes mellitus [57]. Gastroparesis associated with diabetes mellitus may further contribute to malnutrition [58]. Because low-protein diets may help to slow the progression of diabetic nephropathy, inadequate intake in diabetic patients can also be caused by restrictive diets. By the time dialysis is started, these diabetic ESRD patients may be suboptimally nourished due to restrictive diet. Diabetic dialysis patients cannot tolerate dialysis as well as non-diabetic dialysis patients.

In summary, the prevalence of malnutrition is significantly higher in diabetic HD and PD patients compared to non-diabetic patients. This is the major contributory factor to high mortality rate in these patients. Blood pressure control, maintenance of normal blood glucose and body weight, reduction of atherogenic lipid levels and lipoproteins, hyperhomocysteinemia prevention, and nutritional wasting prevention, are important factors for improving survival in diabetic dialysis patients. Diet is the key component for the medical management of dialysis patients with diabetes. Serial monitoring of highly sensitive nutritional parameters such as serum prealbumin level with prompt dietary intervention is essential in maintaining the patients' nutritional status.

ACKNOWLEDGEMENTS

This work was supported in part by grants from the National Kidney Foundation of New York/New Jersey and the Nephrology Foundation of Brooklyn.

REFERENCES

1. King H, Aubert RE, Herman WH. Global burden of diabetes, 1995–2025: prevalence, numerical estimates, and projections. Diabetes Care 1998; 21: 1414–1431.
2. Mokdad AH, Ford ES, Bowman BA, et al. Diabetes trends in the US: 1990–1998. Diabetes Care 2000; 23: 1278–1283.
3. American Diabetes Association: Economic consequences of diabetes mellitus in the US Diabetes Care 1998; 21: 296–309.
4. US Renal Data System: USRDS 2000 Annual Data Report. The National Institutes of Health, National Institute of Diabetes and Digestive and Kidney Diseases, Bethesda, MD 2000.
5. US Renal Data System: USRDS 1999 Annual Data Report. The National Institutes of Health, National Institute of Diabetes and Digestive and Kidney Diseases, Bethesda, MD 1999.
6. Hakim RM, Levin N. Malnutrition in hemodialysis patients. Am J Kidney Dis 1993; 21: 125–137.
7. Bergstrom J. Why are dialysis patients malnourished? Am J Kidney Dis 1995; 26: 229–241.
8. Acchiardo SR, Moore LW, Latour PA. Malnutrition as the main factor in morbidity and mortality of dialysis patients. Kidney Int 1983; 24 (Suppl): S199–S203.
9. Avram MM, Mittman N, Bonomini L, et al. Markers for survival in dialysis: a seven year prospective study. Am J Kidney Dis 1995; 26: 209–219.
10. Lowrie EG, Lew NL. Death risk in hemodialysis: the predictive value of commonly measured variables and an evaluation of death rate differences between facilities. Am J Kidney Dis 1990; 15: 458–463.
11. Avram MM, Goldwasser P, Erroa M, et al. Predictors of survival in continuous ambulatory peritoneal dialysis patients: the importance of prealbumin and other nutritional and metabolic markers. Am J Kidney Dis 1994; 23: 91–98.
12. Avram MM, Sreedhara R, Avram DK, et al. Enrollment parathyroid hormone level is a new marker of survival in hemodialysis and peritoneal dialysis therapy for uremia. Am J Kidney Dis 1996; 28: 924–930.
13. Avram MM, Goldwasser P, Derkatz D, et al. Correlates of long-term survival on hemodialysis. In: Friedman EA (ed), Death on Hemodialysis: Preventable or Inevitable? Kluwer Academic Publishers, 1994, pp 169–176.
14. Avram MM, Sreedhara R, Fein PA, et al. Survival of hemodialysis and peritoneal dialysis over 12 years with emphasis on nutritional parameters. Am J Kidney Dis 2001; 37: S77–S80.
15. Avram MM, Bonomini LV, Sreedhara R, et al. Predictive value of nutritional markers (albumin, creatinine, cholesterol, and hematocrit) for patients on dialysis for up to 30 years. Am J Kidney Dis 1996; 28: 910–917.
16. Avram MM. Long-term survival in end-stage renal disease. Dialysis Transplant 1998; 27: 11–24.
17. Mittman N, Avram MM, Thu T, et al. Predictors of mortality in hemodialysis (HD): analysis of serial data over six years. J Am Soc Nephrol 1996; 7 (Abstract): 1523–1523.
18. Port FK. Worldwide demographics and future trends in end-stage renal disease. Kidney Int 1993; 41: S4–S7.
19. Ritz E, Rychlik I, Lcatelli F, et al. End-stage renal failure in type 2 diabetes: a medical catastrophe of worldwide dimensions. Am J Kidney Dis 1999; 34: 795–808.
20. Rodriguez JA, Cleries M, Vela E. Diabetic patients on renal replacement therapy: analysis of Catalan. Nephrol Dial Transplant 1997; 12: 2501–2509.
21. Avram MM. Proc. Conference on Dialysis as a 'Practical Workshop'. National Dialysis Committee 15, 1966 (Abstract).
22. Avram MM. Dialysis in diabetic patients: three decades of experience of, from 1964 to 1997. In: Friedman EA, L'Esperance Jr FA (eds), Diabetic Renal-Retinal Syndrome. 21st Century Management Now. Kluwer Academic Publishers, 1998, pp 67–78.
23. Kaplan EL, Meier P. Nonparametric estimation from incomplete observations. J Am Stat Assoc 1958; 53: 457–481.

24. Cox DR. Regression models and life tables (with discussion). J R Stat Soc Series B 1972; 34: 187–220.
25. Kelbfleisch J, Prentice R. The proportional hazards model. The statistical analysis of failure time data. Wiley, New York, 1980, pp 70–117.
26. Zimmmerman SW, Oxton LL, Bidwell D, et al. Long-term outcome of diabetic patients receiving peritoneal dialysis. Perit Dial Int 1996; 16: 63–68.
27. Passadakis P, Thodis E, Vargemezis V, et al. Long-term survival on peritoneal dialysis in end-stage renal disease owing to diabetes. Adv Perit Dial 2000; 16: 59–66.
28. Lowrie EG, Lew NL, Huang WH. Race and diabetes as death risk predictors in hemodialysis patients. Kidney Int 1992; 42 (Suppl): S22–S31.
29. Espinosa A, Cueto-Manzano AM, Velazquez-alva C, et al. Prevalence of malnutrition in Mexican CAPD diabetic and non-diabetic patients. Adv Perit Dial 1996; 12: 302–306.
30. Leehey DJ. Hemodialysis in the diabetic patient with end-stage renal disease. Ren Fail 1994; 16: 547–553.
31. Goldwasser P, Mittman N, Antignani A, et al. Predictors of mortality in hemodialysis patients. J Am Soc Nephrol 1993; 3: 1613–1622.
32. Spiegel DM, Anderson M, Campbell U, et al. Serum albumin: a marker for morbidity in peritoneal dialysis patients. Am J Kidney Dis 1993; 21: 26–30.
33. Malhotra D, Tzamaloukas AH, Murata GH, et al. Serum albumin in continuous peritoneal dialysis: its predictors and relationship to urea clearance. Kidney Int 1996; 50: 243–249.
34. Biesenbach G, Debska-Slizien A, Zazgornik J. Nutritional status in type 2 diabetic patients requiring haemodialysis. Nephrol Dial Transplant 1999; 14: 655–658.
35. Scanziani R, Dozio B, Bonforte G, et al. Nutritional parameters in diabetic patients on CAPD. Adv Perit Dial 1996; 12: 280–283.
36. US Renal Data System: USRDS 1992 Annual Data Report. The National Institutes of Health, National Institute of Diabetes and Digestive and Kidney Diseases, Bethesda, MD 1992.
37. Fink JC, Burdick RA, Kurth SJ, et al. Significance of serum creatinine values in new end-stage renal disease patients. Am J Kidney Dis 1999; 34: 694–701.
38. Young GA, Kopple JD, Lindholm B, et al. Nutritional assessment of continuous ambulatory peritoneal dialysis patients: an international study. Am J Kidney Dis 1991; 17: 462–471.
39. Park JS, Jung HH, Yang WS, et al. Protein intake and the nutritional status in patients with pre-dialysis chronic renal failure on unrestricted diet. Korean J Int Med 1997; 12: 115–121.
40. Ingenbleek Y, De Visscher M, De Nayer P. Measurement of prealbumin as index of protein-calorie malnutrition. Lancet 1972; 2: 106–109.
41. Avram MM, Sreedhara R, Mittman N. Prealbumin is a strong predictor of nutritional risk and five year mortality in dialysis patients. ASAIO, 1997 (Abstract).
42. Sreedhara R, Avram MM, Blanco M, et al. Prealbumin is the best nutritional predictor of survival in hemodialysis and peritoneal dialysis. Am J Kidney Dis 1996; 28: 937–942.
43. Goldwasser P, Michel M-A, Collier J, et al. Prealbumin and lipoprotein (a) in hemodialysis: relationships with patient and vascular access survival. Am J Kidney Dis 1993; 22: 215–225.
44. Chertow GM, Ackert K, Lew NL, et al. Prealbumin is as important as albumin in the nutritional assessment of hemodialysis patients. Kidney Int 2000; 58: 2512–2517.
45. Pugh JA, Tuley M, Basu S. Survival among Mexican Americans, non-Hispnic whites, and African Americans with end-stage renal disease: the emergence of a minority pattern of increased incidence and prolonged survival. Am J Kidney Dis 1994; 23: 803–807.
46. Owen WF, Jr. Racial differences in incidence, outcome, and quality of life for African-Americans on hemodialysis. Blood Purif 1996; 14: 278–285.
47. Medina RA, Pough JA, Monterrosa A, et al. Minority advantage in diabetic end-stage renal disease survival on hemodialysis: due to different proportions of diabetic type? Am J Kidney Dis 1996; 28: 226–234.
48. Culp K, Flanigan M, Lowrie EG, et al. Modelling mortality risk in hemodialysis patients using laboratory values as time dependent covariates. Am J Kidney Dis 1996; 5: 741–746.
49. Parker TF, Wingard RL, Ikizler TA, et al. Effect of membrane biocompatibility on nutritional parameters in chronic hemodialysis patients. Kidney Int 1996; 49: 551–556.
50. Llach F, Massry SG. On the mechanism of secondary hyperparathyroidism in moderate renal insufficiency. J Clin Endocrinol Metab 1985; 61: 601–606.
51. Delmez JA, Slatopolsky E. Hyperphosphatemia: Its consequences and treatment in chronic renal failure. Am J Kidney Dis 1992; 19: 303–317.

52. Avram MM. Parathyroid hormone as a predictor of mortality in hemodialysis patients. Nephrology Exchange 1998; 8: 2–6.
53. Chertow GM, Lazarus JM, Lew NL, et al. Intact parathyroid hormone is directly related to mortality in hemodialysis. J Am Soc Nephrol 2000; 11 (Abstract): 573A–573A.
54. Akizawa T, Kinugasa E, Kurihara S, et al. Parathyroid hormone deficiency is an indicator of poor nutritional state and prognosis in dialysis patients. J Am Soc Nephrol 1999; 10 (Abstract): 615A–615A.
55. Heaf JG, Lokkegard H. Parathyroid hormone during maintenance dialysis: influence of low calcium dialysate, plasma albumin and age. J Nephrol 1998; 11: 203–210.
56. Coco M, Rush H. Increased incidence of hip fractures in dialysis patients with low serum parathyroid hormone. Am J Kidney Dis 2000; 36: 1115–1121.
57. Davis M, Comty C, Shapiro F. Dietary management of patients with diabetes treated by hemodialysis. J Am Dietet Assoc 1979; 75: 265–269.
58. Scarpello HB, Barber DC, Hague RV, et al. Gastric emptying of solid meals in diabetics. Br Med J 1976; 2: 671–673.

DAVID A. ANTONETTI AND THE PENN STATE RETINA RESEARCH GROUP

3. New Insights into the Molecular Mechanisms of Vascular Permeability in Diabetes

Editor's Comment: Increased vascular permeability across the 'blood-retinal barrier' is viewed as the first step in development of macular edema, angiogenesis, and proliferative diabetic retinopathy. Antonetti weighs the evidence that various permeabilizing growth factors induce this greater permeability and infers that vascular endothelial growth factor/vascular permeability factor (VEGF/VPF) is the most likely candidate. Noting the 'critical role' for nitric oxide synthase (NO synthase) activity in mediating VEGF/VPF induced vascular permeability, the possibility of blocking VEGF/VPF activation of NO synthase as a therapeutic intervention is raised underscoring the coming strategy of molecular intervention to preempt diabetic complications.

INTRODUCTION

Diabetic retinopathy is a leading cause of blindness in working age people in the United States and contributes significantly to vision loss in the young and elderly. The increasing rate of incidence of diabetes will further impact the visual health of this country unless new treatment modalities are discovered to prevent and cure diabetic retinopathy.

In humans, the metabolic support for the inner retina comes from an arteriole/capillary and post-capillary vascular network that traverses the ganglion cell layer and the outer plexiform layer, while support for the outer retina is achieved by diffusion from the vascularized choroid across the retinal pigmented epithelium (RPE). Together these retinal vessels and the RPE form the blood-retinal barrier (BRB). Breakdown of the BRB is a hallmark of diabetic retinopathy and 20–25% of patients develop macular edema [1–3]. Indeed, vascular permeability remains the pathophysiologic event most closely related to vision loss and precedes angiogenesis that occurs later in the course of the disease. Loss of the BRB also occurs in a number of other diseases including retinal vein occlusion, uveoretinitis, age related macular degeneration, and retinopathy of prematurity. In this review I will discuss some of our current understanding of the molecular mechanisms that mediate changes in vascular permeability. Specifically, I will focus on changes that occur to the tight junctions of the vascular endothelium, which may elicit retinal edema. These changes include a rapid increase in post-translational modifications, redistribution from the cell border to the cell interior and overtime, a decrease in tight junction protein content. Finally, I will discuss possible treatment approaches to reverse these changes to tight junctions and maintain the blood-retinal barrier.

E.A. Friedman and F.A. L'Esperance, Jr. (eds.), Diabetic Renal-Retinal Syndrome,, 23–33.

AGENTS THAT INCREASE VASCULAR PERMEABILITY IN DIABETES

The increase in permeability and angiogenesis that occurs in diabetic retinopathy results, at least in part, from an increase in permeabilizing growth factors. These growth factors include insulin like growth factor 1 and its binding proteins, platelet derived growth factor, fibroblast growth factor and vascular endothelial growth factor/vascular permeability factor (VEGF/VPF) (reviewed in [4–7]). Of these, the potent permeabilizing agent VEGF/VPF has received the most attention. The expression of VEGF/VPF and its receptors increases by 6 months of experimentally induced diabetes [8–10]. In Goto-Kakizaki rats, a model of type II diabetes, the concentration of hormone is significantly elevated over control by 28 weeks. Measurement of VEGF/VPF mRNA by in situ hybridization confirms that VEGF/VPF gene expression occurs in the retinal parenchyma proper and is increased in diabetic rats in the ganglion cell and inner nuclear layers [9]. In addition, VEGF/VPF content increases in patients with proliferative diabetic retinopathy in the vitreous fluid [11, 12] and in epiretinal membranes [13]. Furthermore, several animal studies and two studies in humans reveal that VEGF/VPF expression increases in the retina before proliferative retinopathy, suggesting a specific role in vascular permeability [14, 15]. However, in a study of postmortem control and diabetic retinas with nonproliferative retinopathy, the authors claim that while constitutive VEGF/VPF expression was detectable, there was no significant increase as a result of diabetes [16]. Yet the preponderance of evidence indicates that VEGF/VPF has a role in proliferative diabetic retinopathy and this hormone may also contribute to retinal vascular permeability. It is interesting to note that in a number of these studies VEGF and its receptors were identified in control retinas suggesting a role for this hormone in normal ocular function.

Inhibitor studies and transgenic models have elegantly provided evidence that VEGF/VPF is both necessary and sufficient to cause angiogenesis and vascular permeability in the retina. When mouse pups are placed in a hyperoxic environment and then returned to a normoxic environment, vascular permeability and angiogenesis are stimulated, similar to that observed in retinopathy of prematurity. Treating these mice with a soluble form of the VEGF receptor, injected into the vitreous cavity of the eye, partially blocks angiogenesis [17]. Activation of the VEGF receptor tyrosine kinase increases cytosolic calcium and activates protein kinase C β in bovine aortic endothelial cells [18]. Intraocular injection of VEGF/VPF stimulates vascular permeability to fluorescein in the retina, which is inhibited by oral treatment with the protein kinase C β inhibitor LY333531 [19]. In addition, the more broad based inhibitor, CGP 41251, which blocks PKC as well as VEGF/VPF and PDGF receptor kinase activity, completely abolishes angiogenesis in a mouse model of ischemic retinopathy [20]. Evidence that VEGF/VPF is sufficient to induce retinal angiogenesis derives from the observation that transgenic mice that overexpress VEGF/VPF in the rod cells develop marked vascular growth from the retinal capillary network into the rod layer, which is normally avascular [21, 22].

Physical forces may also contribute to changes in vascular permeability in diabetic retinopathy. Shear stress is defined as the tangential force of fluid on the vessel wall and may profoundly effect vascular permeability. Endothelial cell culture placed under a rotating disk has been used to model step changes in shear stress. Using this model system, a rapid increase in water flux across endothelial monolayers was observed in response to shear stress [23]. Fluid flow and fluid viscosity both effect shear stress and while fluid flow decreases in the retina of patients with diabetes [24, 25] and in animal models of diabetes [26], blood viscosity increases [27], and could contribute to changes in endothelial wall shear stress. Finally, in addition to growth factors produced by the neural retina, permeabilizing agents may be brought to the retina via an inflammatory response. Exciting new work reveals increased leukostasis in diabetic retinopathy that can be inhibited by antibodies to ICAM-1 [28]. Thus, recruitment of leukocytes to retinal vessels and an ensuing inflammation response may also contribute to edema in diabetic retinopathy.

TRANSCELLULAR OR PARACELLULAR PERMEABILITY

Water and solute flux may occur through two basic pathways across a cell monolayer: either the cells remain in close apposition and flux occurs through the cells, i.e. transcellular flux, or the cell junctions change and allow flux between the cells, or para-

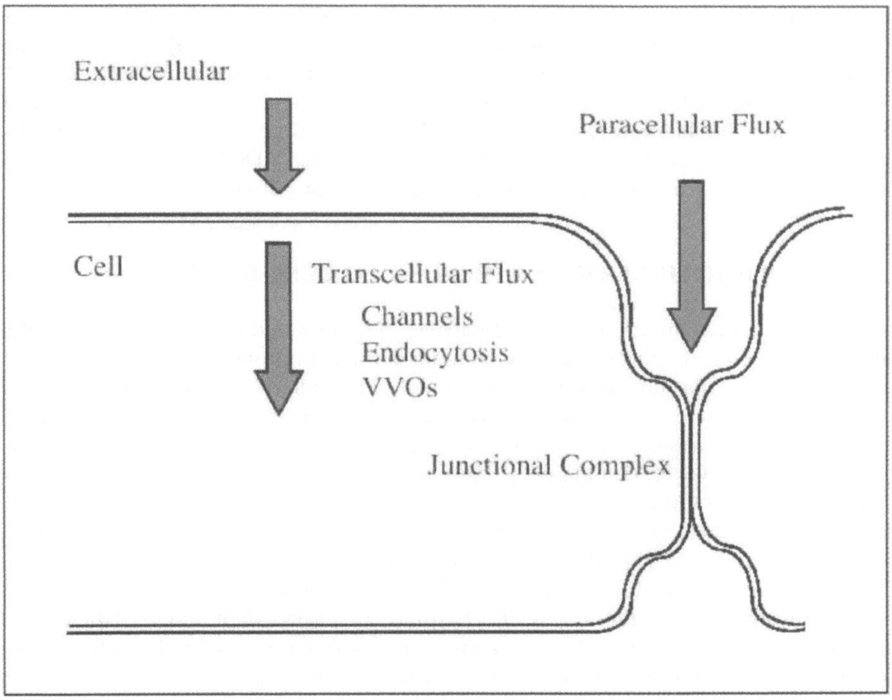

Figure 1. The flow of water and solutes across the endothelial barrier may either occur through the cell (transcellular) or by around the cell (paracellular). Transcellular flux may employ channels, endocytosis or vesiculo-vacuolar organelles (VVOs) while paracellular flux implies a change to the junctional complex that seal cells together.

cellular flux. Transcellular flux may occur through numerous mechanisms. Transporters such as the Glut-1 glucose transporters move molecules across the endothelium in the retina. Alternatively, VEGF-stimulated permeability has been proposed to occur as a result of the formation of vesicle organelle structures that have been termed vesiculo-vacuolar organelles, or VVOs (reviewed in [29–32]). These organelles have been described as grape-like clusters of large vacuoles, which are believed to allow the passage of molecules through the endothelium. However, the number or size of the VVOs does not change in response to VEGF, so if permeability is regulated through VVOs, it must occur through regulation of the interaction between the vacuoles [32]. Recently, Feng et al. proposed that VEGF stimulates transcellular flux across retinal endothelial cells in culture [33]. Using electron microscopy the authors show that VEGF/VPF administration increases horseradish peroxidase entry into cells two-fold. Fluorescent labeling co-localizes the HRP-containing domains with caveolin, providing evidence that caveolae mediate transcytosis. However, the co-localized label appears strongest in a perinuclear location, possibly in the Golgi network. Since caveolin also co-localizes with VEGF receptor, it is unclear whether the HRP-containing vesicles actually represent transcytosis or, alternatively, result from internalization of the VEGF/VPF receptor to the Golgi network. Thus, the debate over the relative contributions of transcellular and paracellular permeability continues. Indeed, it is likely that endothelial cells use multiple mechanisms to regulate flux of various molecules across the BRB. Strong evidence exists in the literature for the role of tight junctions in barrier formation and our laboratory has developed data revealing both acute and chronic changes to tight junctions in response to VEGF/VPF and PDGF commensurate with changes in permeability indicating a role for paracellular permeability.

TIGHT JUNCTIONS GIVE RISE TO CELL BARRIERS

Cell to cell interactions form regulated barriers that allow retinal capillary endothelial cells and pigmented epithelial cells to maintain defined environments within the retina. A complex of cellular junctions bring about cell to cell interactions through a series of structures that includes zonula occludens or tight junctions, zonula adherens or adherens junctions, and desmosomes [34] as well as gap junctions. The endothelial cells of the retinal vessels, however, contain only tight junctions interspersed with desmosomes as observed by electron microscopy [35, 36]. Freeze fracture electron microscopy reveals the zonula occludens as a series of anastamosing fibrils that encircle the most apical region of the cell [34, 37]. The tight junctions create a selective barrier to water and solutes, blocking paracellular flux across the tissue and preventing lipids and proteins from diffusing between the apical and basolateral plasma membranes. Horseradish peroxidase, used as a stain in electron microscopy, diffuses only up to the tight junction in brain cortical capillaries while in other tissues without tight junctions, this marker diffuses out of the vascular lumen [38]. Similar studies in the retina reveal that tight junctions constitute the BRB, preventing solute flux into the retinal parenchyma [35, 36].

Molecular biologists have identified a number of proteins that reside at the tight junction and make important contributions to the barrier properties conferred by these junctions. These proteins include at least 8 structural proteins: Zonula Occludens 1, 2 and 3 (ZO-1, ZO-2, ZO-3), occludin, cingulin, 7H6, symplekin, and claudins (reviewed in [39–42]. Tight junctions also exhibit a range of proteins such as Src, Rab, and PKC, which potentially regulate barrier permeability (Figure 2). Actin filaments bind to ZO-isoforms and occludin, suggesting a role for cytoskeletal regulation of tight junctions.

Several lines of evidence suggest that occludin, a transmembrane protein located exclusively at tight junctions, plays a crucial role in barrier permeability (recently reviewed in [43]). Hydrophobicity plots from several animal species predict that occludin spans the plasma membrane at four regions presenting two extracellular loops [44]. Occludin tissue expression and content correlate well with barrier properties [45]. Over-expression of occludin increases transendothelial electrical resistance in MDCK cells [46] and confers adhesiveness in fibroblasts [47] while microinjection of occludin increases the barrier properties of *Xenopus* oocytes [48]. Furthermore, antisense oligonucleotides to occludin decrease barrier properties to solute flux in arterial endothelial cells [49]. Tight junctions contain other transmembrane proteins in addition to occludin. Genetic ablation of occludin has led to the discovery of a novel tight junction protein family termed claudins [50, 51]. The claudins include at least 16 members and although they are structurally similar to occludin, sequence analysis places occludin in a separate gene family. The diversity of the claudin gene family allows for the possibility that varied claudin expression provides distinct character to that particular tissue's tight junctions, yielding either greater or lesser barrier properties. In support of this model, expression of claudin isoforms in fibroblasts produces cell to cell adhesion between homotypic claudins and particular pairs of hetertypic claudins [52]. Endothelial cells express claudin 5 [53] but any particular junctional properties provided by this claudin remain obscure. The emergence of the claudin gene family has stimulated a great deal

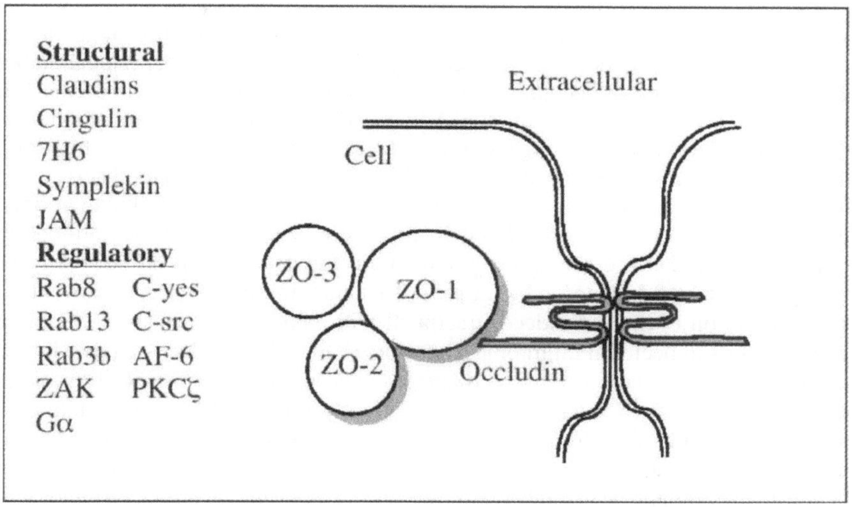

Figure 2. Tight junction. Presented is a schematic of known structural tight junction proteins and regulatory proteins found associated with tight junctions.

of research; however, the functional importance of occludin was recently highlighted by mice with occludin gene deletion [54]. While the mice did maintain an electrical resistance barrier in intestinal epithelial cells they revealed a complex array of pathologies including reduced growth rate, calcification of the brain, and testicular atrophy with male sterility.

The molecular structure of the ZO proteins indicates a functional capacity for multiple protein/protein interactions and this family of proteins may organize tight junctions. ZO-1 belongs to the membrane associated guanylate kinase (MAGuK) family of proteins. Like other MAGuK family members, ZO-1 contains an SH3 domain and three PDZ domains [41, 55, 56]. In vitro binding studies have revealed that ZO-1 can interact directly with occludin [57] as well as ZO-2 and ZO-3 [58] and mutational analysis by Fanning et al. has identified the sites of occludin interaction as the region of homology to guanylate kinase [59]. Other MAGuK family members that utilize PDZ binding domains act as protein nucleation centers, such as the organization of glutamate receptor and potassium channels by the MAGuK family member PSD95 [60]. Because of its multiple protein/protein interaction domains, we and others hypothesize that ZO-1 has a role in organizing tight junctions. In addition, ZO-1 has been reported to bind to connexin 43 of gap junctions [61] as well as α-catenin of adherens junctions [62]. Thus, ZO-1 may in fact, play a role in the organization of the larger junctional complex. In conclusion, tight junctions are complex cell structures that contribute significantly to tissue barrier properties. Here I propose that tight junction composition and organization are both regulated by abnormal levels of growth factors to induce paracellular permeability in diabetic retinopathy.

REGULATION OF TIGHT JUNCTIONS

Vascular endothelial growth factor causes both acute and long term changes to tight junctions that may alter endothelial permeability. Three months of experimental diabetes in rats causes a decrease in occludin content coincident with an increase in permeability to occludin [63]. Furthermore, addition of VEGF/VPF to bovine retinal endothelial cells [63] or brain microvessel endothelial cells reduces occludin content [64]. In addition to the chronic effect of VEGF/VPF on occludin content, we have also observed rapid post-translational modifications to occludin in response to this permeabilizing hormone. Addition of VEGF/VPF to endothelial cells in culture or injection of the hormone into the eye of an anesthetized rat causes a shift in occludin migration on 10% polyacrylamide gels from ~60 kDa to ~62 kDa [65]. This shift in migration is consistent with a phosphorylation event and indeed, treatment with alkaline phosphatase completely ablates this shift in occludin migration. Therefore, VEGF/VPF induces a rapid increase in occludin phosphorylation. Finally, diabetes induces a rapid reorganization of tight junctions; specifically, 4 months of diabetes greatly reduces occludin content in the capillaries of the outer plexiform layer and causes a dramatic reorganization of occludin from the cell border to the cell interior in arterioles within the ganglion cell layer [66]. This reorganization appears to include a punctate cytoplasmic staining and may represent internalization of occludin within specific microdomains. Indeed, others have observed reorganization of occludin from the cell border to the cell cytoplasm in response to VEGF/VPF in human umbilical vein endothelial cell culture [67]. In conclusion, VEGF/VPF induces rapid changes in the occludin phosphorylation state and occludin distribution while over time, VEGF/VPF treatment decreases occludin content. Diabetes, likewise induces redistribution of occludin from the cell border and decreases occludin content, current studies are underway to determine whether diabetes also increases occludin phosphorylation state. It is clear tight junction proteins respond dynamically to VEGF/VPF addition and these alterations may be responsible for the increase in paracellular permeability after hormone addition. Furthermore, VEGF/VPF alone is sufficient to regulate paracellular permeability and induce many of the changes that occur to tight junctions in the retinal blood vessels during diabetes. However, it remains unclear whether VEGF/VPF is solely responsible for these alterations and if VEGF/VPF antagonists will be sufficient to prevent vascular permeability in diabetic retinopathy.

FUTURE STUDIES AND DEVELOPMENT OF TREATMENT OPTIONS

In order to develop new modalities to treat vascular permeability associated with diabetic retinopathy a number of important questions remain to be answered. While changes in occludin phosphorylation, distribution and content are associated with changes in permeability the molecular mechanism of vascular permeability remains obscure. For example, occludin phosphorylation may allow redistribution of occludin from the cell border in a discrete microdomain that eventually fuses with lysosomes leading to occludin degradation. In this model, removal of occludin from the cell border is hypothesized to act as the critical event in allowing paracellular permeability. However, an alternative hypothesis may postulate that occludin phosphorylation causes a conformational change to the protein to form a channel that allows paracellular flux. Such a paracellular channel in tight junctions has, in fact, been shown to exist in the kidney. Paracellin, a member of the claudin gene family is a renal tight junction protein mediating paracellular Mg^{+2} resorption in the thick ascending loop of Henle [68]. This is the first example of a specific paracellular channel through tight junctions and may herald a new understanding of molecule permeability across tight junction-mediated tissue barriers.

Other questions that need to be addressed to develop therapeutic options to treat macular edema relate to the signaling mechanisms that regulate alterations to tight junctions. Towards this end new research has begun to shed light on the signaling pathways employed by VEGF/VPF to regulate vascular permeability. In addition to PKC β activation discussed above, additional signaling pathways have been identified that are essential to regulating permeability. A recent report revealed a unique role for the soluble tyrosine kinase c-Src in mediating vascular permeability in response to VEGF/VPF but not angiogenesis. Mice with c-Src or c-Yes gene deletions no longer responded to VEGF/VPF with increased vascular permeability but did respond with increased angiogenesis [69]. Our research as well as that of others have revealed a critical role for nitric oxide synthase (NO synthase) activity in mediating VEGF/VPF induced vascular permeability [70–72]. Activation of endothelial NO synthase may occur downstream of Src activation. Using the Src kinase specific inhibitor PP2, VEGF/VPF activation of NO synthase was inhibited [73].

In the 'Diabetic Renal-Retinal Syndrome: Proactive not Reactive Intervention' 2001 conference, we present data revealing hydrocortisone directly effects endothelial cell tight junctions and dramatically increases barrier properties of endothelial monolayers. Hydrocortisone has been used for over 35 years to treat patients with brain tumors in order to prevent vascular permeability and stroke. Both diseases represent a number of important similarities regarding endothelial barrier dysfunction. Both brain and retina are neural tissues with vascular endothelial cells containing highly organized tight junctions forming a strong blood-tissue barrier. Both diseases include increased VEGF/VPF associated with vascular permeability and angiogenesis. We have found that hydrocortisone acts in direct opposition to the effect of VEGF/VPF; namely, hydrocortisone increases occludin content, causes the organization of occludin and ZO-1 at the cell border and causes a rapid (4h) dephosphorylation of occludin. Commensurate with these observed changes to tight junctions hydrocortisone dramatically decreases solute and water flux across retinal endothelial cell monolayers. This data strongly supports our hypothesis that changes to occludin mediate changes to paracellular permeability. More importantly, elucidation of the molecular mechanism of hydrocortisone action may allow development of specific drugs that will induce barrier properties of endothelial cells without causing increased hepatic glucose output or inhibition of glucose uptake by muscle and adipose tissue.

The past five years of research in diabetic retinopathy has lead to a dramatic increase in our knowledge of the pathophysiology of this disease. We now know diabetic retinopathy includes a neurodegenerative component [74] (also see chapter in this volume by A. Barber) as well as an inflammatory component x. As discussed in this review, we have greatly advanced our understanding of the mechanisms of vascular permeability and angiogenesis. Future studies will continue to provide novel therapeutic targets so that a panel of effective treatment options can be developed to treat and prevent vision loss that results from diabetes.

REFERENCES

1. Vitale S, Maguire MG, Murphy RP, et al. Clinically significant macular edema in type I diabetes. Incidence and risk factors. Ophthalmology 1995; 102: 1170–1176.
2. Klein R, Klein BEK, Moss SE, Cruscishanks KJ. The Wisconsin epidemiologic study of diabetid retinopathy. XV. The long-term incidence of macular edema. Ophthalmology 1995; 102: 7–16.
3. Anonymous. Progression of retinopathy with intensive versus conventional treatment in the Diabetes Control and Complications Trial. Diabetes Control and Complications Trial Research Group. Ophthalmology 1995; 102: 647–661.
4. Paques M, Massin P, Gaudric A. Growth factors and diabetic retinopathy. Diabetes & Metabolism 1997; 23: 125–130.
5. Aiello LP. Vascular endothelial growth factor and the eye: biochemical mechanisms of action and implications for novel therapies. Ophthalmic Research 1997; 29: 354–362.
6. Miller JW. Vascular endothelial growth factor and ocular neovascularization [comment] [see comments]. American Journal of Pathology 1997; 151: 13–23.
7. Miller JW, Adamis AP, Aiello LP. Vascular endothelial growth factor in ocular neovascularization and proliferative diabetic retinopathy. Diabetes-Metabolism Reviews 1997; 13: 37–50.
8. Murata T, Nakagawa K, Khalil A, Ishibashi T, Inomata, H, Sueishi, K. The relation between expression of vascular endothelial growth factor and breakdown of the blood-retinal barrier in diabetic rat retinas. Laboratory Investigation 1996; 74: 819–825.
9. Gilbert RE, Vranes D, Berka JL, Kelly DJ, Cox A, Wu LL, Stacker SA, Cooper ME. Vascular endothelial growth factor and its receptors in control and diabetic rat eyes. Laboratory Investigation 1998; 78: 1017–1027.

10. Hammes HP, Lin J, Bretzel RG, Brownlee M, Breier G. Upregulation of the vascular endothelial growth factor/vascular endothelial growth factor receptor system in experimental background diabetic retinopathy of the rat. Diabetes 1998; 47: 401–406.
11. Aiello LP, Avery RL, Arrigg PG, Keyt BA, Jampel HD, Shah ST, Pasquale LR, Thieme H, Iwamoto MA, Park, JE, et al. Vascular endothelial growth factor in ocular fluid of patients with diabetic retinopathy and other retinal disorders. N Eng J Med 1994; 331: 1480–1487.
12. Adamis AP, Miller JW, Bernal MT, D'Amico DJ, Folkman J, Yeo TK, Yeo KT. Increased vascular endothelial growth factor levels in the vitreous of eyes with proliferative diabetic retinopathy. American Journal of Ophthalmology 1994; 118: 445–450.
13. Chen YS, Hackett, SF, Schoenfeld CL, Vinores MA, Vinores SA, Campochiaro PA. Localization of vascular endothelial growth factor and its receptors to cells of vascular and avascular epiretinal membranes. British Journal of Ophthalmology 1997; 81: 919–926.
14. Amin RH, Frank RN, Kennedy A, Eliott D, Puklin JE, Abrams GW. Vascular endothelial growth factor is present in glial cells of the retina and optic nerve of human subjects with nonproliferative diabetic retinopathy. Inv Ophthal Vis Sci 1997; 38: 36–47.
15. Lutty GA, McLeod DS, Merges C, Diggs A, Plouét J. Localization of vascular endothelial growth factor in human retina and choroid. Arch Ophthalmol 1996; 114: 971–977.
16. Gerhardinger C, Brown LF, Roy S, Mizutani M, Zucker CL, Lorenzi M. Expression of vascular endothelial growth factor in the human retina and in nonproliferative diabetic retinopathy. American Journal of Pathology 1998; 152: 1453–1462.
17. Aiello LP, Pierce EA, Foley ED, Takagi H, Chen H, Riddle L, Ferrara N, King G.L, Smith LE. Suppression of retinal neovascularization in vivo by inhibition of vascular endothelial growth factor (VEGF) using soluble VEGF-receptor chimeric proteins. Proc Natl Acad Sci 1995; 92: 10457–10461.
18. Xia P, Aiello LP, Ishii H, Jiang ZY, Park DJ, Robinson GS, Takagi H, Newsome WP, Jirousek MR, King GL. Characterization of vascular endothelial growth factor's effect on the activation of protein kinase C, its isoforms, and endothelial cell growth. Journal of Clinical Investigation 1996; 98: 2018–2026.
19. Aiello LP, Bursell SE, Clermont A, Duh E, Ishii H, Takagi C, Mori F, Ciulla TA, Ways K, Jirousek M, et al. Vascular endothelial growth factor-induced retinal permeability is mediated by protein kinase C in vivo and suppressed by an orally effective β-isoform-selective inhibitor. Diabetes 1997; 46: 1473–1480.
20. Seo MS, Kwak N, Ozaki H, Yamada H, Okamoto N, Yamada E, Fabbro D, Hofmann F, Wood JM, Campochiaro PA. Dramatic inhibition of retinal and choroidal neovascularization by oral administration of a kinase inhibitor. American Journal of Pathology 1999; 154: 1743–1753.
21. Okamoto N, Tobe T, Hackett SF, Ozaki H, Vinores MA, LaRochelle W, Zack DJ, Campochiaro PA, Transgenic mice with increased expression of vascular endothelial growth factor in the retina: a new model of intraretinal and subretinal neovascularization [see comments]. American Journal of Pathology 1997; 151: 281–291.
22. Tobe T, Okamoto N, Vinores MA, Derevjanik NL, Vinores SA, Zack, DJ, Campochiaro PA. Evolution of neovascularization in mice with overexpression of vascular endothelial growth factor in photoreceptors. Investigative Ophthalmology & Visual Science 1998; 39: 180–188.
23. Sill HW, Chang YS, Artman JR, Frangos JA, Hollis TM, Tarbell JM. Shear stress increases hydraulic conductivity of cultured endothelial monolayers. Am J Physiol 1995; 268: H535–H543.
24. Cuypers MH, Kasanardjo JS, Polak BC. Retinal blood flow changes in diabetic retinopathy measured with the Heidelberg scanning laser Doppler flowmeter. Graefes Arch Clin Exp Ophthalmol 2000; 238: 935–941.
25. MacKinnon JR, McKillop G, O'Brien C, Swa K, Butt Z, Nelson P. Colour Doppler imaging of the ocular circulation in diabetic retinopathy. Acta Ophthalmol Scand 2000; 78: 386–389.
26. Mori F, King GL, Clermont AC, Bursell DK, Bursell SE. Endothelin-3 regulation of retinal hemodynamics in nondiabetic and diabetic rats. Invest Ophthalmol Vis Sci 2000; 41: 3955–3962.
27. MacRury SM, Small M, MacCuish AC, Lowe GD. Association of hypertension with blood viscosity in diabetes. Diabetic Medicine 1988; 5: 830–834.
28. Miyamoto K, Khosrof S, Bursell SE, Rohan R, Murata T, Clermont AC, Aiello LP, Ogura Y, Adamis A.P. Prevention of leukostasis and vascular leakage in streptozotocin-induced diabetic retinopathy via intercellular adhesion molecule-1 inhibition. Proceedings of the National Academy of Sciences of the United States of America 1999; 96: 10836–10841.
29. Dvorak HF, Nagy JA, Feng D, Brown LF, Dvorak AM. Vascular permeability factor/vascular endothelial growth factor and the significance of microvascular hyperpermeability in angiogenesis. Curr Top Micro Immunol 1999; 237: 97–132.

30. Feng D, Nagy JA, Hipp J, Dvorak HF, Dvorak AM. Vesiculo-vacuolar organelles and the regulation of venule permeability to macromolecules by vascular permeability factor, histamine, and serotonin. Journal of Experimental Medicine 1996; 183: 1981–1986.
31. Feng D, Nagy JA, Hipp J, Pyne K, Dvorak HF, Dvorak AM. Reinterpretation of endothelial cell gaps induced by vasoactive mediators in guinea-pig, mouse and rat: many are transcellular pores. Journal of Physiology 1997; 504: 747–761.
32. Feng D, Nagy JA, Pyne K, Hammel I, Dvorak HF, Dvorak AM. Pathways of macromolecular extravasation across microvascular endothelium in response to VPF/VEGF and other vasoactive mediators. Microcirculation 1999; 6: 23–44.
33. Feng Y, Venema VJ, Venema RC, Tsai N, Behzadian MA, Caldwell RB. VEGF-induced permeability increase is mediated by caveolae. Investigative Ophthalmology & Visual Science 1999; 40: 157–167.
34. Farquhar MG, Palade G. J Cell Biol 1963; 17: 375.
35. Cunha-Vaz JG, Shakib M, Ashton N. Studies on the permeability of the blood-retinal barrier. I. On the existence, development, and site of a blood-retinal barrier. British Journal of Ophthalmology 1966; 50: 441–453.
36. Shakib M, Cunha-Vaz JG. Studies on the permeability of the blood-retinal barrier. IV. Junctional complexes of the retinal vessels and their role in the permeability of the blood-retinal barrier. Experimental Eye Research 1966; 5: 229–234.
37. Staehelin LA. Further observations on the fine structure of freeze-cleaved tight junctions. Journal of Cell Science 1973; 13: 763–786.
38. Reese TS, Karnovsky MJ. Fine structural localization of a blood-brain barrier to exogenous peroxidase. Journal of Cell Biology 1967; 34: 207–217.
39. Citi S, Cordenonsi M. Tight junction proteins. Biochimica and Biophysica Acta 1998; 1448: 1–11.
40. Denker BM, Nigam SK. Molecular structure and assembly of the tight junction. American Journal of Physiology 1998; 274: F1–9.
41. Mitic LL, Anderson JM. Molecular architecture of tight junctions. Annual Review of Physiology 1998; 60: 121–142.
42. Stevenson BR, Keon BH. The tight junction: morphology to molecules. Annu Rev Cell Dev Biol 1998; 14: 89–109.
43. Matter K, Balda MS. Occludin and the functions of tight junctions. International Review of Cytology 1999; 186: 117–146.
44. Ando-Akatsuka Y, Saitou M, Hirase T, Kishi M, Sakakibara A, Itoh M, Yonemura S, Furuse M, Tsukita S. Interspecies diversity of the occludin sequence: cDNA cloning of human, mouse, dog, and rat-kangaroo homologues. J Cell Biology 1996; 133: 43–47.
45. Hirase T, Staddon JM, Saitou M, Anod-Akatsuka Y, Itoh, M Furuse M, Fujimoto K, Tsukita S, Rubin LL. Occludin as a possible determinant of tight junction permeability in endothelial cells. J Cell Sci 1997; 110: 1603–1613.
46. McCarthy KM, Skare I, Stankewich MC, Furuse M, Tsukita S, Rogers RA, Lynch RD, Schneeberger E. Occludin is a functional component of the tight junction. J Cell Sci 1996; 109: 2287–2298.
47. Van Itallie CM, Anderson JM. Occludin confers adhesiveness when expressed fibroblasts. J Cell Sci 1997; 110: 1113–1121.
48. Chen Y, Merzdorf C, Paul DL, Goodenough DA. COOH terminus of occludin is required for tight junction barrier function in early xenopus embryos. J Cell Biol 1997; 138: 891–899.
49. Kevil CG, Okayama N, Trocha SD, Kalogeris TJ, Coe LL, Specian RD, Davis CP, Alexander JS. Expression of zonula occludens and adherens junctional proteins in human venous and arterial endothelial cells: role of occludin in endothelial solute barriers. Microcirculation 1998; 5: 197–210.
50. Furuse M, Fujita K, Hiiragi T, Fujimoto K, Tsukita S. Claudin-1 and -2: novel integral membrane proteins localizing at tight junctions with no sequence similarity to occludin. Journal of Cell Biology 1998; 141: 1539–1550.
51. Morita K, Furuse M, Fujimoto K, Tsukita S. Claudin multigene family encoding four-transmembrane domain protein components of tight junction strands. Proc Natl Acad Sci USA 1999; 96: 511–516.
52. Furuse M, Sasaki H, Tsukita S. Manner of interaction of heterogeneous claudin species within and between tight junction strands. Journal of Cell Biology 1999; 147: 891–903.
53. Morita K, Sasaki H, Furuse M, Tsukita S. Endothelial claudin: Claudin-5/TMVCF constitutes tight junction strands in endothelial cells. Journal of Cell Biology 1999; 147: 185–194.
54. Saitou M, Furuse M, Sasaki H, Schulzke JD, Fromm M, Takano H, Noda T, Tsukita S. Complex

phenotype of mice lacking occludin, a component of tight junction strands [in process citation]. Mol Biol Cell 2000; 11: 4131–4142.
55. Anderson JM, Itallie CMV. Tight junctions and the molecular basis for regulation of paracellular permeability. Am J Physiol 1995; 269: G467–G476.
56. Anderson JM, Fanning AS, Lapierre L, Van Itallie CM. Zonula occludens (ZO)-1 and ZO-2: membrane-associated guanylate kinase homologues (MAGuKs) of the tight junction. Biochemical Society Transactions 1995; 23: 470–475.
57. Furuse M, Itoh M, Hirase T, Nagafuchi S, Yonemura S, Tsukita S, Tsukita S. Direct association of occludin with ZO-1 and its possible involvement in the localization of occludin at tight junctions. J Cell Biol 1994; 127: 1617–1626.
58. Wittchen ES, Haskins J, Stevenson BR. Protein interactions at the tight junction – Actin has multiple binding partners, and ZO-1 forms independent complexes with ZO-2 and ZO-3. Journal of Biological Chemistry 1999; 274: 35179–35185.
59. Fanning AS, Jameson BJ, Jesaitis LA, Anderson JM. The tight junction protein ZO-1 establishes a link between the transmembrane protein occludin and the actin cytoskeleton. Journal of Biological Chemistry 1998; 273: 29745–29753.
60. Schillace RV, Scott JD. Organization of kinases, phosphatases, and receptor signaling complexes. Journal of Clinical Investigation 1999; 103: 761–765.
61. Giepmans BNG, Moolenaar WH. The gap junction protein connexin43 interacts with the second PDZ domain of the zona occludens-1 protein. Curr Biol 1998; 8: 931–934.
62. Itoh M, Nagafuchi A, Moroi S, Tsukita S. Involvement of ZO-1 in cadherin-based cell adhesion through its direct binding to a catenin and actin filaments. J Cell Biol 1997; 138: 181–192.
63. Antonetti D, Barber A, Khin S, Lieth E, Tarbell J, Gardner T, Group, a.t.P.S.R.R. Vascular permeability in experimental diabetes is associated with reduced endothelial occludin content. Diabetes 1998; 47: 1953–1959.
64. Wang W, Dentler, WL, Borchardt RT. 2001. VEGF increases BMEC monolayer permeability by affecting occludin expression and tight junction assembly. Am J Physiol Heart Circ Physiol 1998; 280: H434–440.
65. Antonetti DA, Barber AJ, Hollinger LA, Wolpert EB, Gardner TW. Vascular endothelial growth factor induces rapid phosphorylation of tight junction proteins occludin and zonula occluden 1. A potential mechanism for vascular permeability in diabetic retinopathy and tumors. J Biol Chem 1999; 274: 23463–23467.
66. Barber AJ, Antonetti DA, Gardner TW. Altered expression of retinal occludin and glial fibrillary acidic protein in experimental diabetes. Investigative Ophthalmology & Visual Science 2000; 41: 3561–3568.
67. Kevil CG, Payne DK, Mire E, Alexander JS. Vascular permeability factor/vascular endothelial cell growth factor-mediated permeability occurs through disorganization of endothelial junctional proteins. Journal of Biological Chemistry 1998; 273: 15099–15103.
68. Simon DB, Lu Y, Choate KA, Velazquez H, Al-Sabban E, Praga M, Casari C, Bettinelli A, Colussi C, Rodriguez-Soriano J, et al. Paracellin-1, a renal tight junction protein required for paracellular Mg2+ resorption. Science 1999; 285: 103–106.
69. Eliceiri BP, Paul R, Schwartzberg PL, Hood JD, Leng J, Cheresh DA. Selective requirement for Src kinases during VEGF-induced angiogenesis and vascular permeability. Mol Cell 1999; 4: 915–924.
70. Lakshminarayanan S, Antonetti DA, Gardner TW, Tarbell JM. Effect of VEGF on retinal microvascular endothelial hydraulic conductivity: the role of NO. Invest Ophthalmol Vis Sci 2000; 41: 4256–4261. [MEDLINE record in process].
71. Mayhan WG. VEGF increases permeability of the blood-brain barrier via a nitric oxide synthase/cGMP-dependent pathway. Am J Physiol 1999; 276: C1148–1153.
72. Tilton RG, Chang KC, LeJeune WS, Stephan CC, Brock TA, Williamson JR. Role for nitric oxide in the hyperpermeability and hemodynamic changes induced by intravenous VEGF. Invest Ophthalmol Vis Sci 1999; 40: 689–696.
73. He H, Venema VJ, Gu X, Venema RC, Marrero MB, Caldwell RB. Vascular endothelial growth factor signals endothelial cell production of nitric oxide and prostacyclin through flk-1/KDR activation of c-Src. Journal of Biological Chemistry 1999; 274: 25130–25135.
74. Barber AJ, Lieth E, Khin SA, Antonetti DA, Buchanan AG, Gardner TW. Neural apoptosis in the retina during experimental and human diabetes. Early onset and effect of insulin. Journal of Clinical Investigation 1998; 102: 783–791.

ALISTAIR J. BARBER, MAKOTO NAKAMURA AND
THE PENN STATE RETINA RESEARCH GROUP

4. Apoptosis and Neurodegeneration in Diabetes: Lessons from the Retina

Editor's Comment: Highlighting accelerated death of retinal ganglion cell bodies, Barber suggests that diabetic retinopathy is the result, at least in part, of neurodegenration. Accelerated cell death begins soon after the onset of diabetes resulting in progressive loss of neurons that may induce chronic loss of vision. While the pathogenesis of this apoptosis is uncertain, Barber theorizes that loss of trophic/survival factors (including insulin), leads to hyperglycemia that over stimulates the hexosamine pathway ending in increased vascular permeability, and glutamate excitotoxicity. This line of thinking opens the possibility of molecular intervention to preserve sight in long-term diabetes.

INTRODUCTION

Diabetes compromises vision, leading to diabetic retinopathy. This common complication of diabetes was classified as a vascular disease because of the syndrome of vascular abnormalities that are readily apparent by clinical examination [1, 2]. But it is clear from experimental and histological data that the morphology of the inner (neural) retina is also compromised by diabetes. Our recent findings in the retina of diabetic rats indicate that neuro-degeneration is an important component of diabetic retinopathy. These degenerative changes include increased apoptosis, reduction in the number of retinal ganglion cell bodies, and altered expression of glial fibrillary acidic protein in astrocytes and Müller cells. The fact that diabetes affects many different cell types in the retina should come as no great surprise. Diabetes has a deleterious influence on many different tissues throughout the body. Furthermore, the ultimate loss of vision in diabetic retinopathy must involve the failure of some aspects of neuronal function, since it is the neurons that transmit the visual signal to the brain. But what causes the retinal cell death? Is there something unique about the retina that makes it more sensitive to the stresses of diabetes, or is increased apoptosis a general phenomenon that occurs in other parts of the body during diabetes?

There is an accumulating body of evidence suggesting that the normal homeostasis of the glial cells and neurons of the retina are altered by diabetes. These new data are reminiscent of changes more usually associated with the type of pathology seen in neurodegenerative diseases of the central nervous system. Therefore, we propose a new hypothesis of diabetic retinopathy that embraces all cell types of the retina including the neurons, Müller cells, astrocytes, microglia, and vascular cells. In the following review we describe some of the recent data supporting this general hypothesis.

E.A. Friedman and F.A. L'Esperance, Jr. (eds.), Diabetic Renal-Retinal Syndrome, 35–45.

RETINAL FUNCTION IS IMPAIRED SOON AFTER THE ONSET OF DIABETES

Evidence indicating that retinal function is altered soon after the onset of diabetes has accumulated over a number of years [3]. These changes occur at a time when there is little vascular pathology and may be related to the very earliest increases in vascular permeability.

The electroretinogram (ERG) is a measure of electrophysiological activity in the retina that uses external electrodes to measure changes in field potentials elicited by the entire population of retinal neurons. The ERG has several characteristic components, much like the electroencephalogram, and some of these components are altered in both humans and rats with diabetes. The amplitude of oscillatory potentials was reduced in juvenile individuals who had diabetes for only five years duration, and this has been suggested as a useful method to predict the onset of vascular retinopathy [4, 5]. Oscillatory potentials were also reduced in another group of adolescents with an average duration of 7 years of diabetes. In this study the change in ERG was present in those patients with no vascular retinopathy, again suggesting that the change in retinal function can be measured before the appearance of vascular microaneurysms [6]. More recently it has been suggested that the delay in latency of the oscillatory potential is an even more sensitive indicator of the effects of diabetes on the retina [7, 8].

The effect of diabetes on the electrical activity of the retina has also been measured in animal models of diabetes. Diabetic rats have prolonged peak latencies and in some cases a reduction in the amplitudes of oscillatory potentials on the ERG b-wave [9]. Abnormalities in the rat ERG have been reversed by treatment with heat shock proteins, propionyl- and acetyl-carnitine, aldose reductase inhibitors, and insulin [9–13]. These experimental data suggest that the abnormalities in the ERG are due to early physiological changes that are reversible, and therefore not due to permanent compromise of neuronal function.

The ERG does not correlate directly with conventional measures of visual acuity, but ERG abnormalities are accompanied by deficits in perceptive resolution, such as the ability to discriminate contrast, especially blue-yellow contrast, and night vision. The deficit in contrast sensitivity also appears to be independent of the presence of vascular disease [14]. Patients with non-insulin dependent diabetes also had impaired contrast-sensitivity, which again preceded obvious vascular retinopathy and were worsened in individuals that had developed vascular lesions [15]. Contrast-sensitivity is associated with the degree of vascular retinopathy and correlates more closely with the degree of vascular lesions than with functional tests of color vision and macular recovery time (the ability of the retina to recover visual acuity after exposure to bright light) [16]. Adolescents with diabetes also had impaired contrast-sensitivity in the absence of microvascular lesions. Interestingly the impairment in contrast-sensitivity was worse in those individuals with microalbuminurea, indicating a possible link between the course of retinopathy and that of nephropathy [17]. Reduced contrast-sensitivity was also measured in insulin-resistant obese patients without hyperglycemia, suggesting that this early functional impairment may not be related to serum glucose concentrations [18].

In summary, diabetes causes subtle impairments in vision that are accompanied by alterations in the electrophysiological and psychophysiological measurements of retinal function. These changes appear to precede obvious gross vascular lesions associated with diabetic retinopathy and suggest that diabetes compromises the neurons and glial cells of the inner retina.

DIABETES INCREASES APOPTOSIS IN THE RETINA

Diabetes increases the incidence of cell death by apoptosis in the retina. The first observations of pyknotic cells in human diabetic retinopathy were made 40 years ago [19, 20]. *In situ* TUNEL staining, which detects DNA fragmentation in histological sections, has been used to describe apoptosis in both neurons and vascular cells of the retina [21–23]. In order to quantify apoptosis in the whole retina we applied the *in situ* TUNEL assay to intact retinas that had been dissected and mounted onto microscope slides (flat-mount). We found that the rate of cell death in the rat retina was elevated after one month of streptozotocin-induced diabetes, and remained high for at least a year. By counting cell bodies in paraffin embedded sections we found that retinal ganglion cells are lost as a result of the increase in cell death. The total number of ganglion cell bodies was reduced by 10% after 7.5 months of diabetes compared to age-matched controls, and this was accompanied by a significant reduction in the thickness of the inner retina [24]. This reduction in the number of neurons in the retina was recently corroborated by other investigators using immunohistochemistry for NeuN, a cell-specific marker expressed exclusively in the nuclei of neurons [25]. The loss of retinal ganglion cell bodies is also reflected by a reduction in the number of axons in the optic nerve [26]. Collectively these data suggest that diabetes causes a chronic loss of neurons from the inner retina by increasing the frequency of apoptosis of neurons and possibly glial cells. Importantly, the loss of neurons begins soon after the onset of diabetes. Therefore diabetic retinopathy can be classed as a chronic neurodegenerative disorder of the retina. It is not clear what aspects of diabetes initiate this increase in cell death.

RETINAL NEURONS RESPOND TO INSULIN AND GLUCOSE

From our studies on STZ-diabetic rats it is clear that treatment with insulin, even for 48 hours can reverse a number of the neural and vascular changes described above. Recently we have examined the potential actions of insulin and glucose in a cell culture model of retinal neurons, R28 cells. This cell-line was isolated from postnatal rat retinas and immortalized by transfection of the viral E1A vector [27]. While R28 cells were derived from a mixed retinal population and have been reported to express glial antigens under certain circumstances [28], they have also been noted to express neuron-specific markers [29]. Furthermore, R28 cells undergo caspase-3 dependent apoptosis when exposed to glutamate receptor agonists, a response that is typical of neurons in culture [30]. We found that growing these cells on laminin substrate and adding an absorbable cAMP analogue to the culture medium augmented their neuronal phenotype, causing them to grow neurite-like projections and increase expression of neuron-specific enolase and neurofilament. In contrast, the expression of glial fibrillary acidic protein was undetectable. Therefore, the R28 cell line can be used as a convenient in vitro model of retinal neurons.

First, we examined the potential for insulin to act as a survival factor in R28 cells. Removal of serum from the R28 medium caused apoptosis. After 24 hours of serum-deprivation the number of pyknotic nuclei was significantly increased. Treatment with insulin reduced the amount of apoptosis. The neuroprotective effect of insulin involved the PI-3 kinase/Akt pathway and was blocked by PI-3 kinase inhibitors.[1] In further unpublished work we found that addition of insulin to the medium of healthy R28 cells had no obvious effect on their growth rate or morphology. Also, growing R28 cells in insulin for an extended period of time, followed by its removal, did not cause any significant increase in apoptosis in the way that removal of serum had done. While these data suggest that insulin may act as a trophic factor for neurons in the retina, and that its absence in diabetes may increase the possibility that neurons will succumb to negative factors caused by diabetes, it is still not clear what these factors might be or how insulin protects neurons. In a second study we examined the effect of glucose and glucosamine on apoptosis of R28 cells. While increasing the concentration of glucose in the medium of R28 cells did not increase apoptosis, a protracted incubation with elevated levels of glucose augmented the ability of serum-deprivation to cause apoptosis, suggesting that glucose compromises the survival ability of neurons. Addition of glucosamine, which simulates overload of the hexosamine pathway, induced a large increase in apoptosis. Examination of the expression of the insulin and IGF-1 receptors suggested that glucosamine interfered with receptor processing.[2] This may be due to failure of post-translational glycosylation of the pre-receptor peptides. These data suggest that over-stimulation of the hexosamine pathway in diabetes may compromise the ability of retinal neurons to synthesize appropriate signal transduction apparatus for insulin. This effect may apply to other signal transduction cascades that use glycosylated proteins.

DIABETES ALTERS GLIAL CELLS IN THE RETINA

The retina has two types of macroglial cell. The predominant type is the Müller cell, which is unique to the retina. Müller cells are spindle shaped and span the entire retina from the outer limiting membrane to the retinal ganglion cell layer. The second type is the astrocyte, which migrates into the retina along the optic nerve during development [31, 32]. Astrocytes are less abundant than Müller cells and form a single layer at the inner limiting membrane. Normally retinal astrocytes express the glial specific intermediate filament, glial fibrillary acidic protein (GFAP) while Müller cells do not [33, 34]. Diabetes increases GFAP expression in both the human and rat retina [35–37]. Recently we described dramatic changes in GFAP expression in diabetic rats. While there is increased GFAP expression in Müller cells, this is preceded by reduced expression in astrocytes between two and four months after the onset of streptozotocin-diabetes [38]. Changes in the expression of GFAP occur in many disease states and are a sensitive indicator of central nervous system injury [39]. GFAP also increases in Müller

[1] Barber AJ, Nakamura M, Wolpert EB, Reiter CEN, Antonetti DA, Gardner TW, and the Penn State Retina Research Group. Insulin rescues retinal neurons from apoptosis induced by serum deprivation by activating the PI-3 kinase/akt pathway and NF-κappaB. Submitted to JBC.

[2] Nakamura M, Barber AJ, Reiter CEN, Antonetti DA, Gardner TW, and the PSRRG. High glucose and glucosamine block the neuroprotective effect of insulin in retinal neurons in culture. Submitted to JBC.

cells in response to retinal injury and ischemia and has been termed glial reactivity [40, 41]. GFAP expression is stimulated by glutamate in hippocampal slices, suggesting a mechanism for glial reactivity [42, 43]. These data suggest that the glial cells of the retina react in a way that is similar to their reaction to ischemia or other injury. In further support of this, the abundance of microglia in the retina also increases soon after the onset of diabetes [25]. Taken together these data suggest that the glial cells of the retina take on a reactive state that is associated with neurodegenerative disease.

GLUTAMATE TOXICITY IN DIABETIC RETINOPATHY

An important function of the glial cells in the retina is to clear the extracellular space of glutamate, which the photoreceptors and some of the other neurons use as a neurotransmitter. An over abundance of extracellular glutamate can cause excitotoxicity [41], in which over excitation of neurons can lead to an increase in apoptosis. There have been suggestions that diabetes elevates the concentration of glutamate in the vitreous of humans and in the retina of streptozotocin-diabetic rats [35, 44]. Furthermore, retinas isolated from diabetic rats are less able to convert glutamate to glutamine, compared to control retinas. This is accompanied by a reduction in the activity and content of the glial enzyme, glutamine synthetase [35]. Diabetes also reduces the ability of retinal glial cells to oxidize excess glutamate by conversion to alpha-ketogluterate which is shunted into the tricarboxylic acid cycle [45, 46]. Taken together these data suggest that diabetes compromises the ability of the retina to manage glutamate, which could lead to chronic excitotoxicity by allowing a gradual increase in the levels of extracellular glutamate. In later stages of diabetic retinopathy it is also possible that excess glutamate could originate from serum leaking through the failing blood-retinal barrier. These possibilities are currently being tested in streptozotocin-diabetic rats.

NEUROFILAMENT CHANGES IN DIABETIC RETINOPATHY

Neurofilaments are intermediate filaments found in the cytoskeleton of neurons with large axons, such as the retinal ganglion cells. A recent immunohistochemical study on flat-mount retinas from control and diabetic rats found an increase in the number of axonal swellings containing accumulations of neurofilament protein.[3] The swellings were associated with rare histamine-containing axons called centrifugal fibers, which may originate from cell bodies in the brain [47]. The swellings are similar in structure to corpora amylacea, also described in the retina as hyaline bodies [48–50]. Similar axonal swellings have been observed in peripheral nerve tissue of people with both type I and II diabetes [51]. These abnormalities may be due to a reduced rate of axonal transport which has been measured in both peripheral nerve and the optic nerve of diabetic rats [52–54]. The rate of fast anterograde axonal transport is also reduced in the optic nerve of alloxan-diabetic rabbits, and this may be due to reduced protein synthesis in the retinal ganglion cell bodies [55–57]. Notably, mild glutamate toxicity caused hyperphosphorylation of neurofilament protein in cerebellar granule cells, which may cause a change in the retrograde axon transport rate [58]. Therefore, diabetes reduces axonal transport in both the peripheral and optic nerves, and this may be related to abnormal accumulation of axonal neurofilament. A deficit in retrograde axonal transport of the optic nerve may have significance to the retina in diabetes because this is the mechanism by which certain growth factors, such as brain-derived neurotrophic factor, are transported from the brain to the retina.

FACTORS THAT INDUCE VASCULAR PERMEABILITY MAY ALSO BE NEUROPROTECTIVE

A number of factors that increase vascular permeability have been shown to increase in the eye during diabetes. The most well studied is vascular endothelial growth factor (VEGF) which is elevated in the vitreous of humans with proliferative retinopathy [59, 60], and in diabetic rats [61, 62]. The retinas of humans with diabetes also have increased mRNA for VEGF suggesting that the increased VEGF in the vitreous is derived from the retina [63, 64].

Some of the factors that are likely to induce vascular permeability and angiogenesis may have other effects on the retina. VEGF has been implicated as a survival factor for both vascular endothelial cells [65] and inhibits apoptosis of vascular endothelial cells induced by tumor necrosis factor-alpha [66]. The anti-apoptotic effects of VEGF may also apply to neurons, since it can augment axonal outgrowth and survival of neurons in explant superior cervical ganglion cells and dorsal route ganglion cells [67]. This suggests the intriguing possibility that VEGF has multiple and apparently paradoxical roles, to increase vascular permeability but at the same time protect neurons from apoptosis. Other angiogenic/permeabilizing factors can also act as neuronal survival factors. PDGF and basic fibroblast growth factor (bFGF) cause mitogenesis of Schwann cells [68], while bFGF induces neurite outgrowth in the PC12 cell line [69]. The neuroprotective effects of IGF-1 are better described [70–72] an this factor may also have a role in angiogenesis and vascular permeability [73]. Therefore, some of the growth factors that are increased in the retina by diabetes may have dual, and apparently paradoxical functions of inducing survival mechanisms in neurons while increasing vascular permeability and angiogenesis. It is possible that the primary function of these factors is to protect neurons from the metabolic stresses of diabetes and that their effect on the vasculature is secondary. It is even conceivable that the increase in permeability is a compensatory mechanism that is briefly beneficial to the retina, by increasing the exchange of metabolic substrates between the retina and blood, but is detrimental when invoked for a protracted period of time, as in diabetes.

APOPTOSIS AND DIABETES

Apoptosis in diabetic retinopathy has been the subject of a number of studies. Is this increase in cell death unique to the retina or does it occur in other tissues in diabetes? Apoptosis has been reported to be increased in the kidney of diabetic rats, and the

---

[3] Gastinger MJ, Barber AJ, Khin SA, McRill CS, Gardner TW, Marshak DW. Centrifugal axons are abnormal in streptozotocin-diabetic rat retinas. Submitted to IOVS.

amount of DNA fragmentation was increased four-fold, compared with controls [74]. Apoptosis and neurodegeneration has also been reported to increase in the erectile tissue of streptozotocin rats and may be related to reduced NO synthesis [75, 76]. Apoptosis has also been reported to occur in 30% of the dorsal root ganglion cells in streptozotocin rats, although this unusually high rate of apoptosis is difficult to explain [77]. Interestingly a recent report described a decrease in the amount of DNA fragmentation in rat intestinal mucosa one week after the onset of diabetes. This was accompanied by an increase in the mucosal thickness [78].

These types of studies are made difficult for a number of reasons. Biochemical assays sensitive enough to quantify chronic degenerative cell loss are not yet available. Also, applying TUNEL staining to thin histological sections is not useful for quantifying the total number of dying cells in the tissue. Quantification of the total number of apoptotic cells in the retina was only possible because of the unique properties of this tissue. It may be a number of years before we have a good answer to the question of whether apoptosis due to diabetes occurs in other tissues apart from the retina.

LESSONS FROM THE RETINA

Compromise of functional vision, increased apoptosis, and increased vascular permeability all occur soon after the onset of diabetes. We must take account of all of these components of the disease to fully understand its mechanisms. By assessing changes in the neural cells of the retina we can design studies that examine new pharmacological treatments using protection of visual function as an endpoint. It seems likely that both the vascular and neural components of the retina are intimately associated with each other [79], but it is not clear which is the originator of the pathology. Simple timecourse studies do not clarify the answer to this 'chicken-and-egg' question because the measured onset of each change is determined by the sensitivity of the test used to make the measurement. It is clear that both neural and vascular changes are very early events that may happen coincidentally (see chapter by Antonetti, this volume). In examining how we came to this conclusion, we try to base our general approach to scientific investigation on these lessons:

1. Reexamination of the initial assumptions concerning the etiology of the disease allowed us to make new observations, giving rise to novel hypotheses. Part of this process was to broaden the focus of our research to include the neural elements of the inner retina as well as the vasculature, which enabled us to develop hypotheses about the cellular interactions in the retina. More can be gained by continuing to reexamine our assumptions and preparing to relinquish old hypotheses in the light of new data.
2. Important changes may be difficult to measure. The increase in apoptosis in the retina was significant, but the absolute number of dying cells was not large enough to be detected by biochemical assays or by TUNEL staining in normal 10μm thick histological sections. We adopted a novel approach in which biases due to sampling error were eliminated by quantifying apoptosis in the whole retina, rather than individual histological sections. Significant events such as apoptosis may not have been detected in other tissues because of the insensitivity of the assays used to quantify

them. Therefore we must strive to test our hypotheses in the most stringent way possible, and this includes continuing to develop multiple sensitive assays for cell death, vascular permeability and the other effects of diabetes.

In summary, neurodegeneration by apoptosis is a significant component of diabetic retinopathy. The increase in cell death occurs soon after the onset of diabetes and continues at a steady rate over a long period, resulting in a gradual loss of neurons that may account for the chronic loss of vision. It is still not clear what factor(s) is responsible for this apoptosis but some likely possibilities are: loss of trophic/survival factors (including insulin), elevated glucose and over stimulation of the hexosamine pathway, increased vascular permeability, and glutamate excitotoxicity.

REFERENCES

1. Engerman RL, Kern TS. Retinopathy in animal models of diabetes. Diabetes-Metabolism Reviews 1995; 11(2): 109–120.
2. Kern TS, Engerman RL. Vascular lesions in diabetes are distributed non-uniformly within the retina. Experimental Eye Research 1995; 60(5): 545–549.
3. Lieth E, Gardner TW, Barber AJ, Antonetti DA, The Penn State Retina Research Group. Retinal neurodegeneration: early pathology in diabetes. Clinical and Experimental Ophthalmology 2000; 28: 3–8.
4. Frost-Larsen K, Larsen HW, Simonsen SE. Oscillatory potential and nyctometry in insulin-dependent diabetics. Acta Ophthalmologica 1980; 58(6): 879–888.
5. Simonsen SE. The value of the oscillatory potential in selecting juvenile diabetics at risk of developing proliferative retinopathy. Acta Ophthalmologica 1980; 58(6): 865–878.
6. Juen S, Kieselbach GF. Electrophysiological changes in juvenile diabetics without retinopathy. Archives of Ophthalmology 1990; 108(3): 372–375.
7. Sakai H, Tani Y, Shirasawa E, Shirao Y, Kawasaki K. Development of electroretinographic alterations in streptozotocin-induced diabetes in rats. Ophthalmic Research 1995; 27(1): 57–63.
8. Segawa Y, Shirao Y, Yamagishi S, Higashide T, Kobayashi M, Katsuno K, Iyobe A, Harada H, Sato F, Miyata H, Asai H, Nishimura A, Takahira M, Souno T, Maeda K, Shima K, Mizuno A, Yamamoto H, Kawasaki K. Upregulation of retinal vascular endothelial growth factor mRNAs in spontaneously diabetic rats without ophthalmoscopic retinopathy. A possible participation of advanced glycation end products in the development of the early phase of diabetic retinopathy. Ophthalmic Research 1998; 30(6): 333–339.
9. Segawa M, Hirata Y, Fujimori S, Okada K. The development of electroretinogram abnormalities and the possible role of polyol pathway activity in diabetic hyperglycemia and galactosemia. Metabolism: Clinical & Experimental 1988; 37(5): 454–460.
10. Funada M, Okamoto I, Fujinaga Y, Yamana T. Effects of aldose reductase inhibitor (M79175) on ERG oscillatory potential abnormalities in streptozotocin fructose-induced diabetes in rats. Japanese Journal of Ophthalmology 1987; 31(2): 305–314.
11. Lowitt S, Malone JI, Salem A, Kozak WM, Orfalian Z. Acetyl-L-carnitine corrects electroretinographic deficits in experimental diabetes. Diabetes 1993; 42(8): 1115–1118.
12. Hotta N, Koh N, Sakakibara F, Nakamura J, Hamada Y, Wakao T, Hara T, Mori K, Naruse K, Nakashima E, Sakamoto N. Effect of propionyl-L-carnitine on motor nerve conduction, autonomic cardiac function, and nerve blood flow in rats with streptozotocin-induced diabetes: comparison with an aldose reductase inhibitor. Journal of Pharmacology & Experimental Therapeutics 1996; 276(1): 49–55.
13. Hotta N, Nakamura J, Sakakibara F, Hamada Y, Hara T, Mori K, Nakashima E, Sasaki H, Kasama N, Inukai S, Koh N. Electroretinogram in sucrose-fed diabetic rats treated with an aldose reductase inhibitor or an anticoagulant. American Journal of Physiology 1997; 273(5 Pt 1): E965–971.
14. Della Sala S, Bertoni G, Somazzi L, Stubbe F, Wilkins AJ. Impaired contrast sensitivity in diabetic patients with and without retinopathy: a new technique for rapid assessment. British Journal of Ophthalmology 1985; 69(2): 136–142.
15. Sokol S, Moskowitz A, Skarf B, Evans R, Molitch M, Senior B. Contrast sensitivity in diabetics with and without background retinopathy. Archives of Ophthalmology 1985; 103(1): 51–54.

16. Brinchmann-Hansen O, Bangstad HJ, Hultgren S, Fletcher R, Dahl-Jorgensen K, Hanssen KF, Sandvik L. Psychophysical visual function, retinopathy, and glycemic control in insulin-dependent diabetics with normal visual acuity. Acta Ophthalmologica 1993; 71(2): 230–237.
17. Bangstad HJ, Brinchmann-Hansen O, Hultgren S, Dahl-Jorgensen K, Hanssen KF. Impaired contrast sensitivity in adolescents and young type 1 (insulin-dependent) diabetic patients with microalbuminuria. Acta Ophthalmologica 1994; 72(6): 668–673.
18. Dosso AA, Yenice-Ustun F, Sommerhalder J, Golay A, Morel Y, Leuenberger PM. Contrast sensitivity in obese dyslipidemic patients with insulin resistance. Archives of Ophthalmology 1998; 116(10): 1316–1320.
19. Wolter JR. Diabetic retinopathy. American Journal of Ophthalmology 1961; 51: 1123–1139.
20. Bloodworth JMB. Diabetic Retinopathy. Diabetes 1962; 47: 815–820.
21. Hammes HP, Federoff HJ, Brownlee M. Nerve growth factor prevents both neuroretinal programmed cell death and capillary pathology in experimental diabetes. Molecular Medicine 1995; 1(5): 527–534.
22. Mizutani M, Kern TS, Lorenzi M. Accelerated death of retinal microvascular cells in human and experimental diabetic retinopathy. Journal of Clinical Investigation 1996; 97(12): 2883–2890.
23. Kerrigan LA, Zack DJ, Quigley HA, Smith SD, Pease ME. Tunel-positive ganglion cells in human primary open-angle glaucoma. Archives of Ophthalmology 1997; 115(8): 1031–1035.
24. Barber AJ, Lieth E, Khin SA, Antonetti DA, Buchanan AG, Gardner TW. Neural apoptosis in the retina during experimental and human diabetes. Early onset and effect of insulin. Journal of Clinical Investigation 1998; 102(4): 783–791.
25. Zeng XX, Ng YK, Ling EA. Neuronal and microglial response in the retina of streptozotocin-induced diabetic rats. Visual Neuroscience 2000; 17(3): 463–471.
26. Scott TM, Foote J, Peat B, Galway G. Vascular and neural changes in the rat optic nerve following induction of diabetes with streptozotocin. Journal of Anatomy 1986; 144: 145–152.
27. Seigel GM. Establishment of an E1A-immortalized retinal cell culture [letter]. In Vitro Cellular & Developmental Biology. Animal 1996; 32(2): 66–68.
28. Seigel GM, Mutchler AL, Imperato EL. Expression of glial markers in a retinal precursor cell line. Molecular Vision 1996; 2: 2.
29. Seigel GM. The golden age of retinal cell culture. Molecular Vision 1999; 5: 4.
30. Tezel G, Wax MB. Inhibition of caspase activity in retinal cell apoptosis induced by various stimuli in vitro. Investigative Ophthalmology & Visual Science 1999; 40(11): 2660–2667.
31. Watanabe T, Raff MC. Retinal astrocytes are immigrants from the optic nerve. Nature 1988; 332(6167): 834–837.
32. Ling TL, Mitrofanis J, Stone J. Origin of retinal astrocytes in the rat: evidence of migration from the optic nerve. Journal of Comparative Neurology 1989; 286(3): 345–352.
33. Bignami A, Dahl D. The radial glia of Muller in the rat retina and their response to injury. An immunofluorescence study with antibodies to the glial fibrillary acidic (GFA) protein. Experimental Eye Research 1979; 28(1): 63–69.
34. Stone J, Dreher Z. Relationship between astrocytes, ganglion cells and vasculature of the retina. Journal of Comparative Neurology 1987; 255(1): 35–49.
35. Lieth E, Barber AJ, Xu B, Dice C, Ratz MJ, Tanase D, Strother JM. Glial reactivity and impaired glutamate metabolism in short-term experimental diabetic retinopathy. Diabetes 1998; 47(5): 815–820.
36. Mizutani MC. Gerhardinger and M. Lorenzi. Muller cell changes in human diabetic retinopathy. Diabetes 1998; 47(3): 445–449.
37. Rungger-Brandle E, Dosso AA, Leuenberger PM. Glial reactivity, an early feature of diabetic retinopathy. Investigative Ophthalmology and Visual Science 2000; 41(7): 1971–1980.
38. Barber AJ, Antonetti DA, Gardner TW. Altered expression of retinal occludin and glial fibrillary acidic protein in experimental diabetes. The Penn State Retina Research Group. Investigative Ophthalmology & Visual Science 2000: 41(11): 3561–3568.
39. O'Callaghan JP. Assessment of neurotoxicity: use of glial fibrillary acidic protein as a biomarker. Biomedical & Environmental Sciences 1991; 4(1–2): 197–206.
40. Huxlin KR, Dreher Z, Schulz M, Dreher B. Glial reactivity in the retina of adult rats. Glia 1995; 15(2): 105–118.
41. Osborne NN, Larsen AK. Antigens associated with specific retinal cells are affected by ischaemia caused by raised intraocular pressure: effect of glutamate antagonists. Neurochemistry International 1996; 29(3): 263–270.
42. Wofchuk ST, Rodnight R. Glutamate stimulates the phosphorylation of glial fibrillary acidic protein in

slices of immature rat hippocampus via a metabotropic receptor. Neurochemistry International 1994; 24(6): 517–523.
43. Kommers T, Vinade L, Pereira C, Goncalves CA, Wofchuk S, Rodnight R. Regulation of the phosphorylation of glial fibrillary acidic protein (GFAP) by glutamate and calcium ions in slices of immature rat spinal cord: comparison with immature hippocampus. Neuroscience Letters 1998; 248(2): 141–143.
44. Ambati J, Chalam KV, Chawla DK, D'Angio CT, Guillet EG, Rose SJ, Vanderlinde RE, Ambati BK. Elevated gamma-aminobutyric acid, glutamate, and vascular endothelial growth factor levels in the vitreous of patients with proliferative diabetic retinopathy. Archives of Ophthalmology 1997; 115(9): 1161–1166.
45. Gamberino WC, Berkich DA, Lynch CJ, Xu B, LaNoue KF. Role of pyruvate carboxylase in facilitation of synthesis of glutamate and glutamine in cultured astrocytes. Journal of Neurochemistry 1997; 69(6): 2312–2325.
46. Lieth E, LaNoue KF, Antonetti DA, Ratz M, The Penn State Retina Research Group. Diabetes reduces glutamate oxidation and glutamine synthesis in the retina. Experimental Eye Research 2000; 70(6): 723–730.
47. Wolter JR. Centrifugal nerve fibers in the adult human optic nerve: 16 days after enucleation. Transactions of the American Ophthalmological Society 1978; 76: 140–155.
48. Wolter JR. Hyaline bodies of ganglion-cell origin in the human retina. Archives of Ophthalmology 1959; 61: 127–134.
49. Wolter JR, Liss L. Hyaline bodies of the human optic nerve. Archives of Ophthalmology 1959; 61: 780–788.
50. Woodford B, Tso MO. An ultrastructural study of the corpora amylacea of the optic nerve head and retina. American Journal of Ophthalmology 1980; 90(4): 492–502.
51. Schmidt RE, Beaudet LN, Plurad SB, Dorsey DA. Axonal cytoskeletal pathology in aged and diabetic human sympathetic autonomic ganglia. Brain Research 1997; 769(2): 375–383.
52. Alberghina M. Metabolic correlates and axonal transport in nerves during experimental diabetes. [Review] [61 refs]. Metabolic, Pediatric & Systemic Ophthalmology 1986; 9(2–4): 42–46.
53. Zhang L, Inoue M, Dong K, Yamamoto M. Alterations in retrograde axonal transport in optic nerve of type I and type II diabetic rats. Kobe Journal of Medical Sciences 1998; 44(5–6): 205–215.
54. Zhang LX, Ino-ue M, Dong K, Yamamoto M. Retrograde axonal transport impairment of large- and medium–sized retinal ganglion cells in diabetic rat. Current Eye Research 2000; 20(2): 131–136.
55. Chihara E. Impairment of protein synthesis in the retinal tissue in diabetic rabbits: secondary reduction of fast axonal transport. Journal of Neurochemistry 1981; 37(1): 247–250.
56. Chihara E, Sakugawa M, Entani S. Reduced protein synthesis in diabetic retina and secondary reduction of slow axonal transport. Brain Research 1982; 250(2): 363–366.
57. Tsukada T, Chihara E. Changes in components of fast axonally transported proteins in the optic nerves of diabetic rabbits. Investigative Ophthalmology & Visual Science 1986; 27(7): 1115–1122.
58. Asahara H, Taniwaki T, Ohyagi Y, Yamada T, Kira J. Glutamate enhances phosphorylation of neurofilaments in cerebellar granule cell culture. Journal of the Neurological Sciences 1999; 171(2): 84–87.
59. Adamis AP, Miller JW, Bernal MT, D'Amico DJ, Folkman J, Yeo TK, Yeo KT. Increased vascular endothelial growth factor levels in the vitreous of eyes with proliferative diabetic retinopathy. American Journal of Ophthalmology 1994; 118(4): 445–450.
60. Aiello LP, Avery RL, Arrigg PG, Keyt BA, Jampel HD, Shah ST, Pasquale LR, Thieme H, Iwamoto MA, Park JE. Vascular endothelial growth factor in ocular fluid of patients with diabetic retinopathy and other retinal disorders [see comments]. New England Journal of Medicine 1994; 331(22): 1480–1487.
61. Murata T, Nakagawa K, Khalil A, Ishibashi T, Inomata H, Sueishi K. The relation between expression of vascular endothelial growth factor and breakdown of the blood-retinal barrier in diabetic rat retinas. Laboratory Investigation 1996; 74(4): 819–825.
62. Sone H, Kawakami Y, Okuda Y, Sekine Y, Honmura S, Matsuo K, Segawa T, Suzuki H, Yamashita K. Ocular vascular endothelial growth factor levels in diabetic rats are elevated before observable retinal proliferative changes. Diabetologia 1997; 40(6): 726–730.
63. Lutty GA, McLeod DS, Merges C, Diggs A, Plouet J. Localization of vascular endothelial growth factor in human retina and choroid. Archives of Ophthalmology 1996; 114(8): 971–977.
64. Amin RH, Frank RN, Kennedy A, Eliott D, Puklin JE, Abrams GW. Vascular endothelial growth factor is present in glial cells of the retina and optic nerve of human subjects with nonproliferative diabetic retinopathy. Investigative Ophthalmology & Visual Science 1997; 38(1): 36–47.
65. Alon T, I. Hemo, A. Itin, J. Pe'er, J. Stone and E. Keshet. Vascular endothelial growth factor acts as a

survival factor for newly formed retinal vessels and has implications for retinopathy of prematurity. Nature Medicine 1995; 1(10): 1024–1028.
66. Spyridopoulos I, Brogi E, Kearney M, Sullivan AB, Cetrulo C, Isner JM, Losordo DW. Vascular endothelial growth factor inhibits endothelial cell apoptosis induced by tumor necrosis factor-alpha: balance between growth and death signals [published erratum appears in J Mol Cell Cardiol 1998 Apr;30(4):897]. Journal of Molecular & Cellular Cardiology 1997; 29(5): 1321–1330.
67. Sondell M, Lundborg G, Kanje M. Vascular endothelial growth factor has neurotrophic activity and stimulates axonal outgrowth, enhancing cell survival and Schwann cell proliferation in the peripheral nervous system. Journal of Neuroscience 1999; 19(14): 5731–5740.
68. Davis JB, Stroobant P. Platelet-derived growth factors and fibroblast growth factors are mitogens for rat Schwann cells. Journal of Cell Biology 1990; 110(4): 1353–1360.
69. Rydel RE, Greene LA. Acidic and basic fibroblast growth factors promote stable neurite outgrowth and neuronal differentiation in cultures of PC12 cells. Journal of Neuroscience 1987; 7(11): 3639–3653.
70. Crouch MF, Hendry IA. Co-activation of insulin-like growth factor-I receptors and protein kinase C results in parasympathetic neuronal survival. Journal of Neuroscience Research 1991; 28(1): 115–120.
71. Dudek H, Datta SR, Franke TF, Birnbaum MJ, Yao R, Cooper GM, Segal RA, Kaplan DR, Greenberg ME. Regulation of neuronal survival by the serine-threonine protein kinase Akt. Science 1997; 275(5300): 661–665.
72. Ryu BR, Ko HW, Jou I, Noh JS, Gwag BJ. Phosphatidylinositol 3-kinase-mediated regulation of neuronal apoptosis and necrosis by insulin and IGF-I. Journal of Neurobiology 1999; 39(4): 536–546.
73. Smith LE, Shen W, Perruzzi C, Soker S, Kinose F, Xu X, Robinson G, Driver S, Bischoff J, Zhang B, Schaeffer JM, Senger DR. Regulation of vascular endothelial growth factor-dependent retinal neovascularization by insulin–like growth factor-1 receptor. Nature Medicine 1999; 5(12): 1390–1395.
74. Zhang WP, Khanna P, Chan LL, Campbell G, Ansari NH. Diabetes-induced apoptosis in rat kidney. Biochemical & Molecular Medicine 1997; 61(1): 58–62.
75. Cellek S, Rodrigo J, Lobos E, Fernandez P, Serrano J, Moncada S. Selective nitrergic neurodegeneration in diabetes mellitus – a nitric oxide-dependent phenomenon. British Journal of Pharmacology 1999; 128: 1804–1812.
76. Alici B, Gumustas MK, Ozkara H, Akkus E, Demirel G, Yencilek F, Hattat H. Apoptosis in the erectile tissues of diabetic and healthy rats. BJU International 2000; 85(3): 326–329.
77. Russell JW, Sullivan KA, Windebank AJ, Herrmann DN, Feldman EL. Neurons undergo apoptosis in animal and cell culture models of diabetes. Neurobiology of Disease 1999; 6(5): 347–363.
78. Noda T, Iwakiri R, Fujimoto K, Yoshida T, Utsumi H, Sakata H, Hisatomi A, Aw TY. Suppression of apoptosis is responsible for increased thickness of intestinal mucosa in streptozotocin-induced diabetic rats. Metabolism: Clinical & Experimental 2001; 50(3): 259–264.
79. Gardner TW, Lieth E, Khin SA, Barber AJ, Bonsall DJ, Lesher T, Rice K, Brennan Jr WA. Astrocytes increase barrier properties and ZO-1 expression in retinal vascular endothelial cells. Investigative Ophthalmology & Visual Science 1997; 38(11): 2423–2427.

FREDERICK L. FERRIS III

5. Diabetic Macular Edema

Editor's Comment: As explored by Antonetti in the previous chapter, increased vascular permeability precedes macular edema and is associated with subsequent angiogenesis and proliferative retinopathy. Terming diabetic macular edema 'one of the most difficult challenges for the treating ophthalmologist,' Ferris notes that current management by photocoagulation though effective in preventing further loss of visual acuity, does not improve decreased vision. Noting that new photocoagulation techniques may reduce the amount of tissue destruction while still eliminating the edema, Ferris holds out hope that enhanced medical approaches such as vascular endothelial growth factor inhibitors and growth hormone antagonists may be efficacious.

BACKGROUND

Diabetic retinopathy remains one of the four major causes of blindness in the western world [1, 2]. Blindness from diabetic retinopathy results from consequences of proliferative diabetic retinopathy and from diabetic macular edema. Proliferative diabetic retinopathy is more likely than macular edema to cause rapid and severe loss of vision. Without treatment, eyes that develop proliferative diabetic retinopathy have at least a 50% chance of becoming blind within five years [3, 4]. Appropriate application of treatments that have been developed in the last three decades can reduce this risk of blindness to less than 5% [5]. Although the proportion of persons with diabetic macular edema that become legally blind is less than that for those who develop proliferative diabetic retinopathy, the treatments for diabetic macular edema are not as effective in preventing vision loss in those affected. At best, current treatments reduce the risk of blindness by about 50% [6]. New treatments, both medical and surgical, are needed if we are to reduce blindness from this complication of diabetic retinopathy.

The 10-year incidence of developing diabetic macular edema in persons with Type 1 diabetes is about 14% [7]. and perhaps 42% will develop macular edema during their lifetime [8]. Recent evidence suggests that vascular endothelial growth factor may induce macular edema [9]. Other reports suggest elevated histamines or elevated human growth factor may induce macular edema [10, 11]. Medical treatments have been shown to reduce the frequency of macular edema. Tight blood glucose control can reduce the risk of developing macular edema [12, 13]. Approaches to lower blood pressure may reduce the frequency of macular edema [14], and lowering serum lipids may further reduce the complications of macular edema [15]. Finally surgical approaches, such as vitrectomy in persons with vitreoretinal traction, may dramatically improve visual acuity [16, 17].

E.A. Friedman and F.A. L'Esperance, Jr. (eds.), Diabetic Renal-Retinal Syndrome, 47–58.

NONPROLIFERATIVE DIABETIC RETINOPATHY

The retinal changes in persons with diabetes have been well described and can be divided into five fundamental processes: (1) formation of retinal capillary microaneurysms, (2) excessive vascular permeability, (3) vascular occlusion, (4) proliferation of new blood vessels and accompanying fibrous tissue on the surface of the retina and optic disc and (5) contraction of these fibrovascular proliferations and the vitreous [18, 19]. The first two of these processes, with some contribution from vascular occlusion are responsible for diabetic macular edema. Retinal capillary microaneurysms are the first visible lesions of diabetic retinopathy, but they can also be associated with other retinal vascular diseases, particularly those associated with vascular occlusion. These microaneurysms are hypercellular saccular outpouchings from the capillary wall that are often associated with the development of fluid leakage and macular edema. Although the mechanism for microaneurysm formation is unknown, possible mechanisms include: release of a vasoproliferative factor from dying cells, weakness of the capillary wall (perhaps from loss of pericytes), and increased intraluminal pressure caused by abnormalities of the adjacent retina [20].

Microaneurysms without any of the other components of diabetic retinopathy have no apparent clinical significance except as a marker of the development of diabetic retinopathy. However, when excessive vascular permeability is associated with microaneurysms, vision can be threatened by the development of macular edema [21, 22].

Fluorescein angiography can be used to identify excessive vascular permeability, but fluorescein leakage does not necessarily indicate macular edema, which is defined as retinal thickening from accumulation of fluid and can be best appreciated with slit lamp biomicroscopy or stereoscopic fundus photography. Although retinal thickening is difficult to recognize with the ophthalmoscope, it is often associated with hard exudates. These lipid deposits within the outer retina are easy to recognize as yellow-white deposits, often seen at the border of edematous retina. Edema fluid may come and go within the retina without visual consequence, but lipid deposits, especially when under the center of the macula, are often associated with retinal damage and permanent vision loss [23, 24]. The extent of these lipid deposits in the retina is associated with the degree to which serum lipids are elevated [25, 26].

ETIOLOGY OF MACULAR EDEMA

The abnormal leakage of macromolecules from retinal capillaries into the retina is the probable cause of the development of macular edema. The retinal capillaries are one of the sites in the body where there are tight junctions between endothelial cells. The cells and the tight junctions prevent the free diffusion of molecules from the intravascular space to the extravascular space of the retina. Fluorescein dye, for example, does not diffuse through normal retinal capillaries into the retinal extravascular space. However, in persons with breakdown of the 'blood-retinal barrier' such diffusion is apparent on fluorescein angiography. As large molecules, such as lipoproteins, gain access to the extravascular space they cause an oncotic influx of water into this space resulting in retinal edema (Figure 1). This 'leakage' of macromolecules and ions, when localized to specific areas of microaneurysms or small segments of the retinal

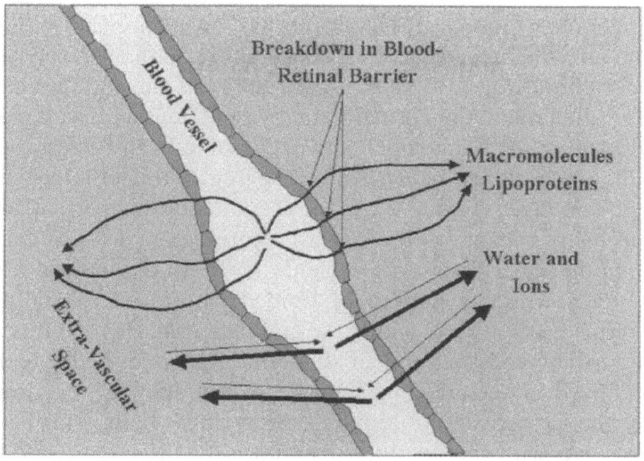

Figure 1. Leakage of large molecules secondary to breakdown of the blood-retinal barrier, leading to macular edema.

capillaries is considered 'focal' in nature. When the leakage occurs from broad areas of the retinal capillaries throughout the posterior retina it is considered 'diffuse.' The leakage can come from individual microaneurysms or from broad areas of abnormal capillaries. How this leakage relates to the histologic changes seen in diabetic retinal vessels, including thickening of the basement membranes of the vascular endothelial cells and loss of pericytes from the vessel walls, has not been clearly elucidated [27–29].

Not all retinal edema comes from retinal vascular leakage. Abnormalities in the barrier or pumping action of the retinal pigment epithelium can lead to macular edema [30]. Abnormalities in the retinal pigment epithelium may theoretically contribute to macular edema by allowing an increased amount of fluid from the choriocapillaris to pass into the sensory retina. Alternatively, and perhaps more probably, defects in the retinal pigment epithelial pumping action could allow edema to accumulate in the retina that would have otherwise been removed.xxxi,xxxii The amount of edema is related to a number of factors. Factors that help to keep fluid out of the retinal extravascular space include: oncotic pressure within the capillaries and pumping of water from the retina to the choroid through the retinal pigment epithelium. Factors that increase the tendency for fluid to accumulate within the retinal extravascular space include: macromolecules (lipoproteins and others) accumulating in the extravascular space, increased blood pressure and decreased oncotic pressure within the capillaries. In addition to edema that is directly a result of leaking microaneurysms or breakdown of the blood retinal barrier, it is clinically apparent that there is edema that is associated with renal failure.

NATURAL COURSE OF DIABETIC MACULAR EDEMA

Edema of the macular region of the retina that is associated with diabetic retinopathy can spontaneously resolve, but frequently it is a chronic process resulting from multiple areas of breakdown of the blood-retinal barrier [33–36]. Classically, this is associated with edema in the immediate area of the 'leakage' sites and the deposition of intraretinal lipid exudates peripheral to these sites. In the areas of the retina where the lipid is deposited, there is often relatively normal retinal vasculature with an intact blood-retinal barrier. The fluid and small ion components of the blood can be resorbed in this area, leaving behind the lipoproteins, which tend to aggregate as lipid deposits, particularly in the outer plexiform layer (Figure 2). Fluid accumulation is particularly pronounced in the macula, probably because of the larger potential extravascular spaces created by the obliquely running nerve fibers in the Henle layer. In the case of a single leakage site, there is often a circinate pattern of lipid deposition surrounding the leakage (Figure 3). The retina is edematous within the circinate ring. Beyond this ring, the retina is often not identifiably edematous. In diabetic retinopathy, the leakage sites are generally not localized to one area and lipid deposition may be scattered in the posterior pole or take the form of multiple intersecting rings. Individual leakage sites can spontaneously resolve with the resultant resolution of the edema and eventually the lipid. Generalized diffuse leakage from the entire capillary bed with development of cystoid macular edema can occur. When the leakage is truly generalized, there are few if any areas of intact blood-retinal barrier and little lipid deposition (Figure 4). More frequently, the process is chronic with new leakage sites replacing the resolved ones, eventuating in persistent edema with multifocal leakage sites (Figure 5). Although perifoveal capillary dropout alone does not cause large losses in visual acuity, it is associated with a poor prognosis, particularly when associated with macular edema [37].

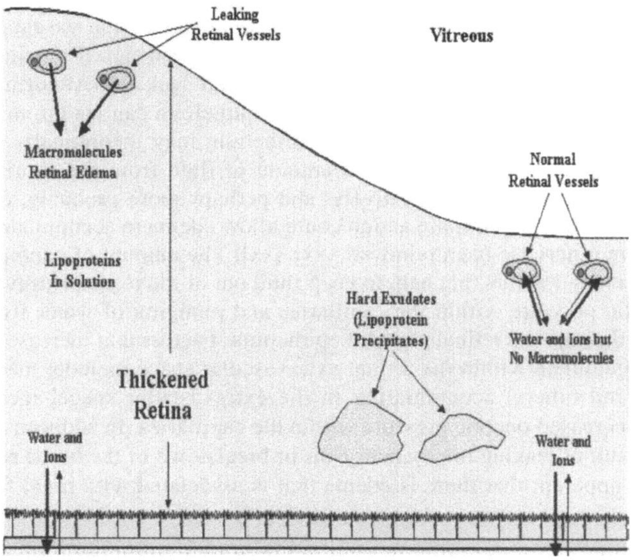

Figure 2. Retina is thickened in areas of breakdown of the blood-retinal barrier with macromolecule leakage. Water is removed in area of intact blood-retinal barrier, leading to lipoprotein precipitates (hard exudates).

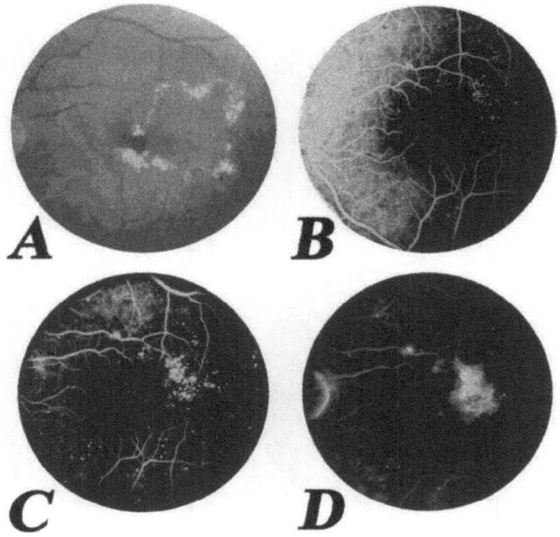

Figure 3. Circinate lipid ring with fluorescein leakage and retinal thickening in the center of the ring. A: Fundus photograph showing white lipid ring in temporal macular area. B: Early phase angiogram with microaneurysms in temporal retinal filling with fluorescein C: Mid phase angiogram with fluorescein now leaking out of microaneurysms and D: Late phase fluorescein angiogram showing accumulation of fluorescein dye in the area of retinal thickening in center of circinate lipid ring.

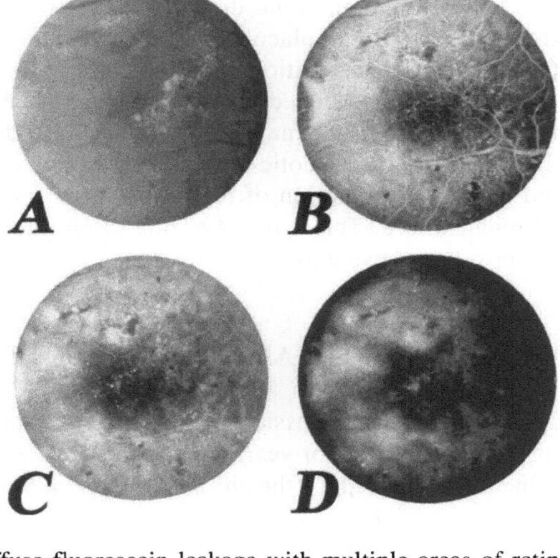

Figure 4. Focal and diffuse fluorescein leakage with multiple areas of retinal thickening and lipid deposits. A: Fundus photograph showing multiple microaneurysms and intraretinal hemorrhages along with multiple areas of yellow lipid deposits. B: Early fluorescein angiogram showing multiple leaking microaneurysms as well as diffuse capillary leakage. C and D: Mid and late phase fluorescein angiograms showing multiple areas of accumulation of fluorescein dye in the retina.

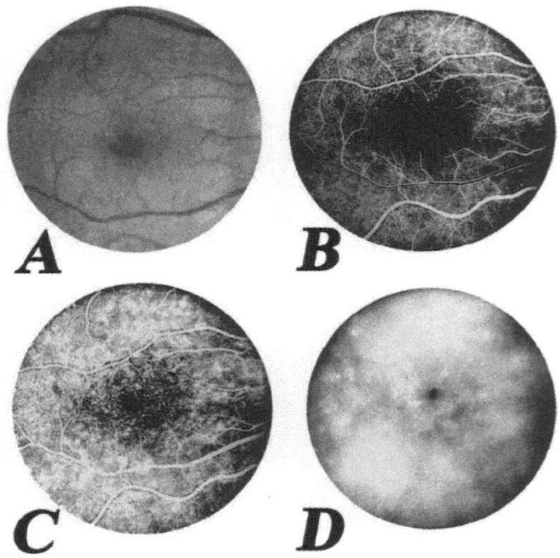

Figure 5. Diffuse fluorescein leakage with retinal thickening throughout the posterior pole. A: fundus photograph showing dilated capillaries with a few small areas of lipid deposits. B and C: Early and mid phase fluorescein angiograms showing diffusely dilated capillary bed with fluorescein leakage from entire vascular tree. D: Late phase fluorescein angiogram showing accumulation of fluorescein throughout the retina in the posterior pole.

Chronic edema fluid may result in some disarray of the retinal segments, but lipid deposits are most destructive. Large placoid lipid deposits in the center of the macula generally indicate little hope of restoration of visual acuity [38, 39]. Finally, there is a type of macular edema that is associated with severe renal disease. Depending on the degree of serum lipids there may or may not be much lipid deposition. This edema may be in part related to decreased oncotic pressure secondary to albuminuria, increased blood pressure and perhaps dysfunction of retinal pigment epithelial pumping related to the generalized renal failure. This type of edema often improves after dialysis but more commonly after renal transplantation.

PREVENTION OF DIABETIC MACULAR EDEMA

Preventing the development or progression of diabetic retinopathy is the most effective approach to preserving vision. For years there was a debate as to whether improved control of blood glucose would reduce the chronic complications of diabetes, including diabetic retinopathy. The Diabetes Control and Complications Trial (DCCT) was initiated to address this important clinical and scientific question [40]. In this study, 1,441 patients with type 1 diabetes (726 with no retinopathy and 715 with mild-to-moderate nonproliferative retinopathy at base line) were randomly assigned to receive either intensive or conventional therapy and were followed for a mean of 6.5 years. There was not only a remarkable reduction in the rate of the development or progression of retinopathy (on the basis of the central grading of fundus photographs with use of a scale that ranged from no retinopathy to proliferative retinopathy) among patients assigned to intensive treatment (Figure 6), but also a reduction in the progression of diabetic nephropathy and neuropathy [12, 41, 42].

A smaller randomized clinical study of 102 patients with type 1 diabetes who were followed for more than seven years also found that intensive treatment reduced all three of the major microvascular complications of diabetes [43]. These results, combined with those of observational studies [44], implicate hyperglycemia in the development of the chronic microvascular complications of diabetes.

Glycemic control has similar beneficial effects on the incidence and progression of microvascular complications in patients with type 2 diabetes as in patients with type 1 diabetes [45], as demonstrated by randomized studies in the United Kingdom and Japan [14, 46, 47].

Better control of hyperglycemia lowers but does not eliminate the risk of retinopathy and other complications of diabetes. Although there was a general reduction of progression of retinopathy in the DCCT, the effect tight blood-glucose control on the development of diabetic macular edema was less pronounced. A more definitive effect of tight blood glucose control on diabetic macular edema was seen in the UKPDS.

The effect of controlling blood pressure on the development of diabetic macular edema was greater than the effect of tight blood glucose control in the UKPDS. Data from this study show that tight control of blood pressure using either the angiotensin-converting enzyme, captopril, or the beta-adrenergic antagonist, atenlolol, slowed the progression of retinopathy compared with more routine blood pressure control. It also slowed the development of diabetic macular edema [48, 49].

Patients with diabetes who have high serum lipid concentrations have an increased risk of both proliferative retinopathy and vision loss from macular edema and associated retinal hard exudates [50, 51]. Reduction of serum lipids to decrease the lipid

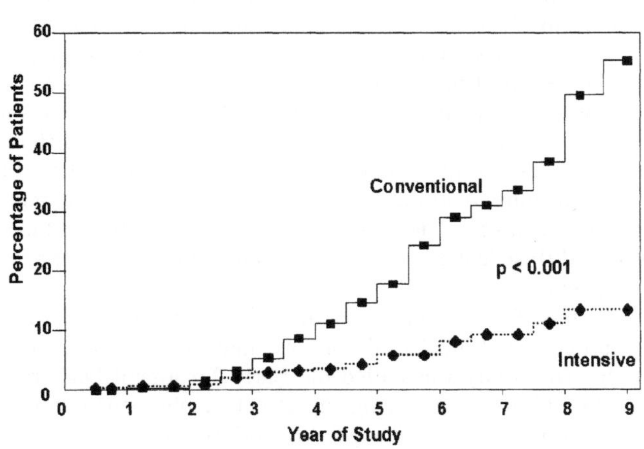

Figure 6. Comparison of 3-step progression of retinopathy rates in DCCT patients with some retinopathy at baseline assigned to conventional control of blood glucose vs. those assigned to intensive control.

accumulations in the retina and their destructive efforts have been attempted. Direct attempts to lower fat intake and reduce circulating cholesterol by substituting unsaturated fats for saturated fats in the diet have resulted in some reduction of the total serum lipids and retinal hard exudates [52]. Pharmacologic agents have a much greater effect in reducing both circulating lipids and retinal hard exudates [53, 54].

Although the clinical trials demonstrate a benefit of reducing serum lipids on the amount of retinal hard exudates there is no clear evidence that this results in either stabilization or improvement of visual acuity. At this point, it is difficult to do a placebo controlled clinical trial to test the hypothesis that lipid lowering will benefit visual acuity, because lowering serum lipids have a proven medical benefit on cardiovascular disease. Trials comparing very intensive lipid lowering versus more standard care are possible and are in the planning phase. These may eventually demonstrate the effect of lipid lowering on diabetic retinopathy and macular edema. In the mean time, it is prudent to reduce hyperlipidemia for cardiovascular reasons and this may also result in a lower risk of vision loss for patients with diabetic macular edema.

Medical approaches to prevent diabetic macular edema should include the control of blood glucose, blood pressure and serum lipids. Additional medical approaches may help control macular edema, but, as of now, none have been demonstrated to be effective. These methods include inhibitors of vascular endothelial growth factor, antagonists of the secretion and action of growth hormone, use of anti-inflammatory agents such as steroids and others. Some clinical trials are underway and others are planned. These will eventually identify which if any of these medical approaches to reduce macular edema are effective.

PHOTOCOAGULATION

The best data demonstrating the effect of photocoagulation comes from randomized clinical trials. Early clinical trials demonstrated some apparent treatment benefit, but the trials were small, admission criteria differed and the treatment protocols were different [21, 55, 56]. Each of these early studies, had a nominally statistically significant difference between treatment and no treatment. Although some eyes in each study had an improvement in visual acuity following treatment, the main difference between treated and untreated eyes was the much larger number of untreated eyes that lost two or more lines of visual acuity over the two year follow-up period. Attempts to evaluate subgroups of patients (with very small sample size) did suggest that initially poor visual acuity and relatively diffuse leakage, as determined in the fluorescein angiograms, were poor prognostic factors.

To better assess the effect of photocoagulation in persons with diabetic macular edema a large multicenter clinical trial was initiated as a follow-up to the Diabetic Retinopathy Study. This study was called the Early Treatment Diabetic Retinopathy Study (ETDRS). In addition to studying the effect of photocoagulation on diabetic macular edema, this study was also designed to address the question of the timing of scatter photocoagulation for proliferative diabetic retinopathy and the effect of aspirin on the progression of diabetic retinopathy. The 3,711 patients enrolled in the study had mild-to-severe nonproliferative or early proliferative diabetic retinopathy, with or without diabetic macular edema. One eye of each patient was randomly assigned to immediate photocoagulation, whereas the other eye was assigned to deferred photocoagulation (i.e., careful follow-up and prompt scatter photocoagulation if high-risk retinopathy developed). If the eye assigned to immediate photocoagulation had macular edema, photocoagulation of areas of edema was also performed, including direct (focal) treatment of leaking microaneurysms and 'grid' treatment with scattered, small laser burns in the areas of diffuse leakage [57]. The ETDRS results demonstrate that focal/grid photocoagulation of patients with diabetic macular edema and mild to moderate nonproliferative diabetic retinopathy reduced the risk of moderate loss of visual acuity (a doubling of the initial visual angle, e.g.. 20/20 to 20/40 or 20/50 to 20/100) by about 50% (Figure 7) [58–61]. However, not all patients with macular edema need immediate treatment. Immediate photocoagulation is beneficial if the edema involves or threatens the center of the macula, but treatment may be deferred if this is not the case. Patients can perceive the scotomas related to the focal laser burns. Careful follow-up with intervention only when retinal thickening or lipid deposits threaten or involve the center of the macula, can reduce the risk of visual loss and limit the need for treatment.

CONCLUSIONS

The best way to prevent vision loss from diabetic macular edema is to maintain the best possible medical care for every patient. Tight control of blood glucose will slow the progression of retinopathy. Tight control of blood pressure can reduce the risk that macular edema will develop. Normalization of elevated serum lipids will be beneficial for the cardiovascular system and may reduce the risk of vision loss from macular edema. Finally, photocoagulation can reduce the risk of vision loss in persons with macular edema that involves or threatens the center of the macula.

Diabetic macular edema remains one of the most difficult challenges for the treating

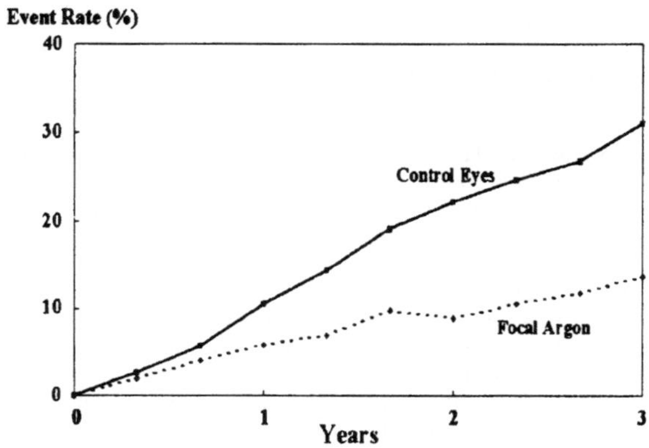

Figure 7. Rates of moderate loss of visual acuity (doubling of the initial visual angle) in ETDRS patients with macular edema and mild to moderate nonproliferative diabetic retinopathy assigned to either immediate argon photocoagulation vs. no photocoagulation for macular edema.

ophthalmologist. While photocoagulation has been demonstrated to be effective in preventing loss of visual acuity, it is a destructive treatment that is not generally effective in improving decreased visual acuity. New photocoagulation techniques are under investigation that may reduce the amount of tissue destruction while still eliminating the edema. In addition, new medical approaches such as vascular endothelial growth factor inhibitors and growth hormone antagonists may prove useful in reducing the edema. Investigations into the effect of immunosuppressants or vitrectomy for difficult macular edema may also lead to more effective means of saving vision.

Clinical trials to evaluate these new treatments for diabetic macular edema are either underway or in the planning stages. Hopefully some of them will demonstrate new and more effective treatments for this difficult condition. In the meantime, careful follow-up, good medical control and photocoagulation when indicated is the standard of care.

REFERENCES

1. Kahn HA, Hiller R. Blindness caused by diabetic retinopathy. Am J Ophthalmol 1974; 78: 58–67.
2. Kahn HA, Bradley RF. Prevalence of diabetic retinopathy: age, sex and duration of diabetes. Br J Ophthalmol 1975; 59: 345–349.
3. Beetham WP. Visual prognosis of proliferating diabetic retinopathy. Br J Ophthalmol 1963; 47: 611–619.
4. Caird FI, Burditt AF, Draper GJ. Diabetic retinopathy; a further study of prognosis for vision. Diabetes 1968; 17: 121–123.
5. Ferris, FL. How effective are treatments for diabetic retinopathy? Commentary. JAMA 1993; 269(10): 1290–1291.
6. The Early Treatment Diabetic Retinopathy Study Research Group. Photocoagulation for diabetic macular edema: Early treatment Diabetic Retinopathy Study Report #1. Arch Ophthalmol 1985; 103: 1796–1806.
7. Klein R, Klein BE, Doss SE, Cruickshanks KJ. The Wisconsin epidemiologic study of diabetic retinopathy XV: the long-term incidence of macular edema. Ophthalmology 1995; 102: 7–16.
8. Javitt JC, Canner JK, Sommer A. Cost effectiveness of current approaches to the control of retinopathy in type 1 diabetes. Ophthalmology 1989; 96: 255–264.
9. Aiello LP, Bersell SE, Clermont A, et al. Diabetes 1997; 46: 1473–1480.
10. Gardener TW and the Penn State Retina Research Group. Vascular permeability in experimental diabetes associated with reduced endothelial occluding content. Diabetes 1998; 47: 1953–1959.
11. Kuijpers RW, Baarsma S, van Hagen PM. Treatment of cystoid macular edema with octreotide. N Engl J Med 1998 (Feb 26); 338(9): 624–626.
12. The DCCT Research Group. Effect of intensive treatment of diabetes on the development and progression of long-term complications in insulin-dependent diabetes mellitus. N Eng J Med 1993; 329: 977–986.
13. The EDICT Research Group. Retinopathy and nephropathy in patients with type 1 diabetes four years after a trial of intensive therapy. The Diabetes Control and Complications Trial/Epidemiology of Diabetes Interventions and Complications Research Group. N Engl J Med 2000; 342(6): 381–389.
14. UK Prospective Diabetes Study (UKPDS) Group. Intensive blood-glucose control with sulphonylureas or insulin compared with conventional treatment and risk of complications in patients with type 2 diabetes (UKPDS 33). Lancet 1998; 352: 837–853.
15. Chew EY, Klein ML, Ferris FL, et al. Association of elevated serum lipid levels with retinal hard exudates in diabetic retinopathy. Arch Ophthalmol 1996; 114: 1079–1084.
16. Lewis H, Abrams GW, Blumenkranz MS, Campo R. Vitrectomy for diabetic macular traction and edema associated with posterior hyaloidal traction. Ophthalmology 1992; 99: 753–759.
17. Harbour JW, Smiddy WE, Flynn HW, Rubsamen PE. Vitrectomy for diabetic macular edema associated with a thickened posterior hyaloid and the inner limiting membrane. Am J Ophthalmol 1996; 121: 405–413.
18. Larson HW. Diabetic Retinopathy. An ophthalmoscopic study with a discussion of the morphologic changes and the pathogenetic factors in this disease. Acta Ophthalmol (Copenh) 1960; 60-1-89.
19. Davis MD. The natural course of diabetic retinopathy. In: Kimura SJ, Caygil WM (eds), Vascular Complications of Diabetes Mellitus. CV Mosby, St. Louis, 1967; pp 139–169.

20. Frank RN. Etiologic mechanisms in diabetic retinopathy. In: Ryan SJ (ed), Retina Vol. 2; Schachat AP, Murphy RP, Patz A (eds) Medical Retina. CV Mosby, St. Louis, 1989; pp 301–326.
21. Patz A, Schatz H, Berkow JW, Gittelsohn AM, Ticho U. Macular edema – an overlooked complication of diabetic retinopathy. Trans Am Acad Ophthalmol Otolaryngol 1977; 77: 34–42.
22. Ferris FL, Patz A. Macular edema. A complication of diabetic retinopathy. Surv Ophthalmol 1984; 28 (suppl): 452–461.
23. King RC. Exudative diabetic retinopathy. Br J Ophthalmol 1963; 47: 666–672.
24. Sigurdsson R, Begg I. Organized macular plaques in exudative diabetic maculaopathy. Br J Ophthmol 1980; 64: 392–397.
25. Klein BEK, Moss SE, Dlein R, Surawicz TX. The Wisconsin Epidemiologic Study of Diabetic Retinopathy, XIII: Relationship of serum cholesterol to retinopathy and hard exudate. Ophthalmol 1991; 98: 1261–1265.
26. Chew EY, Klein ML, Ferris FL, Remaley NA, Murphy RP, Chantry K, Hoogwerf BJ, Miller D; for the ETDRS Research Group. Association of elevated serum lipid levels with retinal hard exudate in progression of diabetic retinopathy. Orch Ophthalmol 1996; 114: 1079–1084.
27. Ashton N. Vascular changes in diabetes with particular reference to the retinal vessels. Br J Ophthalmol 1949; 33: 407–420.
28. Cunha-Vaz JG. Pathophysiology of diabetic retinopathy. Br J Ophthalmol 1978; 62: 351–355.
29. Cogan DG, Toussaint D, Kuwabara T. Retinal vascular patterns. Arch Ophthalmol 1961; 66: 366–378.
30. Whitelocke RAF, Kearns M, Blach RK, Hamilton AM. The diabetic maculopathies. Trans Ophthalmol Soc UK 1979; 99: 314–320.
31. Kaback MB, Tanenbaum HL. The macula in diabetes. Can J Ophthalmol 1974; 9: 202–207.
32. Henkind P, Bellhorn RW, Schall B. Retinal edema: postulated mechanism(s). In: Cunha-Vaz JG (ed), The Blood-retinal Barriers. NATO advanced study institute on the blood-retinal barriers, Expinho, Portugal. NATO advanced study institute series: series A, life sciences, 1980. Plenum Press, New York/London, 32: 251–268.
33. Cheng H, Black RK, Hamilton AM, Kohner EM. Diabetic macular edema. Trans Ophthalmol Soc UK 1972; 92: 407–411.
34. Davis MD. The natural course of diabetic retinopathy. Trans Am Acad Ophthalmol Otolaryngol 1967; 71: 237–240.
35. Dobree JH. Simple diabetic retinopathy. Br J Ophthalmol 1970; 54: 1–10.
36. Patz A, Berkow JW. Visual and systemic prognosis in diabetic retinopathy. Trans Am Acad Ophthalmol Othoaryngol 1967; 71: 253–258.
37. Davis MD, Fisher MR, Gangnon RE, Barton F, Aiello LM, Chew EY, Ferris FL, Knatterud GL. Risk factors for high-risk proliferative diabetic retinopathy and severe visual loss: Early Treatment Diabetic Retinopathy Study Report #18. Invest Ophthalmol Vis Sci 1998 Feb; 39(2): 233–252.
38. King RC. Exudative diabetic retinopathy. Br J Ophthalmol 1963; 47: 666–672.
39. Sigurdsson R, Begg I. Organized macular plaques in exudative diabetic maculopathy. Br J Ophthalmol 1980; 64: 392–397.
40. The DCCT Research Group. The Diabetes Control and Complications Trial: design and methodologic considerations for the feasibility phase. Diabetes 1986; 35: 530–545.
41. The Diabetes Control and Complications Trial (DCCT) Research Group. Effect of intensive therapy on the development and progression of diabetic nephropathy in the Diabetes Control and Complications Trial. Kideny Int 1995; 47: 1703–1720.
42. Idem. Effect of intensive diabetes treatment on nerve conduction in the Diabetes Control and Complications Trial. Ann Neurol 1995; 38: 869–880.
43. Reichard P, Nilsson B-Y, Rosenqvist U. The effect of long-term intensified insulin treatment on the development of microvascular complications of diabetes mellitus. N Engl J Med 1993; 329: 304–309.
44. Klein R, Dlein BEK, Moss SE, Davis MD, DeMets DL. The Wisconsin Epidemiologic Study of Diabetic Retinopathy. Glycosylated hemoglobin predicts the incidence and progression of diabetic retinopathy. JAMA. 1988; 260: 2864–2871.
45. Klein R, Klein BEK, Moss SE. Relation of glycemic control to diabetic microvascular complications in diabetes mellitus. Ann Intern Med 1996; 124: 90–96.
46. UK Prospective Diabetes Study (UKPDS) Group. Effect of intensive glood-glucose control with metformin on complications in overweight patients with type 2 diabetes (UKPDS 34). Lancet 1998; 352: 854–865.
47. Ohkubo Y, Kishikawa H, Araki E, et al. Intensive insulin therapy prevents the progression of diabetic

microvascular complications in Japanese patients with non-insulin dependent diabetes mellitus: a randomized prospective 6-year study. Diabetes Res Clin Pract 1995; 28: 103–117.
48. UK Prospective Diabetes Study Group. Tight blood pressure control and risk of macrovascular and microvascular complications in type 2 diabetes: UKPDS 38. BMJ 1998; 317: 703–713.
49. Idem. Efficacy of atenolol and captopril in reducing risk of macrovascular and microvascular complications in type 2 diabetes: UKPDS 39. BMJ 1998; 317: 313–320.
50. Klein BEK, Moss SE, Klein R, Surawicz TS. The Wisconsin Epidemiologic Study of Diabetic Retinopathy. XIII. Relationship of serum cholesterol to retinopathy and hard exudate. Ophthalmology 1991; 98: 1261–1265.
51. Chew EY, Klein ML, Ferris FL III, et.al. Association of elevated serum lipid levels with retinal hard exudate in diabetic retinopathy: Early Treatment Diabetic Retinopathy Study (ETDRS) report 22. Arch Ophthalmol 1996; 114: 1079–1084.
52. King RC. Exudative diabetic retinopathy. Br J Ophthalmol 1963; 47: 666–672.
53. Duncan L. A three-year trial of atromid therapy in exudative diabetic retinopathy. Diabetes 1968; 17: 458–467.
54. Harold M, Marvin M, Gough K. A double-blind controlled trial of clofibrate in the treatment of diabetic retinopathy., Diabetes 1969; 18: 285–291.
55. Blankenship GW. Diabetic macular edema and laser photocoagulation. Ophthalmology 1979; 86: 69–75
56. Townsend C, Bailey J, Kohner EM. Xenon arc photocoagulation in the treatment of diabetic maculaopathy. Trans Ophthalmol Soc UK 1979; 99: 13–16.
57. Early Treatment Diabetic Retinopathy Study Research Group. Early Treatment Diabetic Study design and baseline characteristics: ETDRS report number 7. Ophthalmology 1991; 98: 741–756.
58. Idem. Photocoagulation for diabatic macular edema: Early Treatment Diabetic Retinopathy Study report number 1. Arch Ophthalmol 1985; 103: 1796–1806.
59. Idem. Treatment techniques and clinical guidelines for photocoagulation of diabetic macular edema: Early Treatment Diabetic Retinopathy Study report number 2. Ophthalmology 1987; 94: 761–774.
60. Idem. Techniques for scatter and local photocoagulation treatment of diabetic retinopathy: Early Treatment Diabetic Retinopathy Study report number 3. Int Ophthalmol Clin 1987; 27: 254–264.
61. Idem. Photocoagulation for diabetic macular edema. Early Treatment Diabetic Retinopathy Study report number 4. Int Ophthalmol Clin 1987; 27: 265–272.

STEVEN T. CHARLES

6. Diabetic Vitrectomy Update

Editor's Comment: estoration of vision in post-hermorrhagic diabetic retinopathy by mechanical removal of clots and fibrous tissue has remarkably extended the potential of eye surgery. Simultaneous replacement of a cataract occluded lens adds to the possible benefit of surgical intervention. Terming current vitrectomy as safer, faster and with better outcomes, Charles advocates performance of the procedures as an outpatient under local anesthesia. By coupling vitrectomy with pre, intra, and post-vitrectomy panretinal photocoagulation aggressive eye intervention has remarkably reduced blindness due to diabetic retinopathy.

There have been considerable recent advances in vitrectomy tools and techniques for treating the complications of diabetic retinopathy. It is essential to recognize that to a substantial extent, a diabetic patient requiring vitrectomy indicates a failure of the diabetic screening process, patient education, medical management, retinopathy screening, and/or laser treatment for diabetic retinopathy. Early detection of clinically significant macular edema and intervention with focal laser is a remarkable success story as is pan-retinal photocoagulation for proliferative retinopathy. Prevention of retinopathy is better for the patient than laser treatment, which is, in turn, far less invasive than vitrectomy.

INDICATIONS FOR VITRECTOMY

A thoughtful approach to the indications for vitrectomy is essential because of potential medical and surgical complications associated with vitrectomy as well as the cost. So-called early vitrectomy tends to be easier with relatively low complication rates in contrast to waiting until epiretinal membranes and traction retinal detachment have occurred. It is important to remember that placing undue emphasis on early vitrectomy can also result in unnecessary surgery as many vitreous hemorrhages clear spontaneously and less than 25% of extra-macular traction retinal detachments progress to include the macula. Surgery for diabetic macular edema or submacular exudate must be very carefully considered because these patients retain ambulatory vision if surgery is not performed and the surgery has significant associated complications.

The decision to operate on a patient with a vitreous hemorrhage is a complex issue based on visual potential of the eye to be operated versus that of the other eye, current status of the other eye, coexistent medical and ocular conditions, and the patient's visual needs. A dense hemorrhage into the vitreous body is much less likely to clear in weeks than diffuse mobile, erythrocytes in the space between the posterior vitreous cortex and the retinal surface. Some patients are more comfortable with surgery on their worst

E.A. Friedman and F.A. L'Esperance, Jr. (eds.), Diabetic Renal-Retinal Syndrome, 59–65.

eye but this may have the unintended consequence of reluctance to have needed surgery on the better eye if the first eye has a poor outcome. It is often difficult to predict which will ultimately be the better or even the only eye.

It is clear that very small extra-macular traction retina detachments have a better surgical prognosis than large macular traction retinal detachments. It is well known that only a small fraction of extra-macular detachments progress to include the macula. Clinical judgment should indicate that a nasal traction detachment or traction detachment located a significant distance from macula are very unlikely to progress to a macula-off detachment.

Hilel Lewis, et al. developed the concept of vitrectomy for macular edema thought to be secondary to traction on the macula from the posterior vitreous cortex. While some surgeons apparently believe that the key effect is reducing tangential traction on the internal limiting membrane; it is more likely that eliminating macular elevation from the retinal pigment epithelium is the key factor. Some surgeons believe that it is not necessary to establish the presence of traction. If this proves to be the case, the mechanism may be that decompartmentalization may result in a reduction of VEGF concentration in the macula as postulated by the author. Another controversial issue is the timing of vitrectomy with respect to focal/grid laser treatment and how much laser should be done before considering surgery. Some observers believe that OCT is useful in determining whether macular elevation due to traction is present.

Ogura developed the concept of removing sub-macular exudates in selected diabetic cases. This procedure remains controversial because of the possibility of pre or intra-surgical photoreceptor and RPE damage as well as the potential for surgical complications such as bleeding, cataract, retinal detachment and neovascular glaucoma. Until further studies are completed it would appear that a well defined 'disk' of seemingly mobile sub-macular exudate that has been given ample time to absorb and is not accompanied by evidence of macular ischemia or RPE damage should be considered for surgery.

MANAGEMENT OF COEXISTENT CATARACT

Many surgeons are advocates of so-called combined procedures including vitrectomy, lens removal and intraocular lens implantation. Although this approach means one less operation for the patient and is an interesting challenge for the surgeon, there are several unintended consequences of this approach. If phakoemulsification is performed prior to vitrectomy, corneal clarity and pupillary dilation may be compromised, negatively impacting surgical visualization. If lens removal is performed after vitrectomy, again surgical visualization may be limited. Lensectomy at the beginning of the case followed by sulcus implantation after the vitrectomy may result in increased bleeding during the obligatory hypotony associated with lens insertion. Patients today expect emmetropia after cataract surgery, which is difficult for the vitreous surgeon to achieve with combined procedures. The author currently prefers sending the patient to the cataract surgeon for phako and posterior chamber lens implantation several weeks prior to vitrectomy if the view is limited. If safe vitrectomy can be performed through the cataract, phako and IOL insertion can be done when the eye is vitreous is clear and all neovascularization has regressed.

VITRECTOMY

Truncation of the posterior vitreous cortex, also described as 3,600 removal of antero-posterior vitreous traction, has been taught almost since the inception of diabetic vitrectomy. This step was has been increasingly safer and faster because of better fluidics and vitreous cutters. The first step in improved fluidics occurred when Conor O'Malley introduced a machine-based aspiration system controlled by the surgeon's foot. The next revolutionary advance was the introduction of proportional suction by the author. Evolutionary design resulted in faster fluidics, which are now 10× faster than the surgeon response time (Alcon Accurus latency is 25 milliseconds). In 2000, the Alcon engineering team in collaboration with the author developed the concept of probe-based flow limiting via high cutting rates. This concept is analogous to the idea of high vacuum, low flow phako introduced by Alcon for cataract surgery. The advantages of high resistance at the orifice in phako surgery are: better anterior chamber stability and reduced capsule rupture and vitreous loss after an occlusion break. Similarly, higher cutting rates limit flow and reduce vitreoretinal traction after sudden elastic deformation of an epiretinal membrane through the port. High cutting rates with InnoVit or axial cutters enable safe removal of many epiretinal membranes without the need for scissors. In addition, high cutting rates produce greater fluidic stability when working near thin, mobile, peripheral retina. Another advantage of high cutting rate is prevention of the movement of uncut vitreous fibers through the port, especially with high vacuum settings and during probe movement.

MANAGEMENT OF EPIRETINAL MEMBRANES

Surgical management of epiretinal began with Robert Machemer's introduction of membrane peeling using a bent needle in 1972. This was followed by Conor O'Malley's development of the rounded membrane pic. Soon thereafter, the author introduced the concept of scissors segmentation of epiretinal membranes. Segmentation means to section the membrane into epicenters in order to relieve tangential traction on the retina and enable reattachment. Scissors delamination of epiretinal membranes was the next revolutionary step developed by the author. Delamination involves the use of scissors in the potential space between the epiretinal membrane and the retina.

The author developed the concept of delamination in broad sheets in an inside-out orientation. This step should be performed before truncation of posterior vitreous cortex, if no vitreous separation is present. Subsequently others introduced the name 'en bloc' but described the inadvisable concept of pulling on the vitreous in an outside-in orientation to assist in removing the ERM. The author's original concept of delamination is now often called 'en bloc'.

Acceptance of delamination was held back by the fear some surgeons had of impaling the retina with the scissors points. The recent introduction of curved delamination scissors by Alcon and DORC have enables much safer and faster delamination.

Dyson Hickingbotham developed the concept of bi-manual surgery that may be better described as 'forceps stabilized epiretinal membrane dissection'. This method has the dual purpose of counteracting tissue dissection forces such as the squeeze-out force produced as scissors close and facilitating visualization of attachment points between the ERM and the retina. The enabling technology is fiber optic forceps.

Although some surgeons advocate visco-dissection this method does not result in shearing or retina-ERM adherence, may lead to higher glial recurrence rates because of retained VEGF, other cytokines, fibronectin and cells and adds cost as well.

HEMOSTASIS

Intra-operative and post-operative bleeding is common in diabetic vitrectomy. At times, surgeons have postulate that the air, other gases, thrombin, viscoelastics, and Factor 13 could reduce post-operative bleeding. Unfortunately, none of these strategies proved to be safe and effective.

McCuen developed the tissue manipulator some years ago in conjunction with Dyson Hickingbotham and Grieshaber. The two-function version included illumination and bipolar diathermy. The three-function version had a lumen for injection or aspiration in addition. The initial implementation had a one-millimeter diameter rather than the twenty-gauge diameter used for all other instruments and produced inadequate illumination. The Alcon EndoZapper version of this concept is ideal for pre-coagulation of vessels to be transected using the small diameter tip for 'focused' diathermy. The author uses the EndoZapper in one hand and the vitreous cutter or curved delamination scissors in the other hand for all diabetic epiretinal membrane dissections. If cutting produces bleeding, the Accurus foot controlled tamponade function is used to quickly raise intraocular pressure and stop the bleeding. The bleeding vessel is then treated with the EndoZapper without the need to remove the dissection tool and introduce a diathermy probe. This approach significantly reduces the development of large, adherent blood clots on the retinal surface.

Stanley Chang introduced the end-aspirating laser probe for use in retinal detachment cases for aspiration of subretinal fluid alternated with laser treatment of retinal breaks, holes, tears, and retinectomies. The author recently introduced the concept of using this probe during diabetic vitrectomy to find and treat all bleeding vascular ERM attachment points.

In addition, all microaneurysms, IRMA, and areas of retinal thickening near the macula can be treated while aspirating residual blood to prevent damage to the nerve fiber layer. Similarly, panretinal photocoagulation can be performed while simultaneously aspirating blood products on the retinal surface to produce a less damage and a more complete PRP.

SUMMARY

The diabetic vitrectomy today is safer, faster and has better outcomes because of tools and techniques such as those described above. The procedures are typically is performed as an outpatient under local anesthesia with an anesthesiologist present. The availability of better glucose, O_2, CO_2, EKG, and electrolyte monitoring and emphasis on normoglycemia during surgery has increased the safety dramatically. Better infusion fluids (BSS Plus) and phakoemulsification and intraocular lens technology has improved the prognosis as well. Pre, intra, and post-vitrectomy panretinal photocoagulation has significantly decreased anterior vitreous cortex fibrovascular proliferation and neovascular glaucoma. It is hoped the screening, wellness programs, better drugs and laser applied by the ETDRS criteria well further reduce avoidable blindness due to diabetic retinopathy.

REFERENCES

1. Chang S. Multifunction endolaser probe. Am J Ophthalmol 1992; 114: 648–649.
2. Charles S. Endophotocoagulation. In: Little H (ed), Diabetic Retinopathy, Chapter 26. Thieme-Stratton, Inc, New York, 1983
3. Charles S. Endophotocoagulation. Ocutome/Fragmatome Newsletter 1980 (September); 5(3): 1.
4. Charles S. Endophotocoagulation. Ophthalmic Laser Therapy 1987; 2(1): 5–12.
5. Charles S. Endophotocoagulation. Ophthalmology Times 1979 (September): 68–69.
6. Charles S. Endophotocoagulation. Retina 1981; 1(2): 117–120.
7. Charles S. Epiretinal membranes: delamination hailed as an improved therapy. Ophthalmology Times 1982 (November 1); 7(16): 52.
8. Charles S. Pars plana vitrectomy for traction retinal detachments. In: Little H (ed), Diabetic Retinopathy, Chapter 25. Thieme-Stratton, Inc, New York, 1983.
9. McCuen BW II, Hickingbotham D. A fiberoptic diathermy tissue manipulator for use in vitreous surgery. Am J Ophthalmol 1984; 98: 803–804.
10. Charles S, White J. Dennison C, Eichenbaum D. Bimanual, bipolar intraocular diathermy. Am J Ophthalmol 1976 (January); 81(1): 101–102.
11. Charles S. Four Papers: Intravenous Fluorescein for Anterior Segment Microsurgery; Bipolar Bimanual Diathermy; Chin Switch; Ocutome. Current Concepts in Cataract Surgery, Selected Proceedings of the Fifth Biennial Cataract Surgical Congress, Jared Emery, Editor, C.V. Mosby Company, 1978.
12. Charles S, Wang C. Linear suction control system. Ocutome/Fragmatome Newsletter 1979 (September); 4(3): 1.
13. Charles S. Vitrectomy for retinal detachment. Trans Ophthal Soc UK 1980; 100(4): 542–549.
14. Charles S. Clinical applications of vitreo-retinal surgery – the Ocutome system. Highlights of Ophthalmology Monthly Letter (Mini-Highlights) 1980; XVI (Chapter 4): 178–216.
15. Charles S, Wang C. A linear suction control for the vitreous cutter (Ocutome). Arch Ophthal 1981; 99: 1613.
16. Charles S. Illuminated intraocular foreign body forceps for vitreous surgery. Arch Ophthal 1981; 99: 1399.
17. Charles S. Bipolar bimanual diathermy (letter). Arch Ophthal 1981; 99; 1868.
18. Charles S, Flinn C. The natural history of diabetic extramacular traction retinal detachment. Arch Ophthal. 1981; 99: 66–68.
19. Charles S. Vitreous Microsurgery. Williams & Wilkins Company, Baltimore, 1981.
20. Charles S, Wang C. Pneumatic intraocular microscissors. Ophthal Digest 1982 (March): 6.
21. Charles S. Diabetic retinopathy. Ophthalmology Times 1982 (June 1).
22. Charles S. Epiretinal membranes: delamination hailed as an improved therapy. Ophthalmology Times 1982 (November 1); 7(16): 52.
23. Charles S. Pars plana vitrectomy for traction retinal detachments. In: Little H (ed), Diabetic Retinopathy, Chapter 25. Thieme-Stratton, Inc, New York, 1983.
24. Charles S. Endophotocoagulation. In: Little H (ed), Diabetic Retinopathy, Chapter 26. Thieme-Stratton, Inc., New York, 1983.
25. Wang CCT, Charles S. Microsurgical instrumentation for vitrectomy, Part 1. Journal of Clinical Engineering 1983 (October-December); 8(4), 321–328.
26. Charles S. The role of intraocular lenses in complex cases of retinal detachment. Personal interview with Boyd B (ed), Highlights of Ophthalmology, 1985.
27. Charles S. The barrier concept. In: Hoffer KJ (ed), Secondary Lens Implantation.
28. Charles S. Pars plana vitrectomy for traction retinal detachments. In: Friedman ER, L'Esperance Jr FA (eds), Diabetic Renal-Retinal Syndrome 3 – Therapy, Grune & Stratton, Inc, Orlando, Florida, 1986, pp 283–295.

29. Charles S. Outpatient vitreoretinal surgery. Outlook, Newsletter of the Outpatient Ophthalmic Surgery Society 1986 (Fall).
30. Blacharski P, Charles S. Thrombin infusion to control bleeding during vitrectomy for Stage V retinopathy of prematurity. Archives of Ophthalmology 1987 (February); 105: 203–205.
31. Charles S. Vitreous Microsurgery, Second Edition. Williams & Wilkins, Baltimore, 1987.
32. Charles S. Laser Endophotocoagulation. In: Gitter K, Schatz H, Yannuzzi L, McDonald R (eds), Laser Photocoagulation of Retinal Disease, Chapter 6. Pacific Medical Press, San Francisco, California, 1988.
33. Charles S. Management of Epiretinal Membranes. Ophthalmic Practice 1988 (December); 6(4): 158, 178.
34. Charles S. Principles and techniques of vitreous surgery. In: Ryan S, Glaser B, Michels R (eds), Retina, Volume III, Chapter 124. C.V. Mosby Company, St. Louis, Missouri, February 1989.
35. Runge P, Charles S. New Instrumentation and Pars Plana Vitrectomy Techniques. In: Ai E, Freeman WR (guest editors), Ophthalmology Clinics of North America, Vol. 3, No. 3, September 1990, pp 517–531.
36. Charles S. Vitreoretinal technology for the 90's. In: Fuller DG (ed), Vitreoretinal Surgery and Technology Newsletter. May/June 1991
37. Charles S. Posterior segment vitrectomy: an overview of indications. In: Wilkinson CP (ed), Consultant for AAO 1991 Focal Points, Clinical Modules for Ophthalmologists, Volume IX, Module 4. San Francisco, California.
38. Varley M, Charles S. Lasers in vitreoretinal surgery. Ophthalmic Practice 1991 (June); 9(3): 118–122.
39. Charles S. Vitreoretinal surgery: new outpatient options. In Morgan K (ed) The Outpatient Ophthalmic Surgery Society Journal, Vol. IV, Winter 1992.
40. Charles S. Periretinal Membranes: Techniques/Economics. Ocular Surgery News, SLACK Incorporated, Thorofare, New Jersey, March 15 & April 1, 1992.
41. Charles S. Lens Management in Vitreoretinal Surgery. Ocular Surgery News, SLACK Incorporated, Thorofare, New Jersey, November 15, 1992.
42. Charles S, Wood B. Vitreoretinal surgical techniques. In: Freeman WR (ed), Practical Atlas of Retinal Disease and Therapy. Raven Press, Ltd, New York, 1993, pp 243–266.
43. Charles S, Sullivan J. Management of the cataract patient with significant retinal disease. In: Lindstrom (ed), Current Opinion in Ophthalmology, vol. 4, no. 1, February 1993.
44. Charles S, Wong R. Decision Making in Vitreoretinal Surgery. Ocular Surgery News, SLACK Incorporated, Thorofare, New Jersey, Vol. 11, No. 10, May 15, 1993 & June 1, 1993.
45. Charles S, Jewart B. Management of the cataract patient with significant retinal disease (update of 1993 article). Current Opinion in Ophthalmology 1994; 5(1): 101–104.
46. Charles S. Simplified Epiretinal Membrane Dissection, Endocapsular Lensectomy and Subretinal Fluid Removal Options. Guest Editor for Vitreoretinal Surgery and Technology Newsletter, Vol 8, No. 2, Spring/Summer 1996.
47. Charles S, Small K. Pathogenesis and Repair of Retinal Detachment. In: Easty DL, Sparrow JM (eds), Oxford Textbook of Ophthalmology, Volume 2, Chapter 3.19. Oxford Medical Publications, Oxford University Press Inc, New York, 1999, pp 1261–1272.
48. Leonard B, Charles S. Diabetic Retinopathy. In: Silverstone B, Lang MA, Rosenthal, B, Faye EE (eds), The Lighthouse Handbook on Vision Impairment and Vision Rehabilitation, Volume 1, Chapter 6. Oxford University Press, New York, New York, 2000, pp 103–127.
49. Zocchi K, Charles S. Management of Cataract in the Diabetic Patient. In: BurattoL, Osher RH, Masket S (eds), Cataract Surgery in Complicated Cases, Chapter 16. Slack, Inc, Thorofare, New Jersey, 2000.
50. Gandorfer A, Messmer EM, Ulbig MW, Kampik A. Resolution of diabetic macular edema after surgical removal of the posterior hyaloid and the inner limiting membrane. Retina 2000; 20: 126–133.
51. Jumper JM, Embabi SN, Toth CA, McCuen BW, II, Hatchell DL. Electron immunocytochemical analysis of posterior hyaloid associated with diabetic macular edema. Retina 2000; 20: 63–68.
52. Lewis H, Abrams GW, Blumenkranz MS, Campo RV. Vitrectomy for diabetic macular traction and edema associated with posterior hyaloidal traction. Ophthalmology 1992; 99: 753–759.
53. Pendergast SD. Vitrectomy for diabetic macular edema associated with a taut premacular posterior hyaloid. Curr Opin Ophthalmol 1998; 9: 71–75.
54. Sakuraba T, Suzuki Y, Mizutani H, Nakazawa M. Visual improvement after removal of submacular exudates in patients with diabetic maculopathy [in process citation]. Ophthalmic Surg Lasers 2000; 31: 287–291.
55. Takagi H, Otani A, Kiryu J, Ogura Y. New surgical approach for removing massive foveal hard exudates in diabetic macular edema. Ophthalmology 1999; 106: 249–256; discussion 256–247.

56. Yang CM. Surgical treatment for severe diabetic macular edema with massive hard exudates. Retina 2000; 20: 121–125.
57. Wu WC, Lin JC. The experience to use a modified en bloc excision technique in vitrectomy for diabetic traction retinal detachment. Kaohsiung J Med Sci 1999; 15: 461–467.
58. Han DP, Murphy ML, Mieler WF. A modified en bloc excision technique during vitrectomy for diabetic traction retinal detachment. Results and complications. Ophthalmology 1994; 101: 803–808.
59. Williams DF, Williams GA, Hartz A, Mieler WF, Abrams GW, Aaberg TM. Results of vitrectomy for diabetic traction retinal detachments using the en bloc excision technique. Ophthalmology 1989; 96: 752–758.

RONALD I. KLEIN AND BARBARA E.K. KLEIN

7. Medical Approaches to Preventing Vision Loss from Diabetic Retinopathy

Editor's Comment: Beginning with the earliest trials of careful metabolic control in diabetic subjects, it has become increasingly evident that striving for euglycemia reduces the rate of microvasculopathy as signaled by proteinuria or vision loss. In this review, Drs. Klein observe that until some means of primary prevention is in hand, it is mandatory to maximize secondary prevention of diabetic eye disease by control of blood pressure and glycemia. Looking forward to promising newer interventions (e.g., reduction of lipids, inhibiting protein kinase C), screening by skilled eye care providers on a regular basis for early detection and treatment, of vision threatening retinopathy with photocoagulation is vital to preservation of sight.

Data from the Diabetic Retinopathy Study [1] and the Early Treatment Diabetic Retinopathy Study [2], demonstrated the efficacy of photocoagulation treatment in preventing visual loss associated with proliferative diabetic retinopathy. However, clinical trial data demonstrating the efficacy of medical interventions for this disease have only become available within the last decade. Nonsurgical interventions prior to the development of severe retinopathy are important because photocoagulation is associated with complications and in some cases may not prevent the progression of retinopathy, resulting in visual loss. The purpose of this review is to examine data from ongoing and completed randomized controlled clinical trials of medical interventions to reduce the progression of retinopathy and loss of vision in persons with diabetes.

GLYCEMIC CONTROL

Historical Perspective

The University Group Diabetes Program was the first prospective clinical trial designed to evaluate the effects of four methods of treatment (oral hypoglycemic agents, insulin in fixed or variable dose, and diet only) on vascular complications and mortality [3]. Data from this trial showed no differences in the incidence or progression or retinopathy in the groups studied at 5 and 12 years after randomization despite differences in levels of glycemic control as measured by fasting blood glucose values. In the 1980's epidemiological data from the population-based Wisconsin Epidemiologic Study of Diabetic Retinopathy (WESDR) showed a strong step-wise relation between glycosylated hemoglobin at baseline and the 4-year incidence of retinopathy, progression of

E.A. Friedman and F.A. L'Esperance, Jr. (eds.), Diabetic Renal-Retinal Syndrome, 67–74.

retinopathy, and the incidence of proliferative retinopathy and macular edema in persons with type 1 and 2 diabetes [4, 5].

Continuous subcutaneous insulin infusion and multidose (3 or 4 times per day) treatment regimens along with newer means of determining glycemic control (e.g., home blood glucose monitoring and glycosylated hemoglobin) led to small randomized therapeutic trials to investigate the relation of glycemic control to the incidence and progression of retinopathy in persons with type 1 diabetes [6–8]. These pilot studies consisted of small numbers of subjects followed for short intervals. While they showed that using intense insulin therapy to achieve good glycemic control was feasible, lowering of glycosylated hemoglobin did not result in differences in progression of retinopathy during the first two to three years of observation.

The Diabetes Control and Complications Trial

The Diabetes Control and Complications Trial (DCCT) was 'designed to compare intensive with conventional diabetes therapy with regard to their effects on the development and progression of the early vascular and neurologic complications of IDDM' [9–13]. The study was designed to make recommendations regarding the benefits and risks associated with intensive insulin therapy. The study had two therapeutic groups, a primary prevention group and a secondary intervention group, the dichotomy based on the presence or absences of retinopathy at study enrollment. A major endpoint in the study and the criterion for being in the primary or secondary prevention arm was based on masked grading of stereoscopic color fundus photographs of the seven standard photographic fields of the ocular fundus. Photographs were taken every six months and graded according to the modified Airlie House Classification scheme and the Early Treatment Diabetic Retinopathy severity scale was applied to these gradings. In 1993, the DCCT reported:

1. a significant reduction in the risk of sustained progression of retinopathy of approximately 50% in the intensive therapy group compared to the conventional therapy group in the primary-prevention cohort after 5 years of follow-up;
2. in the secondary-intervention cohort, the intensive-therapy group had a reduction of progression of retinopathy by 54% during the entire study period compared to those assigned to the conventional therapy group and a 56% reduction in the subsequent rate of laser treatment;
3. an early worsening of retinopathy in the first year of treatment of the intensive therapy group in the secondary-intervention cohort;
4. an absence of a glycemic threshold, with reduction of the risk of progression of retinopathy that was significant across the entire range of glycosylated hemoglobin;
5. that intensive insulin therapy was associated with 50% higher rate of recovery from progression of three or more steps of retinopathy compared to those treated with conventional therapy;
6. that there was a reduction of the incidence and progression of retinopathy for those receiving intensive treatment earlier in the course of diabetes, before the onset of diabetic retinopathy; and
7. that there was a two to threefold increase in severe hypoglycemic reactions in the intensive insulin treatment group compared to the conventional group.

At the conclusion of the DCCT, the patients in the conventional therapy group were offered intensive glycemic therapy and retinopathy was evaluated during the 4th year after the DCCT ended [14]. At that time, the difference in the median glycosylated hemoglobin level between those who had originally been in the intensive therapy group and those who had been in the conventional therapy group had narrowed from 1.9% at the end of the DCCT (9.1% in the conventional and 7.2% in the intensive group) to 0.3% (8.2% in the conventional and 7.9% in the intensive group). Despite this, the risk of progression of retinopathy during the first four years after the end of the DCCT, remained higher in the original conventional therapy group (21%) compared to the original intensive insulin therapy group (6%). There was a 75% reduction in the incidence of proliferative retinopathy, a 69% reduction in the incidence of clinically significant macular edema, and a 58% reduction in the incidence of laser therapy in the intensive therapy group compared to the conventional group four years after the end of the DCCT. These data suggested that the benefits of intensive therapy with near-normal levels of glycosylated hemoglobin over the 7.5 years of the study persists for about four years despite increasing hyperglycemia. Along with the DCCT results, these findings suggest the importance of early intensive insulin therapy after diagnosis of type 1 diabetes.

The United Kingdom Diabetes Prospective Study

The United Kingdom Prospective Diabetes Study (UKPDS) was a randomized controlled clinical trial involving 3,867 newly diagnosed patients with type 2 diabetes [15–17]. After 3 months of diet treatment patients with a mean of two fasting plasma glucose concentrations of 6.1 to 15.0 mmol/L were randomly assigned to an intensive glycemic control group with either a sulfonylurea (chlorpropamide, glibenclamide, or glipizide) or insulin or to a conventional glycemic control group with diet. In addition, metformin was included as one of the treatment arms for 1,704 overweight patients and analyses included comparison of the effect of metformin against conventional therapy in overweight patients. After 12-years of follow-up they reported:

1. a reduction in rate of progression of diabetic retinopathy of 21% and reduction in need for laser photocoagulation of 29% in the intensive vs the conventional treatment group;
2. that there was no difference in reduction in the rate of the retinopathy endpoints among the three agents used in the intensive treatment group (chlorpropamide, glibenclamide, and insulin) but the chlorpropamide treatment group failed to show a reduced rate of retinopathy requiring photocoagulation;
3. that there was no difference between conventional and intensive treatments in the deterioration of visual acuity in 3 years in each group, and the proportion of patients legally blind at 12 years;
4. metformin is preferred as first-line pharmacological therapy in newly diagnosed type 2 diabetic patients who are overweight with a significant (39%) reduction in myocardial infarction compared to the conventional treatment group; however when metformin was added to sulphonlyureas (in both obese and nonobese patients) it was associated with increased diabetes-related (96%) and all-cause mortality (60%) compared to conventional therapy;

5. patients in the intensive treatment group had significantly more major hypoglycemic episodes and weight gain than patients in the conventional group; and
6. that economic analyses of the clinical trial data suggested that intensive glucose control increased treatment costs but substantially reduced complication costs and increased the time free of such complications.

Implications for Clinical Practice

With the results of the UKPDS, the DCCT investigators statement that 'intensive therapy should form the backbone of any health care strategy aimed at reducing the risk of visual loss from diabetic retinopathy' is applicable to persons with both type 1 and 2 diabetes. This has provided further support to the American Diabetes Association guidelines of a target goal of glycosylated hemoglobin of 7.0% for persons with diabetes [18]. However, data from the National Health and Nutrition Examination Survey III [19] and the WESDR [20] suggest that few persons with diabetes reach this targeted level of glycemic control. Until better treatment approaches with minimal complications (e.g., hypoglycemia) to achieve these levels of glycemic control are available, additional medical approaches are needed to prevent retinopathy. One of these includes controlling hypertension.

BLOOD PRESSURE CONTROL

Historical Perspective

Anecdotal observations from clinical studies suggested an association between hypertension and the severity of retinopathy in people with diabetes [21]. While epidemiological data from cross-sectional studies suggested an association of hypertension with diabetic retinopathy, data from cohort studies were inconsistent [22]. Epidemiological analyses from the UKPDS showed that the incidence of retinopathy was associated with systolic blood pressure. For each 10 mm Hg decrease in mean systolic blood pressure a 13% reduction was found for microvascular complications and no threshold was found for any endpoint [23]. In the WESDR, diastolic blood pressure was a significant predictor progression of diabetic retinopathy and the incidence of proliferative diabetic retinopathy in patients with younger-onset type 1 diabetes mellitus [24]. However, neither systolic or diastolic blood pressure or hypertension status at baseline were associated with the incidence and progression of retinopathy in people with older onset type 2 diabetes [22]. Diastolic blood pressure however was found to be associated with a 330% increased 4-year risk of developing macular edema in those with type 1 diabetes and a 210% increased risk in those with type 2 diabetes in the WESDR.

Clinical Trials

The EURODIAB Controlled Trial of Lisinopril in Insulin-Dependent Diabetes Mellitus (EUCLID) Study sought to examine the role of an angiotensin converting enzyme (ACE) inhibitor in reducing the incidence and progression of retinopathy in a group of largely normotensive type 1 diabetic patients of whom 85% did not have microalbuminuria at baseline [25]. This study showed a statistically significant 50% reduction in the progression of retinopathy in those taking Lisinopril over a two-year period after adjustment for glycemic control. Progression to proliferative retinopathy was also reduced. There was no significant interaction with blood glucose control. It was postulated that ACE inhibitors might have an effect independent of blood pressure lowering [26].

The UKPDS also sought to determine whether lower blood pressure achieved with either a beta blocker or an ACE inhibitor was beneficial in reducing macrovascular and microvascular complications associated with type 2 diabetes [27]. One thousand forty-eight patients with hypertension (mean blood pressure 160/94 mm Hg) were randomized to a regimen of tight control with either captopril or atenolol and another 390 patients to less tight control of their blood pressure. The aim in the group randomized to tight control was to achieve blood pressure values < 150/< 85 mm Hg. If these goals were not met with maximal doses of a beta blocker or ACE inhibitor, additional medications were prescribed, including a loop diuretic, a calcium channel blocker, and a vasodilator. The aim in the group randomized to less tight control was to achieve blood pressure values < 180/< 105 mm Hg. Tight blood pressure control resulted in a 35% reduction in retinal photocoagulation compared to conventional control. After 7.5 years of follow-up, there was a 34% reduction in the rate of progression of retinopathy by two or more steps using the modified ETDRS severity scale and a 47% reduction in the deterioration of visual acuity by 3 lines or more using the ETDRS charts (for example, a reduction in vision from 20/30 to 20/60 or worse on a Snellen chart). The effect was largely due to a reduction in the incidence of diabetic macular edema. Atenolol and captopril were equally effective in reducing the risk of developing these microvascular complications. The effects of blood pressure control were independent of those of glycemic control. These findings support the recommendations for tight blood pressure control in patients with type 2 diabetes as a means of preventing visual loss from diabetic retinopathy. Results from other clinical trials that have recently been completed (e.g., ABCD study) [28] or from those that are underway, should provide more information regarding the relative efficacy of blood pressure control and specific antihypertensive medications in reducing the progression of retinopathy in persons with diabetes.

LIPID LOWERING

Historical Prospective

Macular edema is an important cause of loss of vision in people with diabetes [22]. Hard exudate, a lipoprotein deposit, is often associated with macular edema. Data from early clinical studies showed an association of elevated triglycerides and lipids with hard exudate. Clinical trials of clofibrate, showed that treatment with this drug reduced the incidence of hard exudate but failed to restore vision to eyes with established macular edema at the onset of the trial [29]. Clofibrate was associated with liver toxicity and is no longer used. In the WESDR, higher total serum cholesterol was associated with higher prevalence of retinal hard exudates in both the younger- and the older-onset groups taking insulin, but not in those with type 2 diabetes [30]. In the ETDRS, higher levels of serum lipids (triglycerides, low-density lipoproteins and very-low-density lipoproteins) were associated with increased risk of developing hard exudates in the macula and a decrease in visual acuity [31].

Clinical Trials

To our knowledge there are no completed clinical trials showing the efficacy of lipid lowering agents in reducing the progression of retinopathy, the incidence of macular edema, or the loss of vision. Currently there are a number of clinical trials underway or just beginning to examine the efficacy of statins in reducing the risk of progression of retinopathy and visual loss.

OTHER APPROACHES

A clinical trial of aldose reductase inhibitors failed to demonstrate the efficacy of this approach in preventing the incidence and progression of retinopathy in people with diabetes [32]. In addition, data from the ETDRS showed no efficacy of aspirin in preventing the progression of retinopathy. In that trial, treatment with aspirin was associated with a significant reduction in heart attack [33]. It did not increase the risk of vitreous hemorrhage in people with diabetes. Results from this and other studies have led to a recommendation of use of enteric-coated aspirin in low doses of 81 to 325 mg/d in people with diabetes who have a history of angina, myocardial infarction, stroke or transient ischemic attacks, and peripheral vascular disease.

A number of new clinical trials are currently underway to evaluate the efficacy of various new agents including protein kinase C inhibitors, vascular endothelial growth factor inhibitors, and metalloproteinase inhibitors in preventing the incidence or progression of diabetic retinopathy.

CONCLUSIONS

Prevention of diabetes remains an important goal in reducing the complications and costs of this disease. Until approaches for primary prevention become available, clinical trial data have shown that secondary prevention through medical interventions designed to control blood pressure and glycemia will reduce the incidence and progression of retinopathy and loss of vision [9–11, 15, 23, 25]. However, success of these interventions has been limited, in part, due to inability to achieve normalization of blood sugar with current drug delivery systems. While new interventions (e.g., reduction of lipids, inhibiting protein kinase C) may be of further benefit, tertiary prevention of visual loss (screening examination through a dilated pupil by skilled eye care providers on a regular basis for early detection and subsequent treatment, when indicated, of vision threatening retinopathy with photocoagulation) remains an important approach to care for diabetic patients.

AFFILIATIONS/ACKNOWLEDGMENTS

Department of Ophthalmology and Visual Sciences, University of Wisconsin Medical School, Madison, Wisconsin, USA. This research is supported by National Institutes of Health grants EYO3083 (R. Klein, B.E.K. Klein)), EY12198 (R. Klein, B.E.K. Klein) and NL59259 (R. Klein, B.E.K. Klein)).

REFERENCES

1. Diabetic Retinopathy Study Group. Photocoagulation treatment of proliferative diabetic retinopathy: clinical application of Diabetic Retinopathy Study (DRS) findings. DRS Report No. 8. Ophthalmology 1981; 88: 583–600.
2. ETDRS Research Group. Photocoagulation for diabetic macular edema. Arch Ophthalmol 1985; 103: 1796–806.
3. University Group Diabetes Program. Effects of hypoglycemic agents on vascular complications in patients with adult-onset diabetes. VIII. Evaluation of insulin therapy: final report. Diabetes 1982; 31: 1–81.
4. Klein R, Klein BEK, Moss SE, Davis MD, DeMets SL. Glycosylated hemoglobin predicts the incidence and progression of diabetic retinopathy. JAMA 1988; 260: 2864–2871.
5. Klein R, Klein BEK, Moss SE, Davis MD, DeMets DL. The Wisconsin Epidemiologic Study of Diabetic Retinopathy. IV. Diabetic macular edema. Ophthalmology 1984; 91: 1464–1474.
6. The Kroc Collaborative Study Goup. Diabetic retinopathy after two years of intensified insulin treatment: follow-up of The Kroc Collaborative Study. JAMA 1988; 260: 37–41.
7. Lauritzen T, Frost-Larsen K, Larsen HW, Deckert T, Steno Study Group. Two-year experience with continuous subcutaneous insulin infusion in relation to retinopathy and neuropathy. Diabetes 1985; 34: 74–79.
8. Dahl-Jorgensen K, Brinchmann-Hansen O, Hanssen KF, Ganes T, Kierulf P, Smeland E, Sandvik L, Aagenaes O. Effect of near normoglycemia for two years on progression of early diabetic retinopathy, nephropathy, and neuropathy: The Oslo Study. Br Med J 1986; 293: 1995–1999.
9. The Diabetes Control and Complications Trial Research Group. The effect of intensive treatment of diabetes on the development and progression of long-term complications in insulin-dependent diabetes mellitus. N Engl J Med 1993; 329: 977–986.
10. The Diabetes Control and Complications Trial Research Group. The effect of intensive diabetes treatment on the progression of diabetic retinopathy in insulin-dependent diabetes mellitus: The Diabetes control and Complications Trial. Arch Ophthalmol 1995; 113: 36–51.
11 The Diabetes Control and Complications Trial Research Group. Progression of retinopathy with intensive versus conventional treatment in the Diabetes Control and Complications Trial. Ophthalmology 1995; 102: 647–661.
12. The Diabetes Control and Complications Trial Research Group. The absence of a glycemic threshold for the development of long-term complications: The perspective of the Diabetes Control and Complications Trial. Diabetes 1996: 1289–1298.
13. The Diabetes Control and Complications Trial Research Group. Lifetime benefits and costs of intensive therapy as practiced in the Diabetes Control and Complications Trial. JAMA 1996; 276: 1409–1415.
14. The Diabetes Control and Complications Trial/Epidemiology of Diabetes Interventions and Complications Research Group. Retinopathy and nephropathy in patients with type 1 diabetes four years after a trial of intensive therapy. N Engl J Med 2000; 342: 381–389.
15. UK Prospective Diabetes Study Group. Intensive blood-glucose control with sulphonylureas or insulin compared with conventional treatment and risk of complications in patients with type 2 diabetes. UKPDS 33. Lancet 1998; 352: 837–853.
16. UK Prospective Diabetes Study Group. Effective of intensive blood-glucose control with metformin on complications in overweight patients with type 2 diabetes (UKPDS 34). Lancet 1998; 352: 854–865.
17. Gray A, Raikou M, McGuire A, et al. Cost effectiveness of an intensive blood glucose control policy in patients with type 2 diabetes: economic analysis alongside randomised controlled trial (UKPDS 41). BMJ 2000; 320: 1373–1378.

18. American Diabetes Association. Position statement: Standards of medical care for patients with diabetes mellitus. Diabetes Care 1995; 19: 8–15.
19. Harris MI. Health care and health status and outcomes for patients with type 2 diabetes. Diabetes Care 2000; 23: 754–758.
20. Klein R, Klein BEK, Moss SE, Cruickshanks KJ. The medical management of hyperglycemia over a 10-year period in people with diabetes. Diabetes Care 1996; 19: 744–750.
21. Davis MD. Diabetic retinopathy, diabetic control, and blood pressure. Transplant Proc 1986; 18: 1565–1568.
22. Klein R, Klein BEK. Vision disorders in diabetes. In: Harris MI, Cowie CC, Stern MP, Boyko EJ, Reiber GE, Bennett PH (ed), Diabetes in America, 2nd Ed. National Institutes of Health NIH Publication No. 95-1468; 1995, pp 293–338.
23. Adler A, Stratton IM, Neil HAW, et al. Association of systolic blood pressure with macrovascular and microvascular complicaitons of type 2 diabetes (UKPDS 36): prospective observational study. B Med J 2000: 412–419.
24. Klein R, Klein BEK, Moss SE, Cruickshanks KJ. The Wisconsin Epidemiologic Study of Diabetic Retinopathy. XVII. The 14-year incidence and progression of diabetic retinopathy and associated risk factors in Type 1 diabetes. Ophthalmology 1998; 105: 1801–1815.
25. Chaturvedi N, Sjolie, Stephenson JM, et al. Effect of lisinopril on progression of retinopathy in normotensive people with type 1 diabetes: the EUCLID Study Group. EURODIAB Controlled Trial of Lisinopril in Insulin-Dependent Diabetes Mellitus. Lancet 1998; 351: 28–31.
26. Chaturvedi N. Modulation of the renin-angiotensin system and retinopathy. Heart 2000; 84: i29–i31.
27. UK Prospective Diabetes Study Group. Tight blood pressure control and risk of macrovascular and microvascular complications in Type 2 diabetes. UKPDS 38. Br Med J 1998; 317: 703–713.
28. Savage S, Johnson NL, Estacio RO, et al. The ABCD [Appropriate Blood Pressure Control in Diabetes] Trial: rationale and design of a trial of hypertension control (moderate or intensive) in Type II diabetes. Online Journal of Current Clinical Trials, Doc no. 104, 1993.
29. Duncan LJP, Cullen JF, Ireland JT, Nolan J, Oliver MF. A three-year trial of Atromid therapy in exudative diabetic retinopathy. Diabetes 1968; 17: 458–467.
30. Klein BEK, Moss SE, Klein R, Suawicz TS. The Wisconsin Epidemiologic Study of Diabetic Retinopathy. XIII: relationship of serum cholesterol to retinopathy and hard exudate. Ophthalmology 1991; 98: 1261–1265.
31. Chew EY, Klein ML, Ferris FL III, et al. Association of elevated serum lipid levels with retinal hard exudate in diabetic retinopathy. ETDRS Report Number 22. Arch Ophthalmol 1996; 114: 1079–1084.
32. Sorbinil Retinopathy Trial Research Group. A randomized trial of sorbinil, an aldose reductase inhibitor in diabetic retinopathy. Arch Ophthalmol 1990; 108: 1234–1244.
33. Early Treatment Diabetic Retinopathy Study Research Group. Effects of aspirin treatment on diabetic retinopathy. ETDRS Report Number 8. Ophthamology 1991; 98: 757–765.

David H. Berman

8. Ocular Findings at Onset of Uremia

Editor's Comment: The concept of a Renal-Retinal Syndrome in diabetes is starkly supported by Berman's finding, in his initial eye evaluation of 200 diabetic renal failure patients that 78 patients (39%) had visual acuity of 20/40; 64 patients (32%) had visual acuity of 20/50–20/200; 30 patients (15%) had visual acuity of 20/400-CF; 16 patients (8%) had visual acuity of hand motion (HM)/light perception(LP); and 12 patients (6%) had no light perception (NLP). Clearly, appropriate management of kidney failure in diabetes is linked to preservation of remaining vision and perhaps restoration of some sight where possible once kidney failure is diagnosed. The value of careful follow through is illustrated by an retention of useful vision in the study cohort ;20/200 in 137 patients (69%) at baseline, and in 155 patients (78%) at most recent follow-up, an improvement of 13%. Until the molecular manipulation of harmful kinins and nitric oxide is effectively blocked, the application of bread and butter interventions as illustrated by Berman is distinctly beneficial.

INTRODUCTION

The parallel demise of the eye and the kidney remains a serious consequence of long-standing diabetes mellitus (DM). While various studies have reported on the comorbidity of diabetic retinopathy and diabetic nephropathy, few have examined the extent of ocular disease at the onset of uremia. Earlier estimates of diabetic retinopathy in 97%, and visual impairment in 50% of uremic diabetics [1] have been confirmed more recently [2].

It is now widely recognized that, early detection and timely intervention remain key to minimizing the debilitating consequences of blindness and averting its economic drain on society [3]. Since the incidence and severity of diabetic retinopathy are higher in predominantly black and Hispanic populations [4, 5], usually because of a delay in initial diagnosis, aggressive ophthalmologic management is particularly required. Adopting the recommendations of the major National Eye Institute (NEI) studies, aggressive ophthalmologic management has been efficacious in reducing the risk of visual impairment and improving visual outcome in this population, resulting in economically useful vision in 72% of renal transplant recipients, and 64% of patients undergoing hemodialysis [2].The purpose of this report is to review the ophthalmologic findings and course of 200 consecutively referred patients with diabetic end-stage renal disease (ESRD).

E.A. Friedman and F.A. L'Esperance, Jr. (eds.), Diabetic Renal-Retinal Syndrome, 75–88.

METHODS

Between August 1989 and August 1999, a total of 200 patients with advanced diabetic nephropathy were consecutively referred to the Brooklyn Retinal Vascular Center by the renal divisions of University Hospital of Brooklyn, Kings County Hospital, and the Brooklyn Hospital Center. All patients were enrolled in a comprehensive program of continuing eye and kidney care, to be evaluated diagnostically and managed appropriately. There were no exclusion criteria.

Study population parameters included age, sex, race, duration, onset, and type of diabetes, duration of follow-up, and number deceased during the study period (Table 1).

Renal parameters were defined as, chronic renal insufficiency (creatinine clearance < 60 ml/min/1.73 m^2 and/or urine protein excretion > 200 μg [DHBM1]/min), maintenance hemodialysis (at home or at a dialysis center), or renal transplantation (cadaveric or living donor) (Table 2).

Clinical ophthalmologic evaluation at baseline and at most recent follow-up consisted of visual acuity (best-corrected Snellen visual acuity in either eye), applanation tonometry, slit-lamp exam, dilated fundus exam employing slit-lamp bio-microscopy (Volk 78 D lens) and indirect ophthlamoscopy (Volk 2.2 Panretinal lens). Seven-standard field fundus photography and fluorescein angiography were performed sequentially (Zeiss FF-4 fundus camera). Diabetic retinopathy was graded according

Table 1. Study population parameters.

Characteristic	# or years
Subjects, *n*	200
Mean duration F/U (range), in years	4.23 (1–10)
Mean age (range), in years	53.8 (22–77)
Mean duration of DM (range), in years	19.6 (3–57)
Mean age onset of DM, in years	34.3 (3-64)
Onset of DM, *n* = 200	
Younger-onset (%)	53 (26.5)
Older-onset (%)	147 (73.5)
Type of DM, *n* = 200	
I	30 (15)
II	170 (85)
Sex, *n* = 200	
Males (%)	97 (48.5)
Females (%)	103 (51.5)
Race, *n* = 200	
Blacks (%)	139 (69.5)
Hispanics (%)	20 (10)
Asians (%)	15 (7.5)
Whites (%)	26 (13)
Deceased, number (%)	68 (34)

Table 2. Renal parameters.

Renal status, n = 200	# subjects
Renal insufficiency (%)	43 (21.5)
Hemodialysis (%)	135 (67.5)
Transplant recipients (%)	22 (11)

to the Early Treatment Diabetic Retinopathy Study (ETDRS) modified Airlie House scheme [6]. Grades 43–85 were sub-classified as follows: (A) grades 43–53 signify early to severe non-proliferative retinopathy (NPDR); (B) grades 61–81 signify early to severe proliferative retinopathy (PDR); and (C) grade 85 signifies vitreous hemorrhage and/or tractional retinal detachment; (D) 'e' denotes clinically significant macular edema (CSME). Patients who progressed from non-proliferative to a proliferative grade were so classified. Patients with prior pan-retinal photocoagulation for proliferative disease were designated as level 61.

Intervention with focal laser treatment was indicated for clinically significant macular edema as defined by the ETDRS [7], and intervention with panretinal photocoagulation was indicated for high-risk characteristic eyes as defined by the diabetic retinopathy study (DRS), [8]. Vitrectomy surgery was indicated for non-clearing vitreous hemorrhage and/or tractional retinal detachment [9, 10].

Patients with mild to severe lenticular opacities were graded at the slit-lamp and designated as having cataracts. Those requiring cataract extraction or those for whom it was performed prior to the study underwent either intracapsular, extracapsular, or pars-plana extractions, with or without intraocular lens.

Glaucoma was defined as sustained intraocular pressure of ≥ 25, with or without rubeosis.

Visual acuity was further classified into one of five groups: (A) ≥ 20/40; (B) 20/50–20/200; (C) 20/400-CF; (D) HM/LP; ((E) NLP. Groups A and B defined 'economically useful' vision (enabling self-sufficiency); groups A through C defined 'ambulatory' vision (providing the means for patients to 'get around') [11].

RESULTS

Visual Acuity

At baseline evaluation, 78 patients (39%) had visual acuity of ≥ 20/40; 64 patients (32%) had visual acuity of 20/50–20/200; 30 patients (15%) had visual acuity of 20/400-CF; 16 patients (8%) had visual acuity of hand motion (HM)/light perception(LP); and 12 patients (6%) had no light perception (NLP) (Table 3 and see Figure 1).

The prevalence of economically useful vision (≥ 20/200) among diabetics of ≥ 10 years was similar to that in diabetics of < 10 years (70% vs 74%, respectively), although there was greater prevalence of ≥ 20/40 visual acuity in diabetics of < 10 years (48%) compared to diabetics of ≥ 10 years (35%) (Table 4).

Analysis of visual acuity by renal status at baseline revealed that among the 43

Table 3. Visual acuity – baseline/most recent exam.

Visual acuity	Baseline exam, n = 200 (%)	Most recent exam, n = 200 (%)
≥ 20/40	78 (39)	93 (46.5)
20/50–20/200	64 (32)	62 (31)
20/400-CF	30 (15)	20 (10)
HM/LP	16 (8)	10 (5)
NLP	12 (6)	15 (7.5)

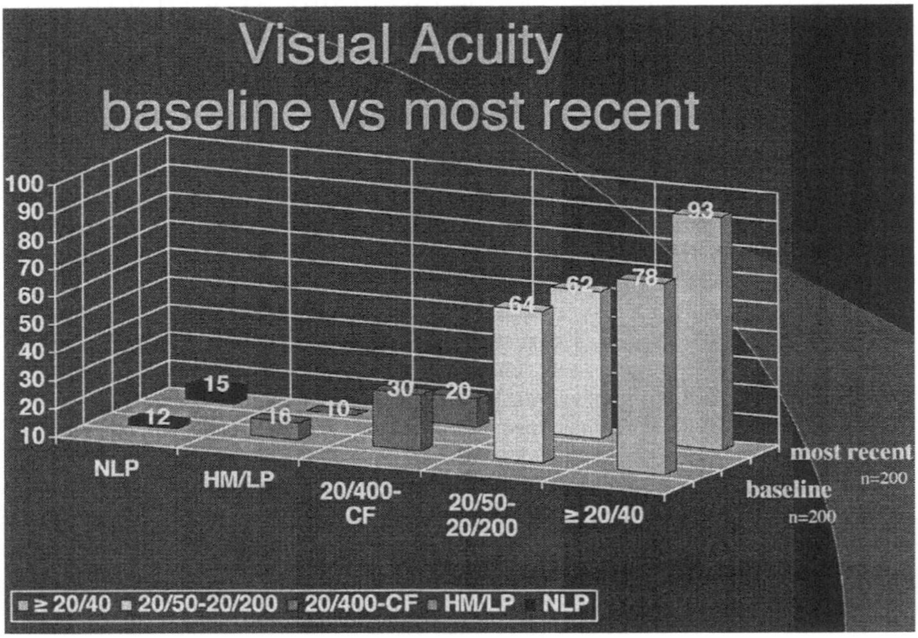

Figure 1.

patients with renal insufficiency, 38 (88%) had visual acuity of ≥ 20/200. Among the 135 patients on maintenance hemodialysis, only 78 (58%) had visual acuity of ≥ 20/200. Visual acuity in the 22 transplant recipients was ≥ 20/200 in 21 (95%)

At most recent follow-up, visual acuity among the renal insufficiency group was ≥ 20/200 in 38 patients (88%), unchanged compared to initial evaluation. Among the patients on hemodialysis, 98 (73%) had visual acuity of ≥ 20/200, reflecting an improvement of 26% with respect to baseline. In the transplant group, visual acuity was ≥ 20/200 in 19 ((86%). Visual acuity in all three renal groups was ≥ 20/200 in 137 patients (69%) at baseline, and in 155 patients (78%) at most recent follow-up, an improvement of 13% (Table 5 and see Figure 2).

Table 4. Visual acuity by duration of DM at baseline exam.

Visual acuity at baseline	Dur DM = 10 yrs, n = 142 (%)	Dur DM < 10 yrs, n = 58 (%)
≥ 20/40	50 (35)	28 (48)
20/50–20/200	49 (35)	15 (26)
20/400-CF	21 (15)	9 (16)
HM/LP	12 (8)	4 (7)
NLP	10 (7)	2 (3)

The pathophysiologic basis of visual acuity ≥ 20/400 at baseline evaluation was attributed to vitreous hemorrhage and/or retinal detachment in 58 patients, whereas the visual acuity at most recent evaluation remained ≥ 20/400 in only 33 patients; this difference was attributed to end-stage tractional retinal detachment in those patients unable benefit from vitrectomy surgery. Of the 94 patients (47%) with visual acuity ≥ 20/200 and active proliferative retinopathy, all had retention or improvement of visual acuity as a result of panretinal photocoagulation. Among the 42 patients who had progressed to proliferative high-risk characteristics, 32 patients (75%) had retention or improvement of visual acuity to ≥ 20/200 (Table 6 and see Figure 3).

Retinopathy

In this cohort of 200 patients, all had some form of retinopathy at baseline evaluation. Non-proliferative retinopathy was present in 94 patients (47%), while 106 patients (53%) manifested proliferative retinopathy including vitreous hemorrhage and/or tractional retinal detachment . Clinically significant macular edema (CSME) was identified in 62 patients (31%) (Table 6). Among the 94 patients with NPDR, 42 (45%) progressed to PDR during the average follow-up period of 3 years (Table 7).

Analyzing the severity of the retinopathy according to duration of diabetes, 142 patients (71%) manifesting any form of retinopathy were diabetic for ≥ 10 years. 95% of patients with PDR had diabetes for ≥ 10 years. Nearly all patients (85%) were type II. Among the 53 patients with younger-onset diabetes (< 30 years of age at the time of diagnosis), 44 (83%) had proliferative disease diagnosed at baseline evaluation, and 6 patients (11%) progressed to PDR during the follow-up period. In the older-onset group (age at diagnosis of diabetes > 30) of 147 patients, 62 (42%) had proliferative disease, a considerably lower incidence compared to the younger-onset group; while 36 (24%) progressed to PDR, more than double compared to the younger-onset group. The incidence of progression to PDR from NPDR was 55% among females, 37% among males. The prevalence of PDR was 61% among females, and 44% among males (Table 7).

The distribution of retinopathy among the three renal groups revealed some distinct differences. At baseline evaluation, CSME was present in 20 patients (47%) with renal insufficiency, in 35 patients (26%) on hemodialysis, and in 4 (18%) of the transplant recipients. Proliferative retinopathy was seen with similar frequency in the renal insuf-

Table 5. Visual acuity by renal status at baseline and most recent exam.

	Renal insufficiency, $n = 43$ (%)	Hemodialysis, $n = 135$ (%)	Transplant, $n = 22$ (%)
Baseline exam			
≥ 20/40	19 (44)	48 (36)	11 (50)
20/50–20/200	19 (44)	35 (26)	10 (45)
20/400-CF	5 (12)	25 (18)	0
HM/LP	0	15 (11)	1 (5)
NLP	0	12 (9)	0
Most recent exam			
≥ 20/40	21 (49)	58 (43)	14 (64)
20/50–20/200	17 (40)	40 ((30)	5 (23)
20/400-CF	5 (11)	15 (11)	0
HM/LP	0	10 (7)	0
NLP	0	12 (9)	3 (13)

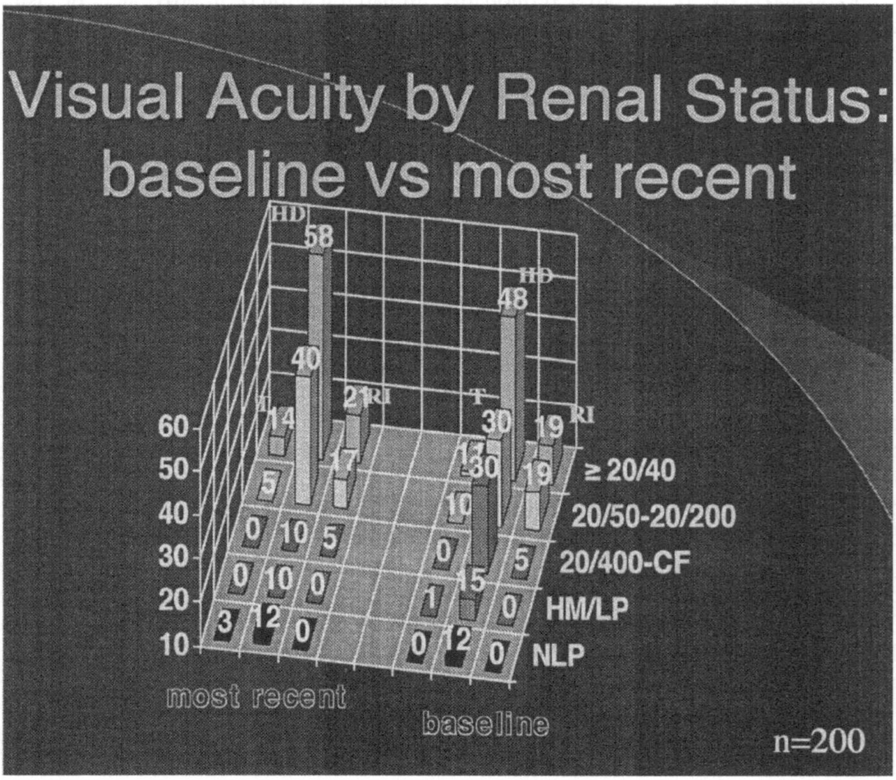

Figure 2.

Table 6. Visual acuity/retinal status at baseline and most recent exam.

Baseline	CSME n = 62 (%)	PDR n = 127 (%)	Prog to PDR n = 42 (%)	VH/RD n = 73 (%)
≥ 20/40	40 (65)	58 (46)	27 (64)	3 (4)
20/50–20/200	20 (32)	36 (28)	13 (31)	12 (16)
20/400-CF	2 (3)	29 (23)	2 (5)	24 (33)
HM/LP	0	4 (3)	0	22 (30)
NLP	0	0	0	12 (16)
Most recent	**Resorbed n = 62 (%)**	**Stable/regress n = 98 (%)**	**Prog w/tx n = 10 (%)**	**VH/RD n = 33 (%)**
≥ 20/40	43 (69)	64 (65)	0	0
20/50–20/200	17 (27)	30 (31)	0	0
20/400-CF	2 (3)	2 (2)	6 (60)	2 (6)
HM/LP	0	2 (2)	3 (30)	16 (48)
NLP	0	0	1 (10)	15 (45)

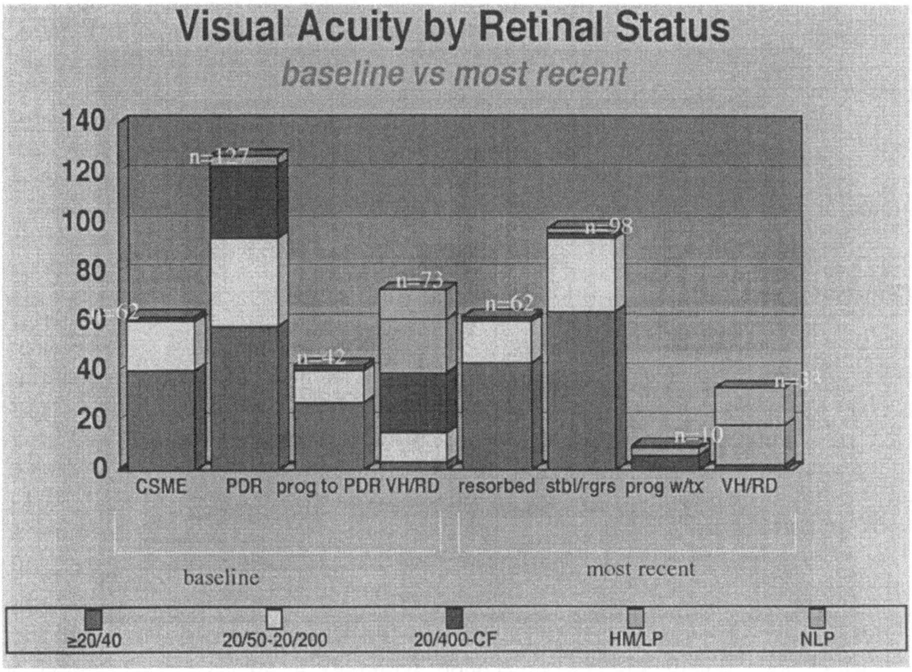

Figure 3.

Table 7. Level of retinopathy [43–85] by duration/onset DM, sex.

	Any rtnp [43–85]	NPDR [43–57]	Prog PDR	PDR [61–85]
Dur. DM				
< 10 yrs, n = 58(%)	58	52 (90)	16 (28)	6 (10)
≥ 10 yrs, n = 142(%)	142	42 (29)	26 (18)	100 (70)
Onset DM				
< 30 yr, n = 53(%)	53	9 (17)	6 (11)	44 (83)
> 30yr, n = 147(%)	147	85 (58)	36 (24)	62 (42)
Sex				
Males, n = 97(%)	97	54 (56)	20 (21)	43 (44)
Females, n = 103(%)	103	40 (39)	22 (21)	63 (61)

ficiency (77%) and hemodialysis groups (80%), but less frequently in the transplant group (59%). The most severe proliferative retinopathy (level 85) was present in 15 patients (35%) with renal insufficiency, 56 patients (42%) undergoing hemodialysis, and in 2 transplant recipients (9%). At most recent follow-up, the renal groups were found to have no change in the level of retinopathy with a frequency range of 32–40%. Level 85 was apparent in similar frequency (18%) among the renal insufficiency and hemodialysis patients, and absent among the transplant recipients. Following appropriate treatment, regression in the level of retinopathy (to a less severe level) was observed in 15 patients (35%) with renal insufficiency, 44 patients (35%) on hemodialysis, and 13 (59%) transplant recipients; CSME was universally absent among the three groups at most recent follow-up (Table 8 and see Figure 4).

Cataracts

The vast majority of patients had cataracts at initial slit-lamp examination (87%). Among the 26 patients without cataracts the average age at diagnosis was 51, less than the average age of the cohort (54); and the duration of diabetes was 15 years, less that the average duration in the cohort (20 years). Considering the renal status, all of the renal transplant recipients had cataracts, 89% of the hemodialysis patients had cataracts, and among those with renal insufficiency, 74% had cataracts. In the cohort, cataract extractions were performed in 43 patients (22%).

Glaucoma

Glaucoma was virtually absent in this cohort, with only 7 patients (2%) diagnosed at baseline evaluation, and all had pre-existing disease. There was no effect on visual function resulting from this condition in any patient.

Table 8. Severity of retinopathy by renal status, at baseline and most recent follow-up.

	Renal insuff, n = 43(%)	Hemodialysis, n = 135(%)	Transplant, n = 22(%)
Retinopathy, baseline			
CSME	20 (47)	35 (26)	4 ((18)
NPDR	7 (16)	22 (16)	9 (41)
PDR	33 (77)	108 (80)	13 (59)
VH/RD	15 (35)	56 (41)	2 (9)
Retinopathy, final			
CSME	0	0	0
No Δ	17 (40)	53 (39)	7 (32)
Progression	7 (16)	38 (28)	2 (9)
Regression	19 (44)	44 (35)	13 (59)
VH/RD	8 (19)	25 (19)	0

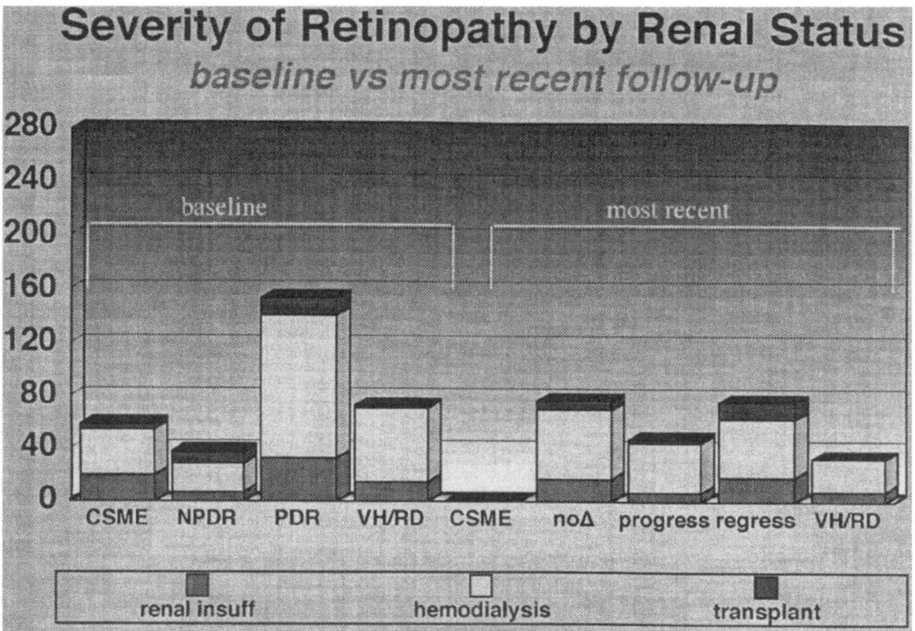

Figure 4.

Renal Status

The distribution of renal disease in the cohort was as follows: 43 patients (22%) had renal insufficiency, 135 patients were on maintenance hemodialysis (67%), and 22 (11%) had received renal transplant. In the renal insufficiency group, the mean creatinine clearance was 38 ml/min, and the mean urine protein excretion was 652 μg/min. Analyzing racial differences among the three renal groups, blacks and hispanics comprised 67% of the renal insufficiency group (blacks only), 88% of the hemodialysis group, and 50% of the renal transplant group. Asians were represented in 5% of the renal insufficiency group, 8% of the hemodialysis group, and 9% of the renal transplant group. Among white patients only 19% of were on maintenance hemodialysis. Transplant recipients were represented equally among blacks and whites (41%), comprising 6% of black subjects, and 35% of white subjects. Comparing younger-onset diabetics to older-onset, renal insufficiency was four times as prevalent in the older-onset group (81%) than in younger-onset (19%); transplant recipients were equally prevalent in younger onset and older-onset subjects; and hemodialysis was three times more prevalent in the older-onset group (75%) compared to younger-onset (25%). When diabetes was present for ≥ 10 years, the prevalence of renal insufficiency was 60%; the prevalence of hemodialysis was 70%; and the prevalence of renal transplantation was 100% (Table 9).

Mortality

During the 10 years that this study was conducted, a mortality rate of 34% was observed (68 patients in the cohort of 200). The mean age of the deceased was 54.7 years, slightly older than the mean age of the cohort (53.8), and the duration of diabetes (19.5 years) was similar in the cohort (19.6 years). 71% of the deceased were older-onset diabetics,

Table 9. Renal Parameters by race, onset/duration DM.

	Renal insuf, n = 43 (%)	HD, n = 135 (%)	Transplant, n = 22 (%)
Race, ethnic freq. (%)			
Black, n = 139 (%)	29 (67) (**21**)	101 (75) (**73**)	9 (41) (**6**)
Hispanic, n = 20(%)	0	18 (13) (**90**)	2 (9) (**10**)
Asian, n = 15 (%)	2 (5) (**13**)	11 (8) (**73**)	2 (9) (**13**)
White, n = 26 (%)	12 (28) (**46**)	5 (4) (**19**)	9 (41) (**35**)
Onset DM			
< 30 yrs	8 (19)	34 (25)	11 (50)
> 30 yrs	35 (81)	101 (75)	12 (50)
Dur. DM			
< 10 yrs	17 (40)	41 (30)	0
= 10 yrs	26 (60)	94 (70)	22 (100)

a finding consistent with earlier reports. Retinopathy severity was particularly prevalent with 84% observed to have had proliferative changes (Table 10).

71% of the patients were on maintenance hemodialysis, 16% had renal insufficiency, and 13% were renal transplant recipients. Looking at inter-racial differences, 84% of the deceased were black/hispanic, compared to 3% and 13% among asians and whites, respectively. Intra-racial mortality rates were highest among hispanics (55%, 11/20), lowest among asians (13%, 2/15), and comparable among blacks and whites (33%, 46/139, and 35%, 9/26, respectively). Diabetes was present for 10 or more years in 84% of the deceased, equally prevalent in among those with renal insufficiency and hemodialysis (80%), and 100% among deceased transplant recipients. Mortality was also higher among older-onset diabetics and females (Table 11).

DISCUSSION

The 100% prevalence of any retinopathy in this inner-city cohort supports data from earlier longitudinal studies [2, 12, 13]. The 40% rate of progression to proliferative retinopathy appears high in comparison to other surveys [4, 12, 14, 15] because of the exclusive composite of advanced renal disease in this population, most of whom (79%) had end-stage renal disease (ESRD), and a retinopathy level of ≥ 43 at baseline, both considered risk factors. Proliferative disease was twice as prevalent in younger-onset diabetics, a finding which is consistent with prior studies [16]. The rate of proliferative disease and the overall severity of retinopathy correlated very strongly to the duration of diabetes. Higher prevalence of CSME in the renal insufficiency group was probably related to more labile hypertension and fluid-electrolyte imbalance seen in the pre-renal patient [17].

A comparison of retinopathy severity at baseline and at most recent assessment revealed stability in approximately one-third of the cohort. A reduction in the rate of level 85 retinopathy was achieved in each of the three renal groups – by almost 50% in the renal insufficiency group, by almost 60% in the hemodialysis group, and eliminated from the transplant recipient group – as a result of vitrectomy surgery. The improvement in overall retinopathy level by 35–59%, and the elimination of CSME as an active problem was a consequence of timely intervention employing DRS, ETDRS, and DRVS protocols. This was reflected in improved and stabilized visual function, validating findings in earlier studies [2, 18].

The high prevalence of cataracts (87%) in this study contratsts considerably to numbers reported in other studies [13, 19], suggesting more deranged aqueous humor

Table 10. Deceased subjects compared to cohort.

	Deceased, n = 68 (%)	Cohort, n = 200 (%)
Mean age	54.7	53.8
Mean dur. DM	19.5	19.6
Onset DM > 30 yrs, number	48 (71)	147 (74)
PDR, number	57 (84)	106 (53)

Table 11. Deceased subjects' renal parameters by race, onset/duration DM, sex.

	Renal insuf, $n = 11$ (%)	HD, $n = 48$ (%)	Transplant, $n = 9$ (%)
Race, ethnic freq., n = 68			
Black, $n = 46$ (%)			
46/68 = 68%	5 (45) (**11**)	37 (77) (**80**)	4 (44) (**9**)
Hispanic, $n = 11$ (%)			
11/68 = 16%	2 (18) (**18**)	8 (17) (**73**)	1 (11) (**9**)
Asian, $n = 2$ (%)			
2/68 = 3%	0	1 (2) (**50**)	1 (11) (**50**)
white, $n = 9$ (%)			
9/68 = 13%	4 (36) (**44**)	2 (4) (**22**)	3 (33) (**33**)
Type DM			
< 30 yrs	4 (36)	14 (29)	2 (22)
> 30 yrs	7 (64)	34 (71)	7 (78)
Dur DM			
< 10 yrs	2 (18)	10 (21)	0
= 10 yrs	9 (72)	38 (79)	9 (100)
Sex			
Males	4 (36)	22 (46)	4 (44)
Females	7 (64)	26 (54)	5 (56)

physiology in the renal population. The majority of patients, however, had subclinical opacities since only 22% had cataract extractions.

Comorbidity of renal disease and diabetes revealed striking contrasts in this multi-racial cohort, particularly weighted in blacks and hispanics. High prevalence of systemic hypertension and aggressive nephropathy in these individuals is well known to the medical community [20, 21]. In this cohort, the most severe retinopathy and the least favorable outcome was seen among the hemodialysis group, comprised of 88% blacks/hispanics. The least severe retinopathy and most favorable outcome were conferred upon the transplant group, comprised of only 50% blacks/hispanics.

The most striking inter-racial difference, however, was the prevalence of black/hispanic ethnicity (84%) among the patients who had died. While socio-economic factors have been cited to play a role in determining the clinical course of diabetes [4], aside from the inner-city origin of this cohort, such a discussion is beyond the scope of this survey. The overall mortality rate, however, appeared lower than figures previously reported [22], in which a 3-year 50% mortality was predicted, compared to 34% in this cohort followed for an average of 3 years. This difference may in part be due to the role of advancing renal care and dialysis technology in improving survival over the last ten years [23].

In spite of ongoing advances in renal, diabetic, and ophthalmologic care, this survey described events such as progression of retinopathy, visual impairment, and mortality. The multiplicity of factors in diabetes mellitus and their multi-organ comorbidity remains highly complex. The role of advanced glycation end-products (AGE's) [24], and systemic hypertension (Mogensen CE, *Renal-Retinal Syndrome*, p. 15) undoubtedly contribute to the accelerated rate of microvasculopathy initiated by the failed or failing kidneys. The elaboration of trophic factors, i.e., vascular endothelial growth factor (VEGF) [25], derangement of nitric-oxide mediated capillary vasodilation [26], and increased activity of the renin-angiotensin axis [27], are among the many areas of ongoing research which will yield further improvement in morbidity and mortality figures, and ultimately to a cure.

REFERENCES

1. Friedman EA. Diabetic renal-retinal syndrome. Archives of Internal Medicine 1980; 140: 1149–1150.
2. Berman DH, Friedman EA, Lundin AP. Aggressive ophthalmological management in diabetic end-stage renal disease: a study of 31 consecutively referred patients. American Journal of Nephrology 1992 (September-October); 12.5: 344–350.
3. Javitt JC, Aiello LP. Cost-effectiveness of detecting and treating diabetic retinopathy. Annals of Internal Medicine 1996; 124.1 part 2: 164–169.
4. Baker RS, Watkins NL, Wilson R, et al. Demographic and clinical characteristics of patients with diabetes presenting to an urban public hospital ophthalmology clinic. Ophthalmology 1998 (August); 105(8): 1373–1379.
5. Cowie CC, Port FK, Wolfe RA, et al. Disparities in incidence of diabetic end-stage renal disease according to race and type of diabetes. New England Journal of Medicine 1989; 321: 1074–1079.
6. Early Treatment Diabetic Retinopathy Study Research Group. Grading diabetic retinopathy from stereoscopic color fundus photographs – an extension of the modified Airlie House classification. ETDRS report number 10. Ophthalmology 1991; 98(suppl): 786–806.
7. Early Treatment Diabetic Retinopathy Study Research Group. Photocoagulation for diabetic macular edema. ETDRS report number 1. Archives of Ophthalmology 1985; 103: 1796–1806.
8. Diabetic Retinopathy Study Research Group. Photocoagulation treatment of proliferative diabetic retinopathy: the second report of Diabetic Retinopathy Study findings. Ophthalmology 1978; 85: 82–106.
9. Diabetic Retinopathy Vitrectomy Study Research Group. Early vitrectomy for severe vitreous hemorrhage in diabetic retinopathy. 2 year results of a randomized trial: DRVS report number 2. Archives of Ophthalmology 1985; 105: 1644–1652.
10. Diabetic Retinopathy Vitrectomy Study Research Group. Early vitrectomy for severe proliferative diabetic retinopathy in eyes with useful vision. Results of a randomized trial: DRVS report number 3. Ophthalmology 1988; 95: 1307–1320.
11. Aiello LM, Rand RI, Sebestyen JG, et al. The eyes and diabetes. In: Marble A, Krall LP, Bradley RF, et al. (eds), Joslin's Diabetes Mellitus, ed 12. Lea & Feibiger, Philadelphia, 1985.
12. Klein R, Klein BEK, Moss SE, Davis MD, DeMets DL. The Wisconsin epidemiologic study of diabetic retinopathy, V: Proteinuria and retinopathy in a population of diabetic persons diagnosed prior to 30 years of age. In: Friedman EA, L'Esperance FA (eds), Renal-Retinal Syndrome, ed 3. Grune & Stratton, Orlando, 1986.
13. Ramsay RC, Cantrill HL, et al. Visual status in diabetic patients following therapy for end-stage nephropathy. In: Friedman EA, L'Esperance FA (eds), Renal -Retinal Syndrome, ed 3. Grune & Stratton, Orlando, 1986.
14. Klein R, Klein BEK, Moss SE, Davis MD, DeMets DL. The Wisconsin epidemiologic study of diabetic retinopathy, IX: four-year incidence and progression of diabetic retinopathy when age at diagnosis is less than 30 years. Archives of Ophthalmology 1989; 107: 237–243.
15. Roy MS. Diabetic retinopathy in African Americans with type I diabetes: the New Jersey 725, I: methodology, population, frequency of retinopathy, and visual impairment. Archives of Ophthalmology 2000; 118: 97–104.
16. Klein R, Klein BEK, Moss SE, Davis MD, DeMets DL. The Wisconsin epidemiologic study of diabetic retinopathy, X: four-year incidence and progression of diabetic retinopathy when age at diagnosis is 30 years or more. Archives of Ophthalmology 1989; 107: 244–249.
17. Nathan DM. Long-term complications of diabetes mellitus. New England Journal of Medicine 1993; 328: 1676–1682.

18. Berman DH, L'Esperance FA, Friedman, EA. Long-term visual outcome of diabetic patients treated with pan-retinal photocoagulation. In: Friedman EA, L'Esperance FA (eds), Renal-Retinal Syndrome, ed 7. Kluwer Academic Publishers, The Netherlands, 1998.
19. Roy MS. Diabetic retinopathy in African Americans with type I diabetes: the New Jersey 725, II: risk factors. Archives of Ophthalmology 2000; 118: 105–115.
20. Brancati FL, Whittle JC, Whelton PK, et al. The excess incidence of diabetic end-stage renal disease among blacks: a population-based study of potential explanatory factors. Journal of the American Medical Association 1992; 268: 3079–3084.
21. Harris M, Klein R, Cowie C, et al. Is the risk of diabetic retinopathy greater in non-hispanic blacks and Mexican Americans than in non-hispanic whites with type 2 diabetes? A US population survey. Diabetes Care 1998; 21: 1230–1235.
22. Shyh TP, Beyer MM, Friedman EA. Treatment of the uremic diabetic. Nephron 1985; 40: 129–138.
23. Miles AM. Heomdialysis in patients with diabetes mellitu. In: Friedman EA, L'Esperance FA (eds), Renal-Retinal Syndrome, ed 7. Kluwer Academic Publishers, The Netherlands, 1998.
24. Makita Z, Radoff S, Rayfield EJ, et al. Advanced glycation end-products in patients with diabetic nephropathy. New England Journal of Medicine 1991; 325: 836–842.
25. Aiello, LP, Avery R, et al. Vascular endothelial growth factor in ocular fluid of patients with diabetic retinopathy and other retinal disorders. New England Journal of Medicine 1994; 331: 1480–1487.
26. Furchgott RF. NO and diabetic complications. In: Friedman EA, L'Esperance FA (eds), Renal-Retinal Syndrome, ed 7. Kluwer Academic Publishers, The Netherlands, 1998.
27. Ravid M, Lang R, Rachmani R, Lishner M. Long-term reno-protective effect of angiotensin-converting enzyme in non-insulin dependent diabetes mellitus. A 7 year follow-up study. Annals of Internal Medicine 1996; 156: 286–289.
28. Diaz-Buxo JA, Burgess WP, Greenman M, et al. Visual function in diabetic patients undergoing dialysis: comparison of peritoneal and hemodialysis. International Journal of Artificial Organs 1984; 7: 257–262.
29. Klein R, Klein BEK. Visual impairment in diabetes. Ophthalmology 1984; 91: 1–9.
30. Klein R, Klein BEK, Moss SE, Cruickshanks KJ. The Wisconsin epidemiologic study of diabetic retinopathy, XIV: ten-year incidence and progression of diabetic retinopathy. Archives of Ophthalmology 1994; 112: 1217–1228.

JONATHAN A. SCHEINDLIN

9. Proliferative Diabetic Retinopathy: Current Treatment Strategies for Progression

Editor's Comment: In what amounts to state-of-the-art 'marching orders,' Scheindlin reviews the options in therapy open to slow vision loss in the diabetic patient. Blending the macroscopic view of a diseased retina with the underlying molecular changes induced by growth factors especially vascular endothelial growth factor, a broad perspective of a virulent disorder is provided. Often neglected in writings for the non-ophthalmologist, the question of 'What to do when proliferation does not regress after panretinal photocoagulation?' is raised and partially answered. Looking toward a future with 'insight into better long term control of diabetes and systemic abnormalities,' Scheindlin explains the what and why of present eye interventions.

BACKGROUND

Diabetes mellitus manifests its detrimental effects on the retina in large part due to development of vascular instability. Visual loss due to effects on the retinal vasculature results from a combination of 1) abnormal permeability leading to deposits within the retina, 2) closure of vessels causing nonperfusion of the retina, and 3) growth of new blood vessels into the vitreous in response to ischemia. When the vessels lose the ability to maintain the blood-retinal barrier, vision is jeopardized when fluid and/or lipid deposit in the fovea. As the vasculature continues to suffer from the effects of abnormal glucose metabolism, the vessels can no longer maintain adequate blood flow to the entire retina. The subsequent ischemia that downstream retina experiences leads to injury of retinal elements and liberation of factors responsible for growth of new blood vessels [1–3]. Proliferation of new vessels, in response to the above process, may progress into the vitreous. The network of new vessels tends to pull the retina towards the center of the eye creating a tractional retinal detachment. In addition, the neovascularization may bleed, dispersing blood into the vitreous cavity and obscuring vision.

The development of proliferative diabetic retinopathy is determined by multiple factors. Although retinal ischemia drives the process of neovascularization, input of other factors fuels the progressive growth of new vessels. The location of these neovascular fronds at the junction of perfused and nonperfused retina, as well as sites relatively distant such as the disc and iris, suggests the presence of diffusible molecules acting on the vasculature to initiate and maintain angiogenesis. Several growth factors identified, such as vascular endothelial growth factor (VEGF), insulin growth factor (IGF), basic fibroblast growth factor (BFGF), and others, have been isolated within ocular tissue of patients with proliferative disease [9–18]. It remains to be seen the exact role each individual factor plays in the process of neovascularization. Recently, much work has focused on VEGF. This has been shown to both induce neovascular-

E.A. Friedman and F.A. L'Esperance, Jr. (eds.), Diabetic Renal-Retinal Syndrome, 89–104.

ization in experimental models and is found in high levels in the aqueous and vitreous of eyes with actively proliferating neovascularization [10]. Further, VEGF levels were found to be reduced in eyes following panretinal photocoagulation suggesting quiescence of the neovascular drive. While VEGF likely participates in the process of neovascularization, the entire chain of events still remains to be identified. It is clear that the complex interaction of numerous events needs to take place for angiogenesis to occur.

Characteristic retinovascular findings implicate regional elaboration of angiogenesis factors. Neovascularization tends to develop in characteristic locations as will be discussed later. Treatment, therefore is aimed at reducing this drive and destruction of the hypoxic cells. Laser photocoagulation has been the mainstay of treatment for proliferative diabetic retinopathy since it's introduction by Meyer-Schwickerath in 1955 8] and later proven benefit in the Diabetic retinopathy study (DRS) [66]. Today, patients with proliferative diabetic retinopathy are treated with panretinal laser photocoagulation as the first step and further treatment is based on the response of the retina. The following is a review of the treatment of proliferative diabetic retinopathy based on clinical studies and options available when there is continued progression of neovascularization after treatment has been initiated.

SYSTEMIC FACTORS

Microcirculation abnormalities in diabetes occur in multiple organ systems including the renal, ocular, and nervous system [19] suggesting systemic factors effect local vascular changes. Cardiovascular atherosclerotic disease is found more frequently among patients with diabetes mellitus [20]. The presence of perfusion deficits and dysfunction in these organs results from interruption of blood flow. The development of abnormal vasculature in these organs, while multifactorial, reflects a systemic milieu that allows progressive degenerative changes within the vessel lumen. We will briefly review some systemic conditions implicated in the process of proliferative disease in the eye.

Much work has focused on systemic abnormalities that would identify those patients at risk for progression of proliferative retinopathy. Several systemic findings have been associated with progression including increased serum levels of fibrinogen, hemoglobin A1C, and cholesterol. Elevated blood pressure, presence of neuropathy, anemia, and increased duration of diabetes have also been shown to be risk factors for the development of proliferative retinopathy in diabetic patients [21–53]. These systemic variables are also related to the development of coronary artery disease. The similarity of intravascular stasis and prothrombotic state in these conditions may represent the results of a systemic environmental threshold that allows local factors to influence structural changes.

Diabetes contributes to a hypercoagulable state and fibrinogen plays a major role in thrombogenesis. Fibrinogen is a high molecular weight protein that circulates in plasma and serves as part of the initial sequence of events leading to thrombosis. Elevated plasma fibrinogen levels have been associated with increased risk of cardiovascular atherosclerotic disease and progression of diabetic retinopathy [20, 22, 23, 34, 40, 42–45]. This relationship suggests that increased blood viscosity in diabetics is due, at least in part, to high fibrinogen levels and disorders in coagulation leads to vascular obstruction in both large and small vessels. With progression of this process, neovascularization results, perhaps in response to local factor influence.

Duration of diabetes has been found to increase the risk of progression to proliferative retinopathy in many but not all studies [21]. More important, it seems, is the presence of other variables such as poor glucose control that increases the risk of progression to proliferative disease [21]. HbA1C has been found consistently to be a strong predictor for progression of retinopathy in both observational studies and clinical trials [21, 24, 26, 28–32, 40]. The reason for this is not completely understood but it may be that elevated glycosylated end-products lead to damaged endothelium and subsequent vessel closure. Not only were measures of glucose control shown to be significant for progression, but also improved glucose management demonstrated less likelihood of developing PDR [6, 21]. An important point is that patients who have had poor control of serum glucose with nonproliferative diabetic retinopathy may be at increased risk for progression during the initial phase of strict glucose control [48]. It is during this time that these patients need to be monitored more frequently to address any progression.

The relationship between hypertension and diabetes is not clearly understood, however, it is clear that the presence of both are common and appear to aggravate the extent of tissue damage from each alone. Numerous studies have looked at both diastolic and systolic pressures to ascertain the effect on progression of proliferative diabetic retinopathy. Some research has found increased risk of proliferative diabetic retinopathy in young patients with elevated diastolic pressure [28, 29, 40]. Still others have found an elevated systolic pressure in patients with long-standing diabetes increases the risk of progression [7, 26, 35]. While the association is still unclear at present, it seems reasonable that the damaging effects to the microcirculation in both conditions exacerbates the effects from each individually. In addition, given the known increased risk for myocardial infarction and stroke in the two conditions, diligent control of both abnormalities has become part of standard clinical management.

Increased levels of cholesterol, particularly triglycerides and low-density lipoproteins (LDL) have been found in patients with progression to PDR [21, 39, 40]. Of the two serum lipids, triglycerides have been more consistently demonstrated in patients with proliferation particularly in the larger studies [21]. Although the mechanism has not been established, it seems serum constituents that lead to atheroma in larger vessels may contribute to vessel closure in the retinal microcirculation as well.

Anemia may play a role in the progression of diabetic retinopathy; however, this relationship has received little notice. It stands to reason that if progression results from local ischemia, systemic decrease in oxygen carrying ability of the blood would accelerate this process. Studies have shown an increase in risk for development of PDR in patients with decreasing hemoglobin [21, 41]. It is likely that anemia contributes to progression when present, but is not a necessary cofactor.

Some authors have suggested that autonomic neuropathy may be a separate predictor of proliferative diabetic retinopathy possibly due to 'decreased responsiveness of the retinal vessels to autonomic humoral mediators' [46]. Other studies as well have supported this theory by also finding nervous system dysfunction associated with PDR [21, 38, 47]. We know the retinal vasculature autoregulates in response to inner retinal oxygen concentration [53], so it is unclear why nervous system dysfunction would have an effect on the retinal vasculature.

Diabetic nephropathy and severe proliferative retinopathy often occur simultaneously but the association between the two has not been clearly defined. Studies have shown a discordance between retinal microvascular changes and glomerular changes [50, 51]. However, in some large epidemiologic studies, proteinuria was significantly associated with proliferative diabetic retinopathy [52]. Clearly the 'Retinal-Renal' syndrome exists but to what extent one causes the other or whether the two occur concomitantly remains to be seen.

CLINICAL COURSE

Proliferative diabetic retinopathy characteristically occurs in the posterior pole and into the midperiphery within 45° of the optic disc [57]. Numerous classification schemes have been proposed to help aid in prognostic, investigational, and communication purposes [4, 56, 57, 59, 60]. What they all have in common is standard configurations of proliferative diabetic retinopathy that occur along a continuum. Description of the various schemes goes beyond the scope of this review, numerous publications have painstakingly described the features of each. Each classification has strengths and weaknesses. Today, the most commonly used classification systems use standard photographs for reference.

When the neovascularization occurs at the junction of perfused and nonperfused retina, the usual location is just outside the posterior pole into the midperipheral retina. There are various size isolated neovascular fronds in this region that range in size from pinpoint up to about 2 disc diameters in size [62] (see Figure 1). These isolated tufts usually do not experience the excessive growth that posterior pole neovascular fronds achieve. The vitreous tends to be firmly adherent to these areas. Although they may not grow in size, they still carry the same risks of traction and rhegmatogenous retinal

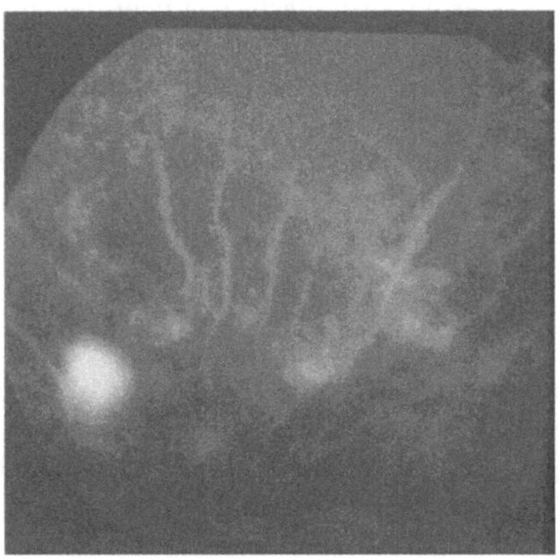

Figure 1. Fluorescein angiography of isolated tufts of neovascularization at the junction of perfused and nonperfused retina. Note the leakage of fluorescein due to the fenestrated nature of the new vessels and the lack of fluorescein in the nonperfused retina just peripheral to the tuft.

detachment as well as vitreous hemorrhage with continued growth along the vitreous. These are often quite responsive to scatter laser photocoagulation of the nonperfused area. Photocoagulation is applied to the peripheral ischemic retina and surrounding the tuft to minimize tractional complications.

More common than midperipheral vessel proliferation is neovascularization that occurs at the disc and along the vascular arcades. The progression of proliferative tissue in the posterior pole occurs in certain characteristic ways. Vessels in and surrounding the optic nerve occur relatively commonly (see Figure 2). The vitreous is firmly adherent to the optic nerve and along the vessels. Growth of the neovascularization extends from the optic nerve into the vitreous and also tangentially along the arcades. We usually see the strongest attachments of the neovascularization at the disc, and along the temporal aspects of the arcades. With continued growth, the florid neovascularization may encompass the entire posterior pole. If the vitreous has not detached from the

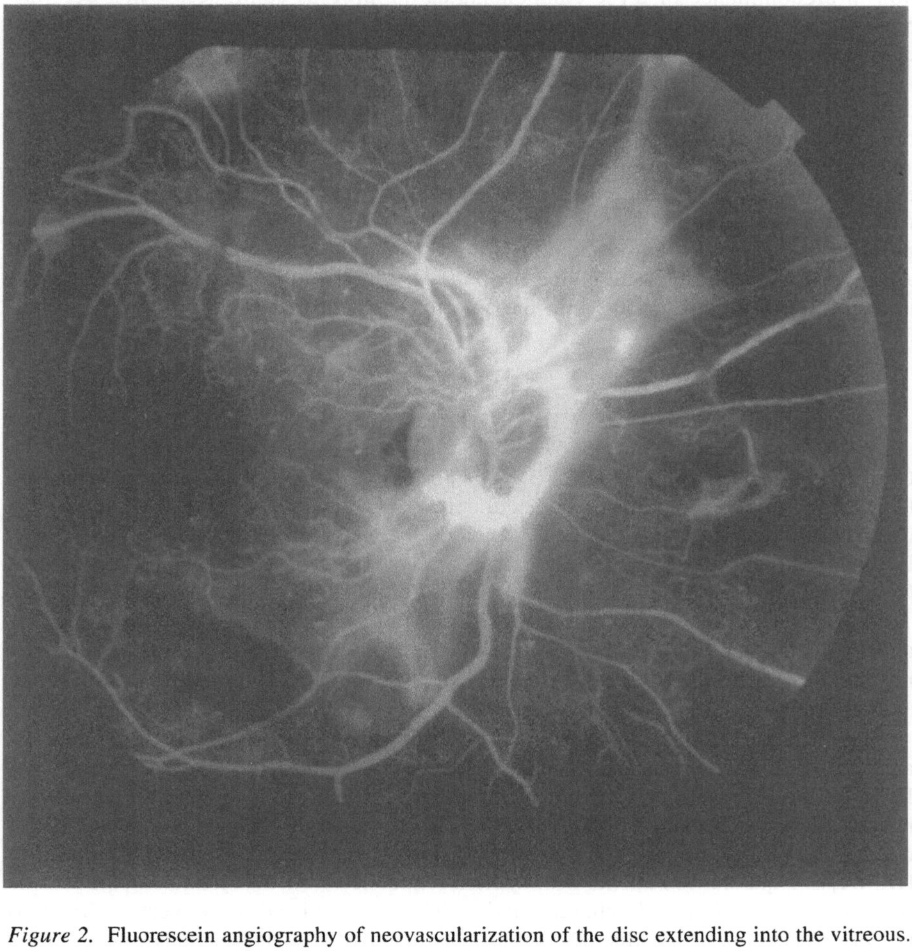

Figure 2. Fluorescein angiography of neovascularization of the disc extending into the vitreous.

posterior pole, we tend to see traction retinal detachments with broad firm adherence points to the retina and complex detachments. At any point along the progressive neovascular process, bleeding into the vitreous and obscuration of vision may occur.

There are two main constituents of proliferative tissue – the vascular component and the gliotic component [61]. The gliotic component likely provides support to the vascularization. These tend to grow along the posterior hyaloid face but may grow into the substance of the vitreous as well. These two components are variable among patients and even between two eyes of the same patient. We have observed in our patient population that depending on which of the two predominate will determine the clinical course. As a general rule, if there is a predominant vascular component, vitreous hemorrhage tends to more likely serve as the cause of visual loss. If the predominant gliotic component progresses, tractional retinal detachment tends to cause visual deterioration. Progression may occur with either or both present and may show variation in predominance of these components as the process continues. When both are present equally, and there is progression, the changes tend to be rapid with severe visual consequences (florid retinopathy) [84].

Neovascular glaucoma secondary to diabetes may occur in patients with varying degrees of proliferative retinopathy. Patients with severe proliferative diabetic retinopathy are at risk for growth of new vessels anteriorly along the iris (rubeosis) and into the angle resulting in scarring and closure of the drainage apparatus of the eye with subsequent elevation in intraocular pressure. It is not uncommon to see patients with seemingly minimal funduscopic changes of the retina that have new vessels growing on the iris and into the angle. Fluorescein angiography typically shows large areas of capillary closure and nonperfusion [61]. The likely cause of this phenomenon is liberation of growth factors into the anterior segment that promotes vessel growth [13].

CLINICAL TRIALS

Current treatment recommendations for proliferative diabetic retinopathy are derived from large randomized clinical studies undertaken in the 1970s and 1980s. Among the most well-known and widely accepted include the Diabetic Retinopathy Study (DRS), Early Treatment Diabetic Retinopathy Study (ETDRS), Diabetes Control and Complications Trial (DCCT) and the Diabetic Retinopathy Vitrectomy Study (DRVS). These were large randomized multicenter clinical studies that looked at diabetic patients and evaluated the role of therapeutic interventions in the progression of diabetic retinopathy. The following is a summary of the findings as related to proliferative diabetic retinopathy.

The DRS sought to establish the natural history of proliferative diabetic retinopathy and establish whether laser photocoagulation reduced the risk of severe visual loss. Inclusion criteria were presence of proliferative diabetic retinopathy (PDR) or severe nonproliferative diabetic retinopathy (NPDR) (extensive retinal hemorrhages in 4 quadrants, venous beading in 2 quadrants, intraretinal microvascular abnormalities [IRMA] in 1 quadrant). Patients with traction retinal detachments threatening the macula were excluded. High-risk characteristics for severe visual loss were defined as neovascularization of the disc [NVD] at least 1/3 disc area and/or neovascularization

(NVD or neovascularization elsewhere [NVE] $\geq$ 1/3 disc area) associated with vitreous or preretinal hemorrhage. In particular, the DRS defined 4 risk factors- presence of new vessels, new vessels located on or within 1 disc diameter of the disc, moderate to severe new vessels and preretinal or vitreous hemorrhage. Presence of three or four risk factors translated into a risk of severe visual loss at 2 years of about 25%. The study showed a decrease in rate of visual loss by about 50% to 11% when scatter laser photocoagulation was performed [64–68].

The ETDRS addressed whether photocoagulation was effective in reducing visual loss or progression to high-risk PDR in patients prior to developing high-risk proliferative retinopathy. Inclusion criteria were presence of mild to severe nonproliferative diabetic retinopathy or early (non high-risk) PDR. NPDR severity was defined according to the degree of retinal abnormalities as defined in the DRS. Non-high risk PDR was defined as proliferative disease not meeting the criteria for high risk PDR. The conclusions of the ETDRS were that up to 50% of patients with mild PDR or severe NPDR progressed to high-risk characteristics during 1 year of follow-up. Early scatter photocoagulation decreased the development of high risk PDR but did not have an effect on visual loss. In addition, many patients with less severe retinopathy did not progress to high-risk retinopathy and were not at risk for severe visual loss. The conclusion to be drawn from this study is that eyes that progress to severe NPDR or worse should be considered for laser treatment 'without delay' prior to developing high-risk retinopathy [69–72].

The DRVS evaluated the role of vitrectomy in patients with vitreous hemorrhage or severe PDR in eyes with useful vision as well as establishing the natural history of severe PDR with conventional management. Patients were included when there was evidence of vitreous hemorrhage accounting for decreased vision or severe proliferating neovascular retinopathy. In both groups, the macula was attached. The study concluded that early vitrectomy was beneficial in improving and maintaining vision for patients with recent severe vitreous hemorrhage. In addition, early vitrectomy was useful in increasing the chance of obtaining 10/20 acuity in patients with proliferating neovascularization. Both results were more pronounced for those patients with Type I DM [73, 74].

The DCCT investigated intensive insulin therapy and its effect on diabetic retinopathy. In particular, this study was designed to determine if aggressive treatment of blood sugar in young diabetics prevented the development or decreased the progression of retinopathy and other systemic complications. The study demonstrated that with intensive insulin treatment in young IDDM patients, development of clinically significant retinopathy was reduced by approximately 50%, and lowered the risk of progression to severe NPDR, PDR and need for photocoagulation by approximately 50%. In addition, there was early worsening of retinopathy within the first year, but over the long term (5 years) there was decreased chance of progression. Systemically, there was decreased albuminuria, proteinuria, and neuropathy in the intensive treatment group. Intensive treatment did, however, lead to increased risk of hypoglycemia events, weight gain and ketoacidosis [32, 33].

It is important to stress that while these studies provide a useful guide to management, there is sufficient variability in proliferative diabetic retinopathy and with technological advancements, strict adherence to the study conclusions may not always be appropriate.

TREATMENT

The goal of treatment is to either prevent the formation of neovascularization or halt the progression once observed. The manner in which this is accomplished relies on destruction of peripheral retinal tissue to decrease the signaling for the neovascular drive. If this is unsuccessful then removal of the 'scaffolding' on which the neovascularization grows may arrest progression. The destruction of peripheral retinal tissue may be achieved with either laser photocoagulation or cryotherapy. Currently, laser photocoagulation is the preferred method by most ophthalmologists mainly for relative ease of use, less inflammation, and patient comfort [75]. Cryotherapy is used in select circumstances, mainly for patients with media opacities and when anterior ocular tissue needs treatment [76–78].

Laser photocoagulation has become the most common mode of treatment for proliferative diabetic retinopathy [80]. There are various techniques that have been described and are effective. No matter the manner in which photocoagulation is applied, the goal of treatment is the same – mainly, the scatter treatment needs to incorporate the midperiphery of the retina, sparing the posterior pole [79] (see Figure 3). In time, the retina demonstrates chorioretinal scarring in the form of well-demarcated hyperpigmented and/or atrophic reaction of the retinal pigment epithelium [85] (See Figure 4).

Usually, an average of 1,200–1,500 burns are placed in multiple sessions. The DRS protocol required from 800–1,600 500µ burns to be scattered from 2 disc diameters from the macula anteriorly to or beyond the midperiphery of the retina at the level of the vortex ampulla [79]. Areas of traction and large retinal vessels are avoided. The

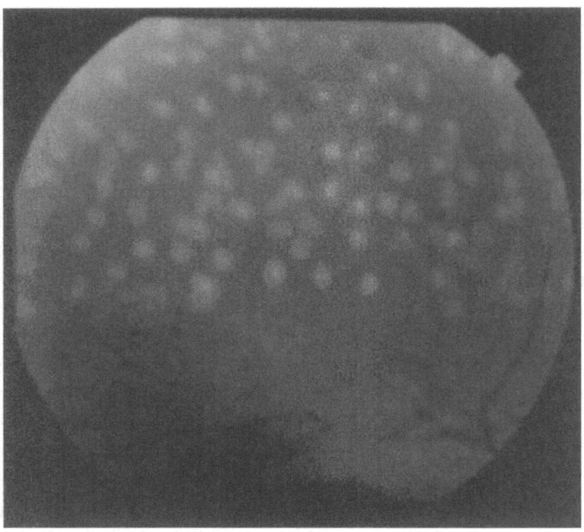

Figure 3. Red-free photograph of panretinal laser photocoagulation immediately after application.

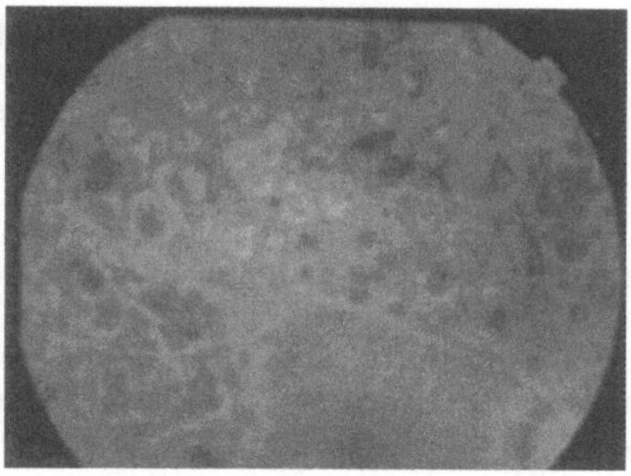

Figure 4. Chorioretinal scar formation of panretinal laser photocoagulation. This generally takes from 10–14 days to form.

DRS called for blue-green argon laser, however, studies have shown various laser wavelengths used for PRP in most patients are effective [80].

Treatment for PDR requires large areas of the retina to receive PRP. The area of the retina is approximately 1,100 mm^2 as has been shown in mathematical models and autopsy eyes [82]. Therefore, to treat the entire peripheral retina out to the ora serrata and leaving the posterior pole untreated (8 disc diameters or 48 mm^2) the total area of treatment is about 1,050 mm^2 (1,100 mm^2–48 mm^2). In other words, to treat the retina with a 'full PRP', 90–95% of the retina is treated. When performing PRP, laser spot sizes are usually 200–500μ in size. There are many variables, but depending on the lens used for application of the laser, the spot size on the retina (500 μ) will be 0.190 mm^2 with a Goldmann lens or 0.509 mm^2 with a Rodenstock lens [82]. To cover the entire retina with treatment, placing the burns exactly end-to-end would take 5,500 burns with a Goldmann lens or 2,050 burns with a Rodenstock lens [82, 85].

Proliferation of new vessels present in a variety of ways, therefore laser photocoagulation also needs to be applied based on the severity of the neovascularization. Patients with extensive progressive neovascularization or neovascular glaucoma need more aggressive PRP in one or two sessions and may require many more burns to be placed [84, 93]. Variations in uptake of photocoagulation spots also at times dictate the need for more sessions. The important point is that treatment needs to be individualized based on the patient's condition.

Response to laser treatment and regression of high-risk characteristics occurs in different forms. One of the earliest signs of regression is narrowing of the retinal vasculature [87, 88, 91]. Complete regression of new vessels is desired but only occurs in some eyes.89 More common is stabilization- vessels still present but without further growth, or a predominant fibrous component remains [90]. Lastly, vitreous and preretinal hemorrhages may resolve with time as the neovascularization stabilizes and the blood is resorbed.

There are various theories as to why photocoagulation works. Destruction of the photoreceptor-RPE complex has been suggested to remove the cells responsible for 2/3 of the oxygen consumption in the retina. Without these cells to consume oxygen, the choroidal circulation may provide oxygen to the inner retina. This results in an autoregulatory response of the retinal vasculature to constrict [91]. Retinal thinning has also been implicated which allows closer approximation of the inner retina to the oxygen-rich choroid. Further, it has been shown that RPE cells secrete an inhibitor of vascular endothelial cell proliferation when treated with photocoagulation in an experimental model [92].

Laser photocoagulation has side effects that need to be considered when treatment is initiated. Heavy treatment applied at one time may cause choroidal effusion with subsequent anterior rotation of the ciliary body and angle-closure glaucoma. In addition, large amounts of intense treatment may cause diffuse retinal swelling, postoperative pain and iritis. PRP may also cause decreases in visual field, contrast sensitivity and night vision. Anteriorly, iris burns, cataract formation and corneal epithelial abrasions have been described [75]. PRP is effective in causing regression of high risk characteristics in the majority of patients (up to 86% show regression in the first 3 weeks after treatment [89]) with standard 'less than complete' PRP. For these reasons, extensive PRP should not be applied to all patients and individualized treatment plans need to be considered.

Pars plana vitrectomy (PPV) is utilized to remove vitreous, relieve traction and clear vitreous hemorrhage. The posterior vitreous cortex and the posterior hyaloid which is adherent to the retinal surface serve as a scaffolding network onto which the neovascularization grows. When the neovascular fronds attach to the vitreous, the complex is firmly adherent and continued growth and scarring lead to contracture and traction on the retina. The goal of surgical vitrectomy is to remove this tissue, or, when not possible, to segment the tissue and isolate the tufts of neovascularization. Most often, laser endophotocoagulation is applied at the same time [58].

PPV is a surgical procedure and therefore has certain inherent limitations. Not all patients are surgical candidates due to coexistent medical problems. In addition, the procedure carries numerous risks including, but not limited to infection, bleeding, cataract formation, and retinal detachment that need to be considered when contemplating treatment. For these reasons, surgical vitrectomy for proliferative diabetic retinopathy is usually reserved for those patients with visual loss secondary to vitreous hemorrhage, traction retinal detachment encroaching on the macula, neovascular glaucoma or proliferation not responsive to photocoagulation.

PROGRESSIVE PROLIFERATION

Treatment approach to patients with severe nonproliferative and proliferative diabetic retinopathy is fairly straightforward in regards to the use of laser photocoagulation. The studies described provide a framework that can be used to individually tailor a treatment strategy. What has not been evaluated is the question as to the best approach to use when proliferation does not regress. After the entire peripheral retina has been treated with photocoagulation, how do we manage patients with progressive neovascularization?

Currently, there are no large-scale studies that adequately answer the above question. Several small-scale studies have addressed progressive proliferation and suggest further fill-in laser is effective in promoting regression [86–88], however, the best approach to patients with 'full' PRP is still in doubt. When the vision declines secondary to effects of proliferation on the macula, most retinal surgeons would agree that pars plana vitrectomy with removal of the posterior hyaloid and fibrovascular tissue causing traction is the appropriate management [58]. What is not so clear is how to approach patients with progression of proliferation when the macula is spared. In 1999, the Vitreous Society surveyed its members for preferences and trends and found vitreoretinal surgeon's views are divided. Of the members surveyed, the following was the response – 6% anterior cryotherapy, 32% further anterior laser with indirect laser ophthalmoscope, 38% further laser with slit lamp delivery, 13% observation, and 13% surgical vitrectomy (Vitreous Society survey, 1999).

Anterior cryotherapy is rarely used today, however has a role in select circumstances. Patients with media opacities such as vitreous hemorrhage, cataract, corneal irregularities, or small pupils may not allow for adequate laser application. Cryotherapy is applied from the outside and treatment may be applied to the retina with less than ideal viewing. Cryotherapy, however, causes inflammation, is painful, and is less exact in its application than laser making its use restricted to select circumstances. In addition, cryotherapy is rarely used in eyes with traction retinal detachments and is considered a contraindication by some [78].

Transscleral diode laser may be used in much the same way as cryotherapy without the severity of inflammation. Due to the practical aspects of treating large areas of retina, this method is rarely used by us for PRP applications.

Application of laser photocoagulation may be accomplished through the pupil with either a slit-lamp delivery system or attached to the headpiece of an indirect ophthalmoscope. When patients have progressive growth of neovascular tissue in the retina or along the iris, often there are areas within the retina that may not have responded adequately to prior laser. When these areas are identified on examination, further fill-in laser is required. However, when there is 'full treatment', sometimes, there may be retina anteriorly which has not been treated and is only found with scleral depression or with wide angle lenses used with a slit lamp. Presumably, the anterior retina may be ischemic and continuing the neovascular drive [94]. Laser treatment to this area serves to 'complete' the PRP. We do not 'retreat' retina that demonstrates adequate response to laser.

When the PRP is 'complete' and there is no more room for further treatment, observation may be considered when the progressive neovascularization does not cause traction on the macula and does not bleed into the vitreous. Progression of only the gliotic component is rare, particularly if the vascular component regressed, and usually occurs slowly. Advancement of the vascular component generally causes vitreous hemorrhage and may recur, necessitating vitrectomy. We have found that if the neovascular frond has both a glial and vascular component, the progression occurs relatively rapidly, within 6–12 months, and these patients usually end up with traction retinal detachments of the macula and/or recurrent vitreous hemorrhages. Often, these are patients with poor control of blood sugar and have coexistent systemic conditions such as infections.

We do not observe these patients for long periods unless medically unstable because the fibrovascular proliferation may detach the retina in a broad sheet making surgery more difficult once the macula is involved. Our approach to patients who fail to regress after full laser PRP and have recurrent vitreous hemorrhage is to consider pars plana vitrectomy. In addition, patients who have had full PRP and demonstrate documented (fundus photographs and fluorescein angiography) progression of fibrovascular proliferation that extends within the vascular arcades are considered for surgical treatment. Many of our patients have extensive traction on the retina on initial presentation, and full PRP is not possible due to the presence of retinal detachment. In these cases, we apply PRP to all areas amenable to treatment and if regression is not seen, vitrectomy with minimal membrane peeling will often halt the progression presumably by the removal of the scaffolding.

Options for patients after 'Full Panretinal Laser Photocoagulation' with Progressive Proliferation

- Anterior Cryotherapy
- Anterior laser photocoagulation with indirect laser
- Anterior laser photocoagulation with slit-lamp laser
- Anterior transscleral diode laser
- Observation

FUTURE DIRECTIONS

Proliferative diabetic retinopathy is presumably a response to reestablish circulation to areas of nonperfusion. At this time, nonperfusion is irreversible, so efforts at prevention of vessel closure would serve to halt or prevent the neovascular drive.(r) This may be accomplished with insight into better long term control of diabetes and systemic abnormalities as previously discussed. Inhibitors of angiogenesis are being studied to inhibit signaling. Substances that create a vitreous detachment through enzymatic means remove the scaffolding onto which the neovascularization grows.(r) These advancements and future therapies will allow diabetics to maintain eyesight and live longer.

REFERENCES

1. Patz A. Studies on retinal neovascularization. Invest Ophthalmol Vis Sci 1980; 19: 1133–1138.
2. Henkind P. Ocular neovascularization. Am J Ophthalmol 1978; 85: 287–301.
3. Jampol LM, Ebroon DA, Goldbaum MH. Peripheral proliferative retinopathies: an update on angiogenesis, etiologies and management. Surv Ophthalmol 1994; 38: 519–540.
4. Friedman EA, L'Esperance FA (eds). Diabetic Retinal-Renal Syndrome. Kluwer Academic Publishers, Great Britain, 1998.
5. Brownlee M, Cerami A, Vlassara H. Advanced glycosylation end products in tissue and the biochemical basis of diabetic complications. N Engl J Med 1988; 318: 1315–1321.
6. Davis MD. Diabetic retinopathy: a clinical overview. Diabetes Metab Rev 1988; 4: 291–322.
7. Janka HU, Warram JH, Rand LI, Krolewski AS. Risk factors for progression of background retinopathy in long-standing IDDM. Diabetes 1989; 38: 460–464.
8. Meyer-Schwickerath G, Schott K. Diabetic retinopathy and photocoagulation. Am J Ophthalmol 1963; 47: 611–619.
9. Aiello LP, Avery R, Arrig P, et al. Vascular endothelial growth factor in ocular fluid of patients with diabetic retinopathy and other retinal disorders. N Engl J Med 1994; 331: 1480–1487.
10. Adamis AP, Miller JW, Bernal MT, et al. Elevated vascular permeability factor/vascular endothelial growth factor levels in the vitreous of eyes with proliferative diabetic retinopathy. Am J Ophthalmol 1994; 118(4): 445–450.

11. Clermont A, Aiello LP, Mori F et al. Vascular endothelial growth factor (VEGF) and severity of nonproliferative diabetic retinopathy (NPDR) mediate retinal hemodynamics in vivo: a potential role for VEGF in the progression of nonproliferative diabetic retinopathy. Am J Ophthalmol 1997; 124: 433–446.
12. Folkman J. Tumor angiogenesis: therapeutic implications. N Engl J Med 1971; 285: 1182–1186.
13. Adamis A, Shima D, Tolentino M, et al. Inhibition of VEGF prevents ocular neovascularization in a non-human primate. Arch Ophthalmol 1996; 114: 66–71.
14. Grant M, Russell B, Fitzgerald C, Merimee TJ. Insulin-like growth factors in vitreous. Studies in control and diabetic subjects with neovascularization. Diabetes 1986; 35: 416–420.
15. Wang Q, Dills D, Klein R, et al. Does insulin-like growth factor 1 predict incidence and progression of diabetic retinopathy. Diabetes 1995; 44: 161–164.
16. Sivalingam A, Kenney J, Brown GC, et al. Basic fibroblast growth factor levels in the vitreous of patients with proliferative diabetic retinopathy. Arch Ophthalmol 1990; 108: 869–872.
17. Smith L, Kopchick J, Chen W, et al. Essential role of growth hormone in ischemia induced retinal neovascularization. Science 1997; 296: 1706–1709.
18. King GL, Johnson S, Wu G. Possible growth modulators involved in the pathogenesis of diabetic proliferative retinopathy. In: Westermark B, Betsholtz C, Hokfelt B (eds), Growth Factors in Health and Disease. Excerpta Medica, New York, 1990.
19. Root H, Pote W, Frehner H. Triopathy of diabetes: sequence of neuropathy, retinopathy and nephropathy. Arch Intern Med 1954; 94: 931–941.
20. Garber A. Attenuating cardiovascular risk factors in patients with type 2 diabetes. Am Fam Physician 2000 (Dec 15); 62(12): 2633–2642.
21. Davis D, Fisher M, Gangnon R, et al. Risk factors for high-risk proliferative diabetic retinopathy and severe visual loss: early treatment diabetic retinopathy study report #18. Invest Ophthalmol Vis Sci 1998; 39: 233–252.
22. De Silva S, Shawe J, Patel H, Cudworth A. Plasma fibrinogen in diabetes mellitus. Diabete Metabolisme 1979; 5: 201–206.
23. Davis T, Moore J, Turner R. Plasma fibronectin, factor VIII-related antigen and fibrinogen concentrations, and diabetic retinopathy. Diabet Metab 1985; 11: 147–151.
24. Klein R, Klein B, Moss S, et al. Relationship of hyperglycemia to the long-term incidence and progression of diabetic retinopathy. Arch Ophthalmol 1994; 112: 1217–1228.
25. Moskalets E, Galstyan G, Starostina E, et al. Association of blindness to intensification of glycemic control in insulin-dependent diabetes mellitus. J Diab Comp 1994; 8: 45–50.
26. Lee E, Zlee V, Lu M, Russell D. Development of proliferative retinopathy in NIDDM: a follow-up study of American Indians in Oklahoma. Diabetes 1992; 41: 359–367.
27. Agardh E, Torffvit O, Agardh C. The prevalence of retinopathy and associated medical risk factors in type I (insulin-dependent) diabetes mellitus. J Intern Med 1989; 226: 47–52.
28. Klein R, Klein B, Moss S, et al. The Wisconsin Epidemiologic Study of Diabetic Retinopathy II. Prevalence and risk of diabetic retinopathy when age at diagnosis is less than 30 years. Arch Ophthalmol 1984; 102: 520–526.
29. Klein R, Klein B, Moss S, et al. The Wisconsin Epidemiologic Study of Diabetic Retinopathy III. Prevalence and risk of diabetic retinopathy when age at diagnosis is 30 or more years. Arch Ophthalmol 1984; 102: 527–532.
30. Moss S, Klein R, Klein B. The incidence of vision loss in a diabetic population. Ophthalmology 1988; 95: 1340–1348.
31. Dwyer MS, Melton J, Ballard DJ, et al. Incidence of diabetic retinopathy and blindness: a population based study in Rochester, Minnesota. Diabetes Care 1985; 8: 316–322.
32. The Diabetes Control and Complications Trial Research Group. The effect of intensive diabetes treatment on the progression of diabetic retinopathy in insulin-dependent diabetes mellitus. Arch Ophthalmol 1995; 113: 36–51.
33. The Diabetes Control and Complications Trial Research Group. The relationship of glycemic exposure (HbA1C) to the risk of development and progression of retinopathy in the Diabetes Control and Complications Trial. Diabetes 1995; 44: 968–983.
34. Fusijawa T, Ikegami H, Yamato E, et al. Association of plasma fibrinogen level and blood pressure with diabetic retinopathy, and renal complications associated with proliferative diabetic retinopathy, in Type 2 diabetes mellitus. Diabetic Medicine 1999; 16: 522–526.
35. Cignarelli M, De Cicco M, Damato A, et al. High systolic blood pressure increases prevalence and severity of retinopathy in NIDDM patients. Diabetes Care 1992; 15: 1002–1008.

36. Marshall G, Garg S, Jackson W, et al. Factors influencing the onset and progression of diabetic retinopathy in subjects with insulin-dependent diabetes mellitus. Ophthalmol 1993; 100: 1133–1139.
37. Teuscher A, Schnell H, Wilson P. Incidence of diabetic retinopathy and relationship to baseline plasma glucose and blood pressure. Diabetes Care 1988; 11: 246–251.
38. UK Prospective Diabetes Study Group. Tight blood pressure control and risk of macrovascular and microvascular complications in type 2 diabetes: UKPDS 38. Br Med J 1998; 317: 703–713.
39. The EURODIAB IDDM Complications Study Group. Fibrinogen and von Willebrand factor in IDDM. Relationships to lipid vascular risk factors, blood pressure, glycaemic control and urinary albumin excretion rate: the EURODIAB IDDM complication study. Diabetologia 1997; 40: 698–705.
40. Lloyd C, Klein R, Maser R, et al. The progression of retinopathy over 2 years: the Pittsburgh Epidemiology of Diabetes Complication (EDC) study. J Diabetes Complications 1995; 9: 140–148.
41. Qiao Q, Keinanen-KiukaanniemiS, Laara E. The relationship between hemoglobin levels and diabetic retinopathy. J Clin Epidemiol 1997; 50: 153–158.
42. Stec J, Silbershatz H, Tofler G, et al. Association of fibrinogen with cardiovascular risk factors and cardiovascular disease in the Framingham offspring population. Circulation 2000 (Oct 3); 102(14): 1634–1638.
43. Asakawa H, Tokunaga K, Kawakami F. Elevation of fibrinogen and thrombin-antithrombin III complex levels of type 2 diabetes mellitus patients with retinopathy and nephropathy. J Diabetes Complications 2000; 14: 121–126.
44. Ceriello A. Coagulation activation in diabetes mellitus: the role of hyperglycaemia and therapeutic prospects. Diabetologia 1993; 36: 1119–1125.
45. Ceriello A, Taboga C, Giacomella R, et al. Fibrinogen plasma levels as a marker of thrombin activation in diabetes. Diabetes 1994; 43: 430–432.
46. Krolewski A, Barzilay J, Warram J, et al. Risk of early-onset proliferative retinopathy in IDDM is closely related to cardiovascular autonomic neuropathy. Diabetes 1992; 41; 430–437.
47. Tesfaye S, Stevens L, Stephenson J, et al. Prevalence of diabetic peripheral neuropathy and its relation to glycemic control and potential risk factors: the EURODIAB IDDM Complications Study. Diabetologia 1996; 39: 1377–1384.
48. The Diabetes Control and Complications Trial Research Group. Progression of retinopathy with intensive versus conventional treatment in the diabetes control and complications trial. Ophthalmology 1995; 102: 647–661.
49. The Diabetes Control and Complications Trial Research Group. The effect of intensive diabetes therapy on the development and progression of neuropathy. Ann Intern Med 1995; 122: 561–568.
50. Chavers B, Mauer S, Ramsay R, Steffes M. Relationship between retinal and glomerular lesions in IDDM patients. Diabetes 1994; 43: 441–446.
51. Bilous R, Viberti G, Sandahl-Christensen J, et al. Dissociation of diabetic complications in insulin-dependent diabetics: a clinical report. Diabetic Nephr 1985; 4: 73–76.
52. Klein R, Moss S, Klein B. Is gross proteinuria a risk factor for the incidence of proliferative diabetic retinopathy? Ophthalmology 1993; 100: 1133–1139.
53. Bill A. Ocular circulation. In: Moses RA (ed), Adler's Physiology of the Eye: Clinical Application, 6th ed. CV Mosby, Saint Louis, 1975, p 222.
54. Taylor E, Dobree J. Proliferative diabetic retinopathy. Site and size of initial lesions. Br J Ophthalmol 1970; 54: 11–18.
55. Dobree J. Proliferative diabetic retinopathy. Evolution of the retinal lesions. Br J Ophthalmol 1964; 48: 637–649.
56. Scuderi G. A classification of diabetic retinopathy. Ann Ophthalmol 1973; 5: 411.
57. The Diabetic Retinopathy Study Research Group: Report 6: Design, methods, and baseline results. Invest Ophthalmol Vis Sci 1981; 21: 149–226.
58. Charles S. Surgical management of proliferative diabetic retinopathy. In: Friedman EA, L'Esperance FA (eds), Diabetic Renal-Retinal Syndrome, Kluwer Academic Publishers, Great Britain; 1998, pp 197–210.
59. Jalkh A, Takahashi M, Topilow H, et al. Prognostic value of vitreous findings in diabetic retinopathy. Arch Ophthalmol 1982; 100: 432–434.
60. Bresnick G, Engerman R, Davis M, et al. Patterns of ischemia in diabetic retinopathy. Trans Am Acad Ophthalmol Otolaryngol 1976; 81: 694–709.
61. McMeel J. Diabetic retinopathy: Fibrotic proliferation and retinal detachment. Trans Am Ophthalmol Soc 1971; 69: 440–493.

62. Shimizu K, Kobayashi Y, Muraoka K. Midperipheral fundus involvement in diabetic retinopathy. Ophthalmology 1981; 88: 601–612.
63. Niki T, Muraoka K, Shimizu K. Distribution of capillary nonperfusion in early stage diabetic retinopathy. Ophthalmol 1981; 88: 601–612.
64. Diabetic Retinopathy Study Research Group. Preliminary reports on effects of photocoagulation therapy. Am J Ophthalmol 1976; 81: 383–396.
65. Diabetic Retinopathy Study Research Group. Photocoagulation treatment of proliferative diabetic retinopathy: the second report of Diabetic Retinopathy Study findings. Ophthalmology 1978; 85: 82–106.
66. Diabetic Retinopathy Study Research Group. Four risk factors for severe visual loss in diabetic retinopathy: the third report from the Diabetic Retinopathy Study. Arch Ophthalmol 1979; 97: 654–655.
67. Diabetic Retinopathy Study Research Group. Photocoagulation treatment of proliferative diabetic retinopathy: relationship of adverse treatment effects to retinopathy severity. Diabetic Retinopathy Study Report No 5. Dev Ophthalmol 1981; 2: 248–261.
68. Diabetic Retinopathy Study Research Group. Photocoagulation treatment of proliferative diabetic retinopathy. Clinical application of DRS findings. DRS report No. 8. Ophthalmology 1981; 88: 583–600.
69. Early Treatment Diabetic Retinopathy Study Research Group. Early photocoagulation for diabetic retinopathy. ETDRS No. 9. Ophthalmol 1991; 98: 766–785.
70. Early Treatment Diabetic Retinopathy Study Research Group. Fundus photographic risk factors for progression of diabetic retinopathy. ETDRS Report No. 12. Ophthalmol 1991; 98: 823–833.
71. Early Treatment Diabetic Retinopathy Study Research Group. Fluorescein angiographic risk factors for progression of diabetic retinopathy. ETDRS Report No. 13. Ophthalmol 1991; 98: 834–840.
72. Early Treatment Diabetic Retinopathy Study Research Group. Techniques for scatter and local photocoagulation treatment of diabetic retinopathy. ETDRS Report No. 3. Int Ophthalmol Clin Winter 1987; 27(4): 254–264.
73. Diabetic Retinopathy Vitrectomy Study Research Group. Early vitrectomy for severe vitreous hemorrhage in diabetic retinopathy. Two-year results of a randomized trial, DRVS Report No. 2. Arch Ophthalmol 1981;103: 1644–1652.
74. Diabetic Retinopathy Vitrectomy Study Research Group. Early vitrectomy for severe proliferative diabetic retinopathy in eyes with useful vision. Results of a randomized trial, DRVS Report No. 3. Ophthalmol 1988; 95: 1307–1320.
75. Quillen D, Gardner T, Blankenship G. Proliferative diabetic retinopathy. In: Guyer D, Yannuzzi L, Chang S, et al (eds), Retina-Vitreous-Macula. W.B. Saunders Co., Philadelphia, 1999, pp 1407–1431.
76. Oosterhuis J, Bijlmer-Gorter H. Cryotreatment in proliferative diabetic retinopathy:long-term results. Ophthalmologica 1980; 87: 306–312.
77. Daily M, Geiser R. Treatment of proliferative diabetic retinopathy with panretinal cryotherapy. Ophthalmic Surg 1984; 15: 741–745.
78. Benedett R, Olk R, Arribas N, et al. Transconjunctival anterior retinal cryotherapy for proliferative diabetic retinopathy. Ophthalmol 1987; 94: 612–619.
79. Blankenship G. A clinical comparison of central and peripheral argon laser panretinal photocoagulation for proliferative diabetic retinopathy. Ophthalmol 1988; 95: 170–177.
80. Seiberth V, Schatanek S, Alexandridis E. Panretinal photocoagulation in diabetic retinopathy: argon versus dye laser coagulation. Graefes Arch Clin Exp Ophthalmol 1993; 231: 318–322.
81. Mcdonald H, Schatz H. Visual loss following panretinal photocoagulation for proliferative diabetic retinopathy. Ophthalmol 1985; 92: 1532–1537.
82. Barr C. Estimation of the maximum number of argon laser burns possible in panretinal photocoagulation. Am J Ophthalmol 1984; 97: 697–703.
83. Vine A. The efficacy of additional argon laser photocoagulation for persistent, severe proliferative diabetic retinopathy. Ophthalmol 1985; 92: 1532–1537.
84. Favard C, Guyot-Argenton C, Assouline M, et al. Full panretinal photocoagulation and early vitrectomy improve prognosis of florid diabetic retinopathy. Ophthalmol 1996; 103: 561–574.
85. Tso M, Wallow I, Elgin S. Experimental photocoagulation of the human retina. I. Correlation of physical, clinical, and pathologic data. Arch Ophthalmol 1977; 95: 1035.
86. Doft B, Blankenship G. Single versus multiple treatment sessions of argon laser panretinal photocoagulation for proliferative diabetic retinopathy. Ophthalmol 1982; 89: 772–779.
87. Doft B, Metz D, Kelsey S. Augmentation laser for proliferative diabetic retinopathy that fails to respond to initial panretinal photocoagulation. Ophthalmol 1992; 99: 1728–1735.

88. Vander J, Duker J, Benson W, et al. Long-term stability and visual outcome after favorable initial response of proliferative diabetic retinopathy to panretinal photocoagulation. Ophthalmol 1991; 98: 1575–1579.
89. Doft B, Blankenship G. Retinopathy risk factor regression after laser panretinal photocoagulation for proliferative diabetic retinopathy. Ophthalmol 1984; 91: 1453–1457.
90. D'Amico D. Diabetic traction retinal detachments threatening the fovea and panretinal argon laser photocoagulation. Semin Ophthalmol 1991; 6: 11–17.
91. Weiter J, Zuckerman R. The influence of the photoreceptor -RPE complex on the inner retina. Ophthalmol 1980; 87: 1133–1139.
92. Yoshimura N, Matsumoto M, Shimizu H, et al. Photocoagulated human retinal pigment epithelial cells produce an inhibitor of vascular endothelial cell proliferation. Invest Ophthalmol Vis Sci 1995; 36: 1686–1691.
93. Reddy V, Zamora R, Olk R. Quantitation of retinal ablation in proliferative diabetic retinopathy. Am J Ophthalmol 1995 (June); 119(6): 760–766.
94. Terasaki H, Miyake Y, Awaya S. Fluorescein angiography of peripheral retina and pars plana during vitrectomy for proliferative diabetic retinopathy. Am J Ophthalmol 1997 (March); 123(3): 370–376.

Carl Erik Mogensen

10. Microalbuminuria in Perspectives

Editor's Comment: No investigator has furthered our understanding of the progression of diabetic nephropathy more than Mogensen who initially documented the predictive value of the detection of microalbuminuria in both type 1 and type 2 diabetes. In this 'all you wanted to know about microalbuminuria' essay, Mogensen starts with the history of its discovery and application and then provides a grasp for screening and management if discovered. Also proffered is a consensus perspective of the place of angiotensin converting enzyme inhibition as a key therapeutic intervention for both type 1 and type 2 diabetes. By collecting key studies of the effect of treatment with angiotensin converting enzyme inhibitors on structural changes in diabetic subjects with microalbuminuria as in Table 6, Mogensen has provided the scholarship that facilitates understanding of a key advance in diabetes.

1. INTRODUCTION

The development of immuno-assays (including radio-immuno-assays) in the beginning of the 1960s allowed measurement of previously undetectable amounts of albumin in the urine [1, 2]. Initially procedures were complicated, slow, and in general expensive. However, there has been a rapid development in this area, so by now easy, quick and clinically readily applicable procedures are available for routine use in daily diabetes care as well as elsewhere [2–7].

This is important because microalbuminuria is now documented and widely recognised as an important marker, not only of early diabetic renal disease, but also risk marker of early vascular complications including a very strong predictor of early mortality [1, 8–9]. This is the case in diabetic patients, but also in hypertensive patients and other patients. Even more importantly, the identification of microalbuminuria as the risk marker or factor was followed by many studies on the clinical course of microalbuminuria and associated abnormalities and finally by clear treatment strategies, based upon many intervention trials. Thus there is now consensus to treat diabetic patients with microalbuminuria with early ACE-inhibition or alternative blood pressure lowering treatment, even in the presence of so-called normal blood pressure [1, 10–13]. Glycemic control is obviously always important [14, 15]. However, diabetic patients have a unique vulnerability to hypertensive injury [16] first documented in the kidney [17, 18].

In addition, studies now document that renal structure [19–22] is better preserved by antihypertensive treatment and that renal function [23] is better maintained in long-term studies. This can be described as a hat trick in clinical medicine, 1) identification of the risk factor and risk marker, 2) defining the clinical course, 3) and implementation of intervention strategies that are now well accepted by the medical community.

E.A. Friedman and F.A. L'Esperance, Jr. (eds.), Diabetic Renal-Retinal Syndrome, 105–120.

Therefore screening for microalbuminuria and follow-up of patients is now general practise in most countries with advanced diabetes care, but not everywhere [24].

Generally accepted diagnostic definitions of normo- micro- and macroalbuminuria is shown in Table 1, but it should be emphasised that albuminuria is a continuum; for practical reasons it is however, important to stratify patients since this yields a definitively more precise prognosis in the subsequent years for such patients in the clinic. Borderline cases are always problematic. In this review various perspectives of microalbuminuria will be discussed, bearing in mind that this is an area where the number of publications is increasing steadily. Some proposed diagnostic and treatment strategies are indicated in Table 2 [3].

2. THE GEOGRAPHICAL PERSPECTIVES (US/EUROPE)

The first papers on the predictive use of microalbuminuria came from the UK and Denmark and these studies have largely been confirmed although it should be noted that improved treatment of diabetes with respect to glycemia and blood pressure may have changed the clinical course, meaning that the initial studies may not reflect the present situation [24]; still the newly published follow-up from the DCCT [14] clearly showed progression of the same magnitude as our own study published in 1984 in conventionally treated patients [25] (Table 3).

Interestingly, the intervention studies in microalbuminuria patients were mainly conducted in Europe and Australia with only a few studies in the US, where there has been some reluctance to accept microalbuminuria as a predictor of diabetic nephropathy. Thus screening in the US is not as widespread as in Europe, which may in part explain the greater number of overt nephropathy in the US compared to Europe [24].

Obviously, microalbuminuria should be monitored continuously and it is clear that a single or a few measurements of albumin excretion rate are unable to define long-term risks of diabetic nephropathy as shown in initial studies [1, 5, 6, 17, 18]. However, this is possible with continuous follow-up of patients where the main risk factors for

Table 1. Diagnostic definitions of normo-, micro- and macroalbuminuria.

Condition	24-h urinary albumin excretion rate	Overnight urinary albumin excretion rate	Albumin: creatinine ratio*
Macroalbuminuria (overt nephropathy)	> 300 mg/day	> 200 μg/min	> 25 mg/mmol
Microalbuminuria	30–300 mg/day	20–200 μg/min	2.5–25.0 mg/mmol (for men) 3.5–25.0 (for women)
Normoalbuminuria	< 30 mg/day	< 20 μg/min	< 2.5 mg/mmol (for men) < 3.5 mg/mmol (for women)

* Usually, but not necessarily measured in the first morning urine sample.

Table 2. New strategies in diagnosis and treatment of diabetic renal disease.

Strategy	Helpful in diagnosis	Helpful in treatment
Analysis of genetic factors [27]	No, but studies are still needed	At risk patients cannot yet be found.
Familial factors	Early diagnosis of hypertension(?)	Probably not (?)
Antiglycemic treatment [14, 28]	Yes (HbA1c monitoring)	Yes, clearly demanding and sometimes not feasible
Various types of AGE-inhibitors [1]	No	Some renal studies stopped, other studies started (?)
Aldose reductase inhibitor [1]	No	No, renal studies stopped (?)
Growth factor inhibition [1]	No	Needs investigation
Protein kinase C inhibitors [1]	No	Needs investigation
Antihypertensive treatment [1, 10, 29]	Yes, along with abnormal albuminuria and/or high BP	Yes, mainly ACE-I as basis for combination therapy
ACE-inhibitors [10, 12, 30, 31] or receptor-blockers [1, 25, 32]	No (genotyping not useful)	Yes, profoundly, especially in microalbuminuric patients
Lipid lowering [1]	Yes, dyslipidemia	Probably, but needs further confirmation
Aspirine [1]	No	Only for macrovascular diseases
Endothelial and endopeptidase inhibitors [1]	No	Under investigation
Glycosaminoglycans [33]	No	Under investigation
Metalloproteinase [1]	No	Under evaluation

Table 3. NEJM: IDDM prediction of worsening nephropathy.

Publication:	Mogensen (1984) [25]	Post-DCCT (2000) [14]
Initial Micro % to proteinuria	14 patients 8.5%/year	64 patients 8%/year*

* 2% with intensive therapy.

progression in renal disease, from the very beginning is glycemic control and blood pressure level, and possibly blood lipids, according to ideas of Bradford Hill [26].

Provided continuos follow-up, albuminuria is clearly able to define patients who are safe or who are at risk and it has a sufficient clinical validity to define adequate clinical decision making and also it is very useful in clinical trials. So far it has not been possible to define other markers [24].

It may be easy, but not fair to criticise the early studies for small sample sizes, post-hoc analysis and using variable microalbuminuria definitions. Such critics will not lead to any progress in the area, in particular in view of the obvious lack of relevant alternatives [24]. In fact, initial data has been clearly confirmed (Table 3).

3. MICROALBUMINURIA: A HISTORICAL PERSPECTIVE

Diabetic kidney disease is defined as the presence of clinical proteinuria or dipstick positive proteinuria. In the definition of microalbuminuria excretion is lower than proteinuria but higher than normal excretion rate (Table 1). The predictive values of microalbuminuria in the various studies differ which is not surprising considering various follow-ups and different patient numbers as well as different proposals for discriminate albumin excretion rate. However, there is by now very much general consensus in the area, where the limits in Table 1 are generally accepted [1, 3].

On the other hand, it is clear that albuminuria is a continuum not only in range but also in time. Initially virtually all patients with type 1 diabetes will have an albumin excretion rate in the normal range unless severe ketosis [1]. In type 2 diabetes, on the other hand, the disease may have been present for several years at clinical presentation, thus patients may also present with microalbuminuria, also due to elevated BP. This is contrast to patients with type 1 diabetes where microalbuminuria is not found in the early years although microalbuminuria may be present in cases of very poor metabolic control, a reversible abnormality. Several studies are now and have been conducted on the significance of upper-normal albuminuria which certainly represent a risk [34–42].

Microalbuminuria has emerged as a strong and highly significant predictor of renal and of cardiovascular disease in diabetes and also in the general population and in essential hypertension. However it should be stressed that to some extent the definition is arbitrarily defined in the original consensus paper [1]. Some of the authors used a higher threshold for albuminuria e.g. 70 μg/min but this author was inclined to use a lower value (15 μg/min); the consensus was 20–200 μg/min, a level that has proven quite relevant and useful over the years and has gained significant general acceptance. For unknown reasons, American authors have often used a different threshold, e.g. in the DCCT [14].

An important study has recently been published from Ravid's group in Israel [39], who for many years has followed patients with diabetes and various degrees of albuminuria and also has published very important intervention studies on the positive effect of ACE-inhibition [43–45]. The main purpose of screening for microalbuminuria is to identify patients at risk and if possible to treat them with early intervention strategies, such as antihypertensive treatment and optimised glycemic control. Rachmani and Ravid studied 599 normoalbuminuric patients over 8 years and divided them into 3 categories, Group 1 from 0–10 mg/24 h, Group 2 from 10–20 mg/24 h and finally Group 3 from 20–30 mg/24 h. With the common definition of microalbuminuria based upon 24 h samples the excretion rate for microalbuminuria is 30–300 mg/24 h. It was clear from this study that during the 8-year follow-up there was an increasing risk of progression to microalbuminuria according to the level of baseline albuminuria, an observation that was highly significant also from a clinical point of view. Interestingly, in the excretion rates between 20–30 mg/24 h there was also a tendency to a decline in GFR with time. The correlation between baseline albumin excretion rate and subsequent decline was exponential without a clear-cut threshold value. Interestingly also odds ratio for cardiovascular end-points such as death, non-fatal myocardial infarction etc. were higher in Group 3. Therefore the authors rightly conclude that albumin excretion rates between 20–30 mg/24 h have accelerated risk for decline in GFR and development of cardiovascular events compared to those below 20 mg/24 h. Also, the authors correctly point out that albuminuria is a continuos variable and that one should consider a redefinition of the threshold not only in research work but also in clinical practise. In this writers mind, however, it is preferential to stick to the original definition of microalbuminuria but certainly underscore that albuminuria is a continuos variable and that patients in the upper normal range are at greater risk of progression, not only regarding renal disease but also development of cardiovascular disease (that is 'upper normal' normoalbuminuria) [38, 40, 41]. Microalbuminuria is an important indicator of patients at risk and certainly studies now show that improved metabolic control is important, e.g. in the follow-up study from the DCCT. Treatment with ACE-inhibitors is now commonplace and has proven beneficial in the long run with respect to preservation of GFR [1, 23] and also reduction in cardiovascular events as shown in the HOPE Study [11–13].

In practical terms it has now been proposed to screen for microalbuminuria by using an early morning urine sample or a random urine sample and to measure albumin excretion ratio [3]. The normal value is usually indicated as being below 2.5 mg/mmol of albumin over creatinine. Clearly, clinicians should be aware of the findings in Rachmani's and Ravid's study and more carefully consider follow-up clinical intervention in patients with albuminuria in the so-called upper normal range. A new study indicates a positive effect of ACE-inhibition, even in normoalbuminuric patients [46].

It is equally clear that excretion rates in the upper microalbuminuric range signify greater risk for progression to overt proteinuria. However, it is also important to bear in mind that abnormal albuminuria may reverse or decrease with better metabolic as well as BP control as seen in several studies [1].

Hyperfiltration precedes microalbuminuria [6], an abnormality that can be ameliorated by better metabolic control and by treatment with Losartan, an angiotensin 2 receptor-blocker [31].

4. MICROALBUMINURIA AND THE MORTALITY PERSPECTIVE

In a follow-up study first communicated in 1983–84 it was shown that microalbuminuria in Type 2 diabetes is strongly predictive of an increased mortality [1, 47–48]. This observation has since been confirmed in numerous studies [8]. Other factors could be associated with poor prognosis such as long-term hyperglycaemia and high blood pressure. Microalbuminuria seems though to be the strongest overall predictor of mortality as has also been found in non-diabetic population-based studies [37]. Interestingly, poor metabolic control has recently been shown to predict microalbuminuria, and could also be a risk factor for its progression, whereas high blood pressure can be more important in more advanced disease. Poor glycaemic control correlates with typical diabetic glomerular lesions. New studies have shown structural abnormalities in microalbuminuric Type 2 diabetic patients [49]. Among the many risk factors analysed hyperglycaemia and hypertension seem to be of particular importance, as documented in initial studies on the relevance of higher albumin excretion rates in relation to other risk factors. Thus again it appears that there is a requirement for two risk factors; hyperglycaemia and high blood pressure factors must combine to produce clinically important disease. Since microalbuminuria in Type 2 diabetes is a better predictor of cardiovascular than of microvascular disease other macrovascular disease factors must be taken into consideration. Microalbuminuria as a predictor for early mortality was recently confirmed in a US follow-up study where patients with risk of retinopathy were examined [48]. Very similar observations are thus now available both for Europe and the US. Data from the UKPDS is still awaited [50–54].

5. SCREENING

Several agencies, like the ADA and the BDA, now propose screening for microalbuminuria with the perspective of identifying patients at risk and initiating early treatment [55].

Many studies in the area of hypertension and in population studies clearly confirm microalbuminuria as a risk marker, and many studies are now ongoing to use the presence of microalbuminuria as an intervention marker.

The requirement for a risk factor marker according to Bradford Hill [26] is a clearly fulfilled for microalbuminuria. The same is the case for WHO-criteria for screening (Table 4).

6. NEW PERSPECTIVES: ALTERNATIVES TO MICROALBUMINURIA

Microalbuminuria has emerged as a very powerful clinical predictor of overt renal and cardiovascular disease. This was recently confirmed in the post DCCT-study [14] and the HOPE-study [11, 12]. Although the correlation to structural changes may not be perfect, especially in type 2 patients, microalbuminuria certainly predicts clinical proteinuria in both types of diabetes, and in type 2 diabetes, also cardiovascular mortality. The power of microalbuminuria is strengthened by continuos follow-up in the diabetes clinic. It should be mentioned that ACE-inhibition and other antihypertensive agents may reduce or even normalise microalbuminuria so evaluation of the actual clinical situation should be done after discontinuation of therapy for one or two months [23]. Early treatment has also proven to improve prognosis. However, it would be useful to have alternatives to microalbuminuria, especially in evaluation of the long-term fate of patients, but so far this has not been possible, as discussed elsewhere [1, 6, 24].

New technologies have also failed to replace microalbuminuria or to add accurately to the predictions of future disease. Identification of genes has so far been clinically too unreliable and in some studies no association is found. Other substances to be measured in blood and urine (like tubular markers [42]) have not added any new progress. It has been proposed that extra-cellular-matrix molecules or products of glycation could be of importance, but this needs still to be confirmed. Measurements of cellular function in skin and lymphocytes were interesting research tools, but

Table 4. Screening for microalbuminuria in NIDDM (middle-aged).

	Generally accepted criteria for screening (most often in population based screening)	Criteria fulfilled	Modifications
1	The condition sought should be an an important health problem	Yes	Predict cardiovascular disease and renal mortality
2	There should be an accepted treatment for patients with recognised disease	Yes	Best medical treatment should be offered plus blocking the RAS
3	Facilities for diagnosis and treatment should be available	Yes	Available in most countries
4	There should be a recognisable latent or early symptomatic stage	Yes	Related both to cardiovascular and renal disease
5	There should be a suitable test or examination	Yes	Related to other risk factors and complications
6	The test should be acceptable to the population	Yes	Urine samples usually convenient
7	The natural progression of the condition, including the development from latent to declared disease, should be adequately understood	Yes	Relationship to cardiovascular disease poorly understood
8	There should be an agreed policy about whom to treat as patients	Yes	(best medical treatment should be offered)
9	The cost of case finding should be economically balanced in relation to possible expenditure on medical care as a whole	Yes	Difficult to calculate
10	Case finding should be a continuing process and not a 'once-and-for-all' project	Clearly	Generally accepted. (With some exceptions [24])

(Screening in type 1 diabetes even more indicated).

presently not useful in clinical evaluation [2]. Also new imaging technologies, such as Positron-emission tomography and magnetic resonance imaging have not been useful although new studies may be required.

It has been proposed that increasing serum prorenin precedes the onset of microalbuminuria, a highly interesting area that needs further investigation. So far the overlap between non-progressors and progressors is too large [56, 57], and also it is too demanding to evaluate serum prorenin in the diabetes clinic (with variable assays results). However, the observation is very interesting, also because dual blockade of the renal angiotensin system seems to be useful in clinical practise [32].

It can thus be concluded today that there are no alternatives in the present situation. Microalbuminuria quite accurately predicts renal disease, especially with careful follow-up and measurements of albumin excretion rate in the clinics. Patients with normal albumin excretion or even microalbuminuria have normal or supra-normal GFR [58, 59]. Some American investigators, however, remain curiously inclined to use much more invasive techniques such as renal biopsies [24]; as evidenced by the title of a paper: 'Can the insulin-dependent diabetic patients be managed without kidney biopsy'. A more realistic approch is outlined in Figure 1.

Table 5. Proposed alternatives (partly out-dated and not realistic) [24].

1.	Genes
2.	'Substances' in blood and urine
3.	Tubular function
4.	Cellular functions (Skin fibroblasts)
5.	Fine needle aspiration from the kidney
6.	PET, MRI
7.	Prorenin
8.	Sodium-lithium Countransport in RBC
9.	Lipid level
10.	Smoking
11.	Family history

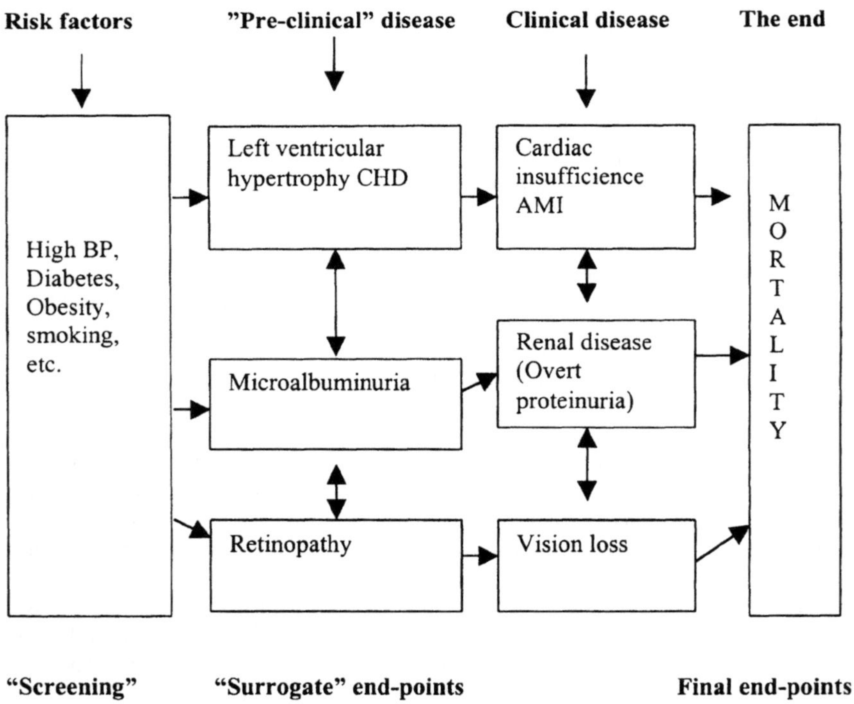

Figure 1.

7. EARLY ACE-INHIBITOR TREATMENT IN MICROALBUMINURIC PATIENTS

Two points must be stressed here. Firstly, blood pressure, like albuminuria, is a continuous variable; therefore, unless long-term studies with detailed BP-recordings have been carried out to that effect, it is impossible to define cut-off points corresponding to high or pathological blood pressure levels at which cardiovascular and renal lesions start to develop in specific types of patients or populations, whether this concerns type 1 or 2 diabetes or the general population. Secondly, diabetic patients are more susceptible to blood pressure-related renal damage than nondiabetics or patients with essential hypertension [16]. The first study to take these points into consideration was an early clinical intervention study with a self-control approach, which enrolled microalbuminuric type 1 diabetic patients with so-called 'normal' blood pressure [60]. This study showed that antihypertensive treatment with β-blockers could lead to regression of microalbuminuria. Further studies showed that antihypertensive combination treatment in such patients was associated with a decrease in albuminuria and a slower progression of early diabetic renal disease (no decline in GFR) [61, 62]. This therapeutic benefit was confirmed in studies using ACE inhibitors in patients with microalbuminuria [10]. Numerous studies [63, 64] have shown a regression of microalbuminuria with these agents in type 1 diabetes. Interestingly, this effect is reported even at a very early stage or renal involvement in diabetes, i.e., with mild microalbuminuria between 20 and 70 μg/min. Thus, in a 2-year controlled clinical trial of ACE-inhibitor treatment in type 1 diabetic subjects, microalbuminuria increased, as expected, in the untreated control group where a mean increase rate between 15% and 20% was observed. In contrast, a clear-cut reduction in albuminuria was reported in patients on ACE-inhibitor therapy [63]. Thus, it appears that ACE inhibitor treatment is effective right from the onset of microalbuminuria. This study also indicated a correlation between the decrease in albuminuria and that in filtration fraction, confirming the specific renal impact of ACE inhibition [63]. The long-term aim, however, is not regression or microalbuminuria per se, although this could be useful as a surrogate marker, but primarily preservation of renal function (GFR) – and perhaps structure – which, to be assessed, requires long-term studies of up to 6 to 8 years. Mathiesen and co-workers [23] showed that the effect of ACE inhibition is long-lasting: microalbuminuric patients treated with ACE inhibitors and diuretics over the previous 8 years maintained an adequate GFR even after a pause in treatment, whereas a clear-cut decline in GFR was reported in those who did not receive antihypertensive treatment. This was the first study to document long-term preservation of GFR in type 1 diabetes. Rudberg et al. documented better preservation of renal structure [21]. Earlier long-term studies had demonstrated preservation of GFR along with reduction in microalbuminuria in type 2 diabetes [43, 45, 64]. Table II gives the principal details of 10 controlled studies (of 2 years' duration or longer) in type 1 diabetic patients with microalbuminuria [10, 21, 23, 34, 63–70]

8. EFFECT OF ACE INHIBITION ON RENAL STRUCTURAL DAMAGE IN PATIENTS WITH EARLY DIABETIC NEPHROPATHY

Several studies suggest a beneficial effect of ACE inhibitors and β-blockers on renal structure in patients with microalbuminuria. Rudberg et al. [21] documented a benefi-

Table 6. 10 Controlled Studies of IDDM Microalbuminuric patients (duration of study ≥ 2y) (745 pts) (modified from [1]).

Study	Drugs	Base-line no of pts	Mean BP	Dur of study, y	Mean or median UAE μg/min	Effect on UAE of drug	Effect on BP	Note
European Captopril Study (1994)	Cap/Pla	46/46	124/77	2	55	↓	↓	European Arm
North American Captopril Study (1995)	Cap/Pla	70/73	120/77	2	62	↓	↓	Northamerican arm
EUCLID (1997)	Lis/Pla	32/37	122/80	2	–42	↓	↓	Little/No effect on normoalb.
Bojestig et al. (2001)	Ram/Ram/Pla 1,25/5,0/-	18/19/18	?	2	61	No	No	HbA_{1C} = 7.4 Long duration of DM
Italian Micro-albuminuria (1998)	Lis/Nif/Pla	33/26/34	129/83	3	71	Lis ↓↓/ Nif ↓	L ↓↓/ Nif ↓	Effect with Nif
Mathiesen, Steno (1999)	Cap+Diu/ Con	21/23	126/77	8	93	↓	?	Preservation of GFR by ACE-I+Diu
ATLAN-TIS 2001	Ram/Ram/Pla 1,25/5,0/-	44/44/46	132/76	2	53	Ram 1,25/5 mg ↓	↓	Not dose-dependent HbA_{1c} = 11.0
Melbourne DNSG	Per/Nif/Pla	13/10/10	132/77	2-1/2	62	Per ↓/Nif-	Tenden-cy by Per ↓/Nif ↓	No effect with Nif
Padua/Aarhus (Low grade micro) (2001)	Lis/Pla	32/28	124/83 131/81	2	36 (range 20–70)	↓	↓ (24h)	Effect in low micro-albuminuria related to FF
Rudberg (1999)	Ena/Meto/Ref	7/6/9	125/81	~3	~31	E ↓/M ↓/ ref.-	No	Preservation of structure by Ena/Meto
Europe 9/ North-Am. 1	Mostly ACE-I	745 pts	127/79	2,8		Mainly ACE-I ↓	Mostly ↓	

Cap = Captopril, Lis = Lisinopril, Ram = Ramipril, Per = Perindopril, Ena = Enalapril, Meto = Metoprolol, Nif = Nifidipine, Con = Control group, Pla = Placebo, Ref: Reference group,

Mean age of patients: 32 years. Duration of Diabetes in mean: 16 years.

cial effect in patients with microalbuminuria, together with a reduction in albumin excretion rate. This is an important observation, since it seems that reduction in microalbuminuria also translates into beneficial effects on renal structure. In a follow-up study, the same authors reported a beneficial effect on glomerular dimensions and cortical interstitial expansion in microalbuminuric patients, which may be partly explained by blood pressure reduction [21–22]. A French study showed that perindopril was able to arrest further interstitial expansion in hypertensive patients with diabetes and biopsy-documented glomerulopathy [19]. This study highlighted the role of angiotensin II in the progression of interstitial changes, suggesting that ACE-inhibitor treatment would be effective. An Australian study [20] in, which in patients with microalbuminuria were given perindopril, also documented a long-term beneficial effect on glomerular structural damage.

In contrast, in the European Study for the Prevention of Renal disease In type 1diabetes (ESPRIT) [71], a low dose of an ACE inhibitor was used in a mixed group of patients with slow progression of renal disease. No effect on structural parameters could be documented [71].

9. THE GFR PERSPECTIVE

Many studies have documented the early hyperfiltration in especially type 1 diabetes, an abnormality that may be predictive of late nephropathy [6, 17]. Hyperfiltration is most often found in long-term diabetes with normo- or microalbuminuria. However, increasing albuminuria indicate future decline in GFR [30, 39]; clearly normoalbuminuria (and even microalbuminuria) ensures well-preserved GFR when measured with reliable techniques [58, 59]. Microalbuminuria is a very early feature in the development of renal disease in diabetes [72]. Figure 2 shows data on GFR in type 1 diabetes (female patients) with long- and short-term duration. GFR is clearly well-preserved in all patients. This is in contrast to data presented by M. Mauer, but in accordance with other studies.

10. CONFOUNDING ELEMENTS

In spite of the fact that several factors may influence UAER, they are usually easy to evaluate in a diabetes clinic [5]. The most important confounder is the inherent variability of UAER, which clearly makes several tests advisable in the continuous follow-up of patients. Factors such as urinary tract infection, cardiac decompensation, hypertension, exercise, and poor metabolic control with ketosis can easily be controlled for [5]. These conditions usually provoke increases only when they are clearly apparent or clinically symptomatic. New immunoturbidimetric procedures are certainly cost effective and may be recommended in conjunction with measurement of creatinine using the UACR as an important parameter [3]. However, studies suggest that even an early morning urine concentration of albumin is highly predictive of complications and mortality, at least in NIDDM patients [8, 9].

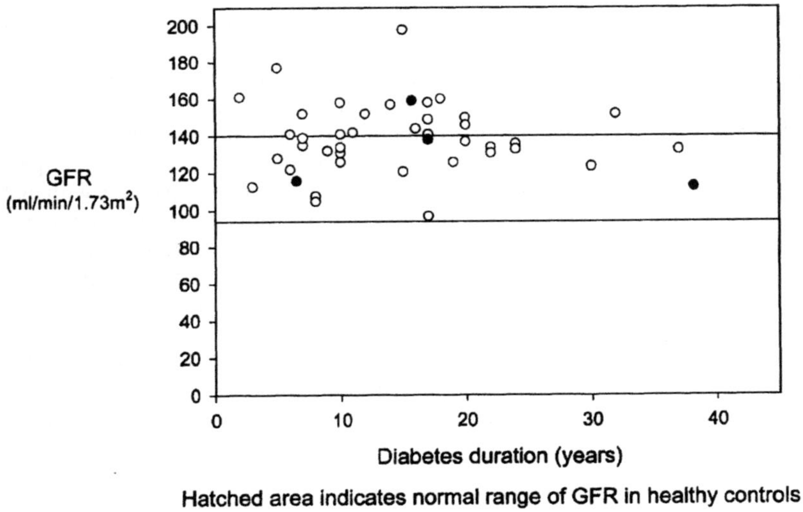

Figure 2.

11. SUMMARY AND CONCLUSION

Microalbuminuria is associated with significant glomerular pathology and increased renal as well as cardiovascular risks. There are many associated systemic alterations, which are amenable to effective treatment. Recent strong evidence supports the benefit of intervention at the phase of microalbuminuria in type 1 diabetes in particular. In type 2 diabetes treatment of those with microalbuminuria remains of equally great importance. It remains to be seen whether treatment of at risk patients with type 2 diabetes (and type 1) and a normal albumin excretion is as cost effective and beneficial as treatment of patients with microalbuminuria [73, 74]. Combined normoglycemia and low blood pressure are key elements in the prevention of micro- and macrovascular disease in diabetes.

REFERENCES

1. Mogensen CE. Microalbuminuria, blood pressure and diabetic renal disease: origin and development of ideas. In: Mogensen CE (ed), The Kidney and Hypertension in Diabetes Mellitus. Kluwer Academic Publ, 2000, pp 655–706.
2. Thomas SM, Viberti GC. Is it possible to predict diabetic kidney disease (review). J Endo Inv 2000; 23: 44–53.
3. Baenziger JC. Reporting units for albumin/creatinine ratio. Laboratory Med 2000; 31: 597.
4. Marshall SM. Blood pressure control, microalbuminuria and cardiovascular risk in type 2 diabetes mellitus. Diabetic Med 1999; 16: 358–372.
5. Mogensen CE, Vestbo E, Poulsen PL, Christiansen C, Damsgaard EM, Eiskjær H, Frøland A, Hansen KW, Nielsen S, Pedersen MM. Microalbuminuria and potential confounders. Diabetes Care 1995; 18: 572–581.

6. Mogensen CE. Prediction of clinical diabetic nephropathy in IDDM patients. Alternatives to microalbuminuria? Diabetes 1990; 37: 761–767.
7. Thomas SM. What should we do about microalbuminuria? In: Gill GV, Pickup JC, Williams G (eds), Difficult Diabetes. Blackwell Science Ltd, Oxford, 2001.
8. Dinneen S, Gerstein HC. The association of microalbuminuria and mortality in non-insulin-dependent diabetes mellitus. A systematic overview of the literature. Arch Intern Med 1997; 157: 1413–1418.
9. Schmitz A, Vaeth M. Microalbuminuria: a major risk factor in non-insulin-dependent diabetes: a 10-year follow-up study of 503 patients. Diabetic Med 1988; 5: 126–134.
10. Chaturvedi N. Should all type 1 diabetic microalbuminuric patients receive ACE inhibitors? A mega regression analysis. Ann Intern Med 2001, in press.
11. Heart Outcomes Prevention Evaluation (HOPE) Study Investigators. Effects of ramipril on cardiovascular and microvascular outcomes in people with diabetes mellitus: results of the HOPE study and MICRO-HOPE substudy. Lancet 2000; 355: 253–259.
12. The Heart Outcomes Prevention Evaluation Study Investigators. Effects of angiotensin-converting-enzyme inhibitor, ramipril, on death from cardiovascular causes, myorcardial infarction and stroke in high-risk patients. N Engl J Med 2000; 342: 145–153.
13. Gerstein HC. Cardiovascular and metabolic benefits of ACE inhibition. Diabetes Care 2000; 23: 882.
14. The Diabetes Control and Complications Trial/Epidemiology of Diabetes Interventions and Complications Research Group. Retinopathy and nephropathy in patients with type 1 diabetes four years after a trial of intensive therapy. N Engl J Med 2000; 342: 381–389.
15. Warram JH, Scott LJ, Hanna LS, Wantman M, Cohen SE, Laffel LMB, Ryan L, Krolewski AS. Progression of microalbuminuria to proteinuria in type 1 diabetes. Nonlinear relationship with hyperglycemia. Diabetes 2000; 49: 94–100.
16. Williams B. The unique vulnerability of diabetic subjects to hypertensive injury. Journal of Human Hypertension 1999; 13(S2): S3–S8.
17. Mogensen CE, Damsgaard EM, Froland A, Hansen KW, Nielsen S, Pedersen MM, Schmitz A, Thuesen L, Osterby R. Reduced glomerular filtration rate and cardiovasc ular damage in diabetes: a key role for abnormal albuminuria. Acta Diabetol 1992; 29: 201–213.
18. Mogensen CE, Hansen KW, Osterby R, Damsgaard EM. Blood pressure elevation versus abnormal albuminuria in the genesis and prediction of renal disease in diabetes. Diabetes Care 1992; 15: 1192–1204.
19. Cordonnier DJ, Pinel N, Barro C, et al. For the DIABIOPSIES GROUP. Expansion of the cortical interstitium is limited by converting enzyme inhibition in type 2 diabetic patients with glomerulossclerosis. J Am Soc Nephrol 1999; 10: 1253–1263.
20. Nankervis A, Nicholls K, Kilmartin G, Allen P, Ratnaike S, Martin FIR. Effects of perindopril on renal histomorphometry in diabetic subjects with microalbuminuria: a 3-year placebo-controlled biopsy study. Metabolism 1998; 47(suppl 1): 12–15.
21. Rudberg S, Østerby R, Bangstad H-J, Dahlquiest G, Persson B. Effect of angiotensin-converting enzyme inhibitor or beta-blocker on glomerular structural changes in young microalbuminuric patients with type 1 (insulin-dependent) diabetes mellitus. Diabetologia 1999; 42: 589–595.
22. Østerby R, Bangstad H-J, Rudberg S. Follow-up study of glomerular dimensions and cortical interstitium in microalbuminuric type 1 diabetic patients with or without antihypertensive treatment. Nephrol Dial Transplant 2000; 15: 1609–1616.
23. Mathiesen ER, Hommel E, Hansen HP, Smidt UM, Parving H-H. Randomised controlled trial of long-term efficacy of captopril on preservation of kidney function in normotensive patients with insulin-dependent diabetes and microalbuminuria. BMJ 1999; 319: 24–25.
24. Caramori ML, Fioretto P, Mauer M. The need for early predictors of diabetic nephropathy risk. Is albumin excretion rate sufficient? Diabetes 2000; 49: 1399–1408.
25. Mogensen CE. ACE-inhibition and antihypertensive treatment in diabetes: focus on microalbuminuria and macrovascular disease. JRAAS 2000; 1: 234–239.
26. Bradford HA. The environment and disease: association or causation? Proc R Soc Med 1965; 58: 295–300.
27. Bain SC, Chowdhur TA. Genetics of diabetic nephropathy and microalbuminuria (review). J Roy S Med 2000; 93: 62–66.
28. Schichiri M, Kishikawa H, Ohkubo Y, Wake N. Long-term results of the Kumamoto Study on optimal diabetes control in type 2 diabetic patients. Diabetes Care 2000; 23(s2): B21–B30.
29. Adler AI, Stratton IM, Neil HAW, Yudkin JS, Matthews DR, Cull CA, Wright AD, Turner RC, Holman RR on behalf of the UK Prospective Diabetes Study Group. Association of systolic blood pressure with

macrovascular and microvascular complications of type 2 diabetes (UKPDS 36): a prospective observational study. BMJ 2000; 321: 412–419.
30. Lemley KV, Abdullah I, Myers BD, Meyer TW, Blouch K, Smith WE, Bennett PH, Nelson RG. Evolution of incipient nephropathy in type 2 diabetes mellitus. Kidney Int 2000; 58: 1228–1237.
31. Nielsen S, Hove KY, Dollerup J, Poulsen PL, Christiansen JS, Schmitz O, Mogensen CE. Losartan modifies glomerular hyperfiltration and insulin sensitivity in type 1 diabetes. Diabetes, Obesity & Metabolism 2001, in press.
32. Mogensen CE. Neldam S, Tikkanen I, Oren S, Viskoper R, Watts RW, Cooper ME for the CALM Study Group. Randomised controlled trial of dual blockade of renin-angiotensin system in patients with hypertension, microalbuminuria, and non-insulin dependent diabetes: the candesartan and lisinopril microalbuminuria (CALM) study. BMJ 2000; 321: 1440–1444.
33. Raats CJ, Vandenbo J, Berden JHM. Glomerular heparan-sulfate alterations – mechanisms and relevance for proteinuria (review) Kidney Int 2000; 57: 385–400.
34. Bojestig M, Karlberg BE, Lindström T, Nystrom FH. Reduction of ACE activity is insufficient to lower micro-albuminuria in normotensive patients with type 1 diabetes. Diabetes Care 2001, in press.
35. Forsblom CM, Groop P-H, Ekstrand A, et al. Predictors of progression from normoalbuminuria to microalbuminuria in NIDDM. Diabetes Care 1998; 21: 1932–1938.
36. Nielsen S, Schmitz A, Poulsen PL, Hansen KW, Mogensen CE. Albuminuria and 24-h ambulatory blood pressure in normoalbuminuric and microalbuminuric NIDDM patients: a longitudinal study. Diabetes Care 1995; 18: 1434–1441.
37. Damsgaard EM, Frøland A, Jørgensen OD, Mogensen CE. Microalbuminuria as predictor of increased mortality in elderly people. BMJ 1990; 300: 297–300.
38. Poulsen PL, Hansen KW, Mogensen CE. Ambulatory blood pressure in the transition from normo- to microalbuminuria. A longitudinal study in IDDM patients. Diabetes 1994; 43: 1248–1253.
39. Rachmani R, Levi Z, Lidar M, et al. Considerations about the threshold value of microalbuminuria in patients with diabetes mellitus. Lessons from an 8-year follow-up study of 599 patients. Diabetes Res Clin Prac 2000; 49: 187–194.
40. Royal College of Physicians of Edinburgh Diabetes Register Group. Near-normal urinary albumin concentrations predict progression to diabetic nephropathy in type 1 diabetes mellitus. Diabetic Med 2000; 17: 782–791.
41. Schultz CJ, Neil HAW, Dalton RN et al. Risk of nephropathy can be detected before the onset of microalbuminuria during the early years after diagnosis of type 1 diabetes. Diabetes Care 2000; 23: 1811–1815.
42. Schultz CJ, Dalton RN, Neil HAW, Konopelska-Bahu T, Dunger DB on behalf of the Oxford Regional Prospective Study Group. Markers of renal tubular dysfunction measured annually do not predict risk of microalbuminuria in the first few years after diagnosis of type 1 diabetes. Diabetologia 2001; 44: 224–229.
43. Ravid M, Brosh D, Levi Z, Bar-Dayan V, Ravid D, Rachmani R. Use of enalapril to attenuate decline in renal function in normotensive, normoalbuminuric patients with type 2 diabetes mellitus. Ann Intern Med 1998; 128: 982–988.
44. Ravid M, Brosh D, Ravid-Safran D, Levy Z, Rachmani R. Main risk factors for nephropathy in type 2 diabetes mellitus are plasma cholesterol levels, mean blood pressure and hyperglycemia. Arch Intern Med 1998; 158: 998–1004.
45. Ravid M, Savin H, Jutrin I, Bental T, Katz B, Lishner M. Long-term stabilizing effect of angiotensin-converting enzyme inhibition on plasma creatinine and on proteinuria in normotensive type II diabetic patients. Ann Intern Med 1993; 118: 577–581.
46. Kvetny J, Gregersen G, Pedersen RS. Randomised placebo-controlled trial of perindopril in normotensive, normoalbuminuric patients with type 1 diabetes mellitus. QJM 2001; In press.
47. Mogensen CE. Microalbuminuria predicts clinical proteinuria and early mortality in maturity-onset diabetes. N Engl J Med 1984; 310: 356–360.
48. Valamadrid CT, Klein R, Moss SE, Klein BEK. The risk of cardiovascular disease mortality associated with microalbuminuria and gross proteinuria in persons with older-onset diabetes mellitus. Arch Intern Med 2000; 160: 1093–1100.
49. Fioretto P, Vestra MD, Saller A, Mauer M. Renal Structure in type 2 diabetic patients with micro-albumuria. In CE Mogensen (Ed.) The Kidney and Hypertension in Diabetes Mellitus. Kluwer Academic Publishers, 2000, pp 225–236.
50. Holman R, Turner R, Stratton I, et al. for United Kingdom Prospective Diabetes Study Group. Efficacy

of atenolol and captopril in reducing risk of macrovascular and microvascular complications in type 2 diabetes: United Kingdom prospective diabetes study 39. BMJ 1998; 317: 713–720.
51. Stratton IM, Adler AI, Neil HAW, Matthews DR, Manley SE, Cull CA, Hadden D, Turner RC, Holman RR on behalf of the UK Prospective Diabetes Study Group. Association of glycaemia with macrovascular and microvascular complications of type 2 diabetes (UKPDS 35): prospective observational study. BMJ 2000; 321: 405–412.
52. Stratton IM, Kohner EM, Aldington SJ, Turner RC, Holman RR, Manley SE, Matthews DR for the UKPDS Group. UKPDS 50: Risk factors for incidence and progression of retinopathy in type 2 diabetes over 6 years from diagnosis. Diabetologia 2001; 44: 156–163.
53. Turner R, Holman R, Stratton 1 et al. for United Kingdom Prospective Diabetes Study Group. Tight blood pressure control and risk of macrovascular and microvascular complications in type 2 diabetes: United Kingdom prospective diabetes study 38. BMJ 1998; 317: 703–713.
54. UKPDS 33. An intensive blood glucose control policy with sulphonylureas or insulin reduces the risk of diabetic complications in patients with Type 2 diabetes. Lancet 1998; 352: 837–853.
55. American Diabetes Association. Diabetes Nephropathy. Diabetes Care 2001; 24(suppl): S69–S72.
56. Deinum J, Rønn B, Mathiesen E, Derkx FHM, Hop WCJ, Schalekamp MADH. Increase in serum prorenin precedes onset of microalbuminuria in patients with insulin-dependent diabetes mellitus. Diabetologia 1999; 42: 1006–1010.
57. Bojestig M, Nystrom FH, Arnqvist HJ, Ludvigsson J, Karlberg BE. The renin-angiotensin-aldosterone system is suppressed in adults with type 1 diabetes. JRAAS 2000; 1: 353–356.
58. Hansen KW, Pedersen MM, Christensen CK, Schmitz A, Christiansen JS, Mogensen CE. Normoalbuminuria ensures no reduction of renal function in type 1 (insulin-dependent) diabetic patients. Journal of Internal Med 1992; 232: 161–167.
59. Sackmann H, Tran-Van T, Tack I, Hanaire-Broutin H, Tauber J-P, Ader J-L. Contrasting renal functional reserve in very long-term type 1 diabetic patients with and without nephropathy. Diabetologia 2000; 43: 227–230.
60. Christensen CK, Mogensen CE. Effect of antihypertensive treatment on progression of disease in incipient diabetic nephropathy. Hypertension 1985; 7: II109–II113.
61. Pedersen MM, Christensen CK, Hansen KW, Christiansen JS, Mogensen CE. ACE-Inhibition and renoprotection in early diabetic nephropathy. Response to enalapril acutely and in long-lerm combination with conventional antihypertensive treatment. Clin Invest Med 1991; 14: 642–651.
62. Pedersen MM, Hansen KW, Schmitz A, Sorensen K, Christensen CK, Mogensen CE. Effects of ACE inhibition supplementary to beta blockers and diuretics in early diabetic nephropathy. Kidney Int 1992; 41: 883–890.
63. Poulsen PL, Ebbehøj E, Nosadini R, et al. Early ACE-I intervention in microalbuminuric patients with type 1 diabetes: effects on albumin excretion, 24:h ambulatory blood pressure and renal function. Diabetes Metab 2001, in press.
64. Cooper ME, McNally P, Boner C. Antihypertensive treatment in NIDDM with special reference to abnormal albuminuria. In: Mogensen CE (ed), The Kidney and Hypertension in Diabetes Mellitus, 5th ed. Kluwer Academic Publishers, Boston, MA, 2000.
65. Poulsen PL, Ebbehøj E, Mogensen CE. Lisinopril reduces albuminuria during exercise in low grade microalbuminuric type 1 diabetic patients: a double blind randomised study. J Intern Med 2001, in press.
66. Laffel LMB, McCill JB, Cans DJ, on behalf of the North American Microalbuminuria Study Group. The beneficial effect of angiotensin-converting enzyrne inhibition wilh captopril on diabetic nephropathy in normotensive IDDM patients with microalbuminuria. Am J Med 1995; 99: 497–504.
67. Crepaldi C, Carta Q, Deferrari G, et al. Effects of lisinopril and niredipine on the progression to overt albuminuria in IDDM patients with incipient nephropathy and normal blood pressure. The Italian Microalbuminuria Study Group in IDDM. Diabetes Care 1998; 21: 104–110.
68. The ATLANTIS study group. Low-dose ramipril reduces microalbuminuria in type 1 diabetic patients without hypertension: results of a randomized controlled trial. Diabetes Care 2000; 23: 1823–1829.
69. Jerums C on behalf of the Melbourne Diabetic Nephropathy Study Group, Melbourne, Australia. ACE inhibition vs calcium channel blockade in normotensive type 1 and type 2 diabetic patients with microalbuminuria. Nephrol Dial Transplant 1998; 13: 1065–1066.
70. The EUCLID study group. Randomised placebo-controlled trial of lisinopril in normotensive patients with insulin-dependent diabetes and normalbuminuria or microalbuminuria. Lancet 1997; 349: 1787–1792.

71. The European Study for the Prevention of Renal disease in Type 1 diabetes (ESPRIT). Three years of antihypertensive therapy has no detectable effect on renal structure in type 1 diabetic patients with albuminuria. Diabetes, 2001. In press.
72. Thomas W. Shen Y, Molitch ME, Steffes MW. Rise in albuminuria and blood pressure in patients who progressed to diabetic nephropathy in the diabetes control and complications trial. J Am Soc Nephrol 2001; 12(2): 333–340.
73. Borch-Johnsen K, Wenzel H, Viberti GC, Mogensen CE. Is screening and intervention for microalbuminuria worthwhile in patients with insulin dependent diabetes. BMJ 1993; 306: 1722–1725.
74. The ACE-Inhibitors in Diabetic Nephropathy Trialist Group. Should all patients with type 1 diabetes mellitus and microalbuminuria receive angiotensin-converting enzyme inhibitors? A meta-analysis of individual patient data. Ann Intern Med 2001; 134: 370–379.

S. MICHAEL MAUER, MARIA LUIZA AVANCINI CARAMORI AND PAOLA FIORETTO

11. Which Comes First: Renal Injury or Microalbuminuria?

Editor's Comment: Expounding thinking countrary to the growing establishment view that microalbuminuria is an early (?earliest) sign of diabetic nephropathy portending approaching renal deterioration, Mauer et al. provide carefully drawn evidence that glomerular injury – at least in some diabetic individuals – may precede development of microalbuminuria. In Maurer et al.'s cohort of 125 long-standing type 1diabetic patients (including 72 of female gender) some with normalbuminuria expressed glomerular lesions of diabetic nephropathy. Furthermore, there was a disconnect between reduced glomerular filtration rate and proteinuria in one subset of 14 women. According to these and other studies by the Mauer team, glomerular lesions of diabetes may precede increases in albumin excretion. The lesson here is that our confidence in the sequence of changes in diabetic nephropathy requires reflection as hypertension and a reduced glomerular filtration rate may precede proteinuria as the first clinical indicator of nephropathy. As summarized by the authors, the natural history of nephropathy in type 1 and type 2 diabetes 'is far from completely described and continued longitudinal renal structural and functional observations are needed.'

INTRODUCTION

Diabetic nephropathy (DN) is the single most important cause of renal failure in the Western world and is consequent to the advanced progression of a constellation of renal lesions including widening of extracellular basement membranes, mesangial expansion, arteriolar hyalinosis, interstitial fibrosis, tubular atrophy, and global glomerular sclerosis (GS) [1]. This review argues that renal lesions precede the development of increased albumin excretion rates (AER) in patients with type 1 diabetes mellitus (type 1 D) but that the pathologic picture in type 2 diabetes mellitus (type 2 D) may be more complex and includes subsets of patients with increased AER but without significant diabetic glomerulopathy changes whose natural history, albeit incompletely described, is probably more benign.

DIABETIC NEPHROPATHY IN TYPE 1 D

Increase in glomerular basement membrane (GBM) width occurs early in type 1 D [2] and is paralleled by widening of the tubular basement membrane (TBM) [3]. In the GBM this is associated with an increase in the density of $\alpha 3$, $\alpha 4$ chains of type IV collagen in the lamina densa and a decrease in the density of $\alpha 1$, $\alpha 2$ chains of type IV [4] and type VI [5] collagen in the lamina rara interna. At more advanced stages of

E.A. Friedman and F.A. L'Esperance, Jr. (eds.), Diabetic Renal-Retinal Syndrome, 121–128.

disease there is a decrease in the density of heparan sulfate proteoglycan charge sites in the lamina rara externa and interna of the GBM [6]. Increase in the fraction of glomerular volume which is mesangium {mesangial fractional volume or [Vv(Mes/glom)]} can also be measured within 3–5 years after onset of type 1 D [2]. This increase is mainly due to expansion in the fraction of the glomerulus occupied by mesangial matrix [Vv(MM/glom)] although a lesser expansion of mesangial cell fractional volume [Vv(MC/glom)], also occurs [7]. The nature of the accumulating mesangial matrix material is not fully understood. The major component of MM, the classical chains of type IV collagen, decrease in density as the disease progresses [4], and the same is true for type VI collagen [5]. Quantitative immunohistochemical analyses of other MM molecules have not yet been done. Whatever its composition, the increase in the Vv(Mes/glom) eventually results in a decrease in relative peripheral GBM filtration surface [Sv(PGBM/glom)] [8]. Enlargement of glomerular volume (GV) can delay the decline of filtration surface per glomerulus (S/G) but, ultimately, as mesangial expansion progresses S/G declines [9]. Most patients developing diabetic renal disease have diffuse mesangial expansion but many also have nodular (Kimmelstiel-Wilson) lesions which appear to be consequent to the development of capillary microaneurisms [10]. These nodules are seen in about one-half of proteinuric (P) type 1 D patients (Mauer M, Fioretto P, Huang C, unpublished observations).

Expansion of the fraction of renal cortex occupied by interstitium [Vv(Int/cortex)] occurs in most type 1 D patients, even in relatively normal-appearing areas of renal cortex [areas without tubular atrophy, marked TBM thickening and reduplication, or global GS [11]. More severe interstitial expansion, however, occurs in areas of advanced tubular injury and glomerular scarring [11]. The initial stages of interstitial expansion are mainly due to increases in the cellular compartment of the interstitium with increase in the fibrillar collagen component of the interstitium seen only when Vv(Int/cortex) is about twice normal and GFR is already reduced [12]. The molecular nature of this increase in interstitial extracellular matrix (ECM) has not been carefully studied, but it appears that, initially, there are increases in type I and III collagen while in the late stages of disease type IV collagen accumulates [13].

Hyalinosis of afferent and efferent arterioles, i.e. the replacement of smooth muscle cells by waxy, homogeneous, PAS positive material [1, 8, 14], which appears to be consequent to an exudate of plasma proteins [15], is a characteristic diabetic renal lesion. Advanced arteriolar hyalinosis lesions, where one or more vessels on a renal biopsy are completely or nearly completely replaced by hyaline material, is associated with an increased percent of GS [14].

Although renal lesions in type 1 D tend to develop in parallel, there are still considerable variations within a given patient in the severity of the individual lesions. Thus, for example, although highly statistically significant, the relationship of GBM thickness and Vv(Mes/glom) is not precise ($r = 0.56$, $P \leq 0.0005$) [8]. These levels of structural interrelationships are present at both the earlier and the later stages of development of the renal lesions of diabetes, with the exception that the relationships of Vv(Int/cortex) with glomerular parameters and with TBM width are much weaker or not statistically significant at the earlier stages of the disease [3], and become significant only when large numbers of patients with more advanced nephropathology and renal dysfunction are included in these analyses [11, 16].

Renal Structural Functional Relationships in Type 1 D Diabetes

Vv(Mes/glom) is the structural measure which, on a cross-sectional analysis, is most closely (and inversely) related to glomerular filtration rate (GFR) [8]. GBM [8] and TBM width [3] are also inversely related to GFR, but less precisely so. Vv(Int/cortex) is not correlated with GFR, unless more advanced cases of DN are included [3, 11]. The increase in Vv(Int/cortex) in these cases is in association with lesions of marked TBM thickening and reduplication, tubular atrophy, and global GS [11]. Thus, it is more likely that interstitial expansion in diabetes is, at least in part, consequent to advanced glomerular, tubular and vascular injury and interstitial fibrosis as a driving force for declining GFR may only become important at the late stages of DN.

Increase in Vv(Mes/glom) is also the lesion most closely associated with increasing AER in type 1 D [17]. This conclusion is based on sequential biopsies performed five years apart, in a group of long-standing type 1 D patients whose AER was increasing over this time [17]. Several patients were in transition from normal to microalbuminuria (MA), or from MA to overt nephropathy. The single structural variable which correlated with the increase in AER over the five years of this study was the increase in the Vv(Mes/glom) over this same time. GBM width and Vv(Int/cortex) did not change significantly over this time [17].

It is difficult to sort out whether diabetic renal lesions are the cause or the consequence of elevated systemic blood pressure (BP) in type 1 D patients. Our earlier studies suggested that rapid development of mesangial expansion can be occurring in patients with systemic BP which are entirely normal and, in fact, are identical to those with patients with very slow rates of development of mesangial expansion [18]. Thus, systemic hypertension is not a necessary precondition for rapid mesangial expansion. However, this does not answer the question as to whether systemic hypertension can accelerate the rate at which the crucial lesions of DN develop. Since hypertension is frequently the consequence of advanced glomerular and tubulointerstitial lesions, early and longitudinal studies with serial measures of BP and renal structure are necessary in order to answer this question. Preliminary data from studies of the natural history of DN in young type 1 D patients show no significant relationships of carefully measured clinic BP values with renal biopsy findings. However, 24-hour ambulatory BP measurements in these normotensive type 1 D patients were linearly related to GBM width and Vv(Mes/glom) was related to 24-hour ambulatory BP measurements, but this relationship was not linear and was seen only in patients with BPs in the upper part of the normal range (Sharma, A and the International Diabetic Nephropathy Study Group, unpublished data). Thus, BP could be one determinant of the rate of development of early DN lesions. In cross-sectional studies at later stages of disease, mesangial expansion was a stronger statistical predictor of GBM width, interstitial fibrosis, and GS than was hypertension (clinic BP), whereas hypertension was more strongly related to arteriolar hyalinosis [18].

In summary, increase in Vv(Mes/glom), primarily due to an increase in Vv(MM/glom), is closely associated with all of the clinical manifestations of DN including increasing AER, declining GFR, and rising systemic BP. Although other structural variables are also related to these functional alterations, Vv(Mes/glom) is the strongest structural predictor of renal dysfunction in type 1 D patients and, in longitudinal studies of a small group of patients, the only correlate of increasing AER.

DIABETIC NEPHROPATHY IN TYPE 2 DIABETES

The first electron microscopic morphometric analysis of glomeruli in Japanese type 2 D patients revealed structural interrelationships and structural-functional relationships similar to those described for type 1 D patients [19]. There was striking inverse correlation between Vv(Mes/glom) and Sv(PGBM). Further, Vv(Mes/glom) was closely and directly related to urinary protein excretion and inversely related creatinine clearance whereas relationships of these functional parameters to GBM thickness were weaker or nonexistent [19]. However, the picture of type 2 D patients is probably more complex than suggested by this study. Thus, Østerby et al. found that type 2 D patients tended to have less marked glomerular changes than type 1 D patients with the same degree of P [20]. Also, several type 2 D patients, despite proteinuria, had glomerular parameters in the normal range. However, in this comparison, the type 1 D patients had lower levels of GFR when compared to type 2 D patients with the same degree of P [20]. One possible explanation of these findings could be that by the much larger glomerular volumes in the type 2 D patients, results in preservation of filtration surface. However, the explanation for the P in these type 2 D patients was somewhat obscure. These authors noted that the type 2 D patients were more heterogeneous in their pathology, although this heterogeneity was diminished when patients with both P and retinopathy were considered. Nonetheless, these authors found a significant correlation between S/G and GFR, which ranged from 24–146 ml/min/1.73 m^2 in these patients [20]. In a study performed in type 2 D Pima Indians, Vv(Mes/glom) increased progressively from early diabetes, to long-term diabetes with normoalbuminuria (NA), to MA, and to clinical nephropathy [21]. There was no relationship between GFR and S/G in these various functional subgroups, but the range of GFRs in this patient population was not as great as in the Danish studies of Østerby et al. [20]. Global GS was considered an important correlation of reduced GFR in this study of Pima Indian type 2 D patients [21].

The looser association between electron microscopic morphometric analysis of glomerular structure and renal function in type 2 D patients compared to type 1 D patients noted in the above-mentioned studies may in part be explained by observations suggesting more complex patterns of renal injury in type 2 D patients with MA [22]. Thirty-four Caucasian northern Italian type 2 D patients with MA were biopsied for research purposes. Three categories of renal structure were discerned by light microscopic analysis. In category I [(CI), $n = 10$] renal structure by light microscopy was normal or near normal showing only mild mesangial expansion, tubulointerstitial changes, or arteriolar hyalinosis, in any combination. In category II [(CII), $n = 10$] patients had typical diabetic nephropathology with balanced severity of glomerular, tubulointerstitial, and arteriolar changes, more typical of what is seen in type 1 D patients. Category III [(CIII), n-14] patients had atypical patterns of renal injury. These patients had absent or only mild glomerular diabetic changes with disproportionately severe renal structural lesions including tubulointerstitial injury, advanced glomerular arteriolar hyalinosis, and global GS exceeding 25%. That these lesions in CIII patients are of diabetic origin is suggested by the fact that hemoglobin A_1C was more elevated in this group and in CII than in CI patients. However, CIII patients differed from CII patients in having lower HbA_1C, higher body mass index and a lower incidence of proliferative retinopathy. Thus, the CII patients in this study may be similar to the type 2 D patients with retinopathy in the Danish study referred to above [20]. Findings in the CIII patients suggest that the kidney may react differently to hyperglycemia in different subpopulations with type 2 D. This might reflect the heterogeneous nature of type 2 D *per se*. Alternately, lesions in CIII patients may be, at least in part, non-diabetic in origin.

Recent unpublished observations suggest that these patterns are also seen in P type 2 D patients, although a lower proportion of these patients are in CI compared to normo-or MA patients; still, about 15% of type 2 D patients with P had minimal renal lesions (Fioretto P, unpublished data). A substantial proportion of P type 2 D patients were in CIII, similar to that in MA patients. Importantly, the course of GFR loss was also heterogeneous in these patients, and in type 2 D patients progressive GFR loss is related to mesangial expansion and does not occur over 4–5 years of follow-up, in patients who do not have typical DN lesions [23].

In summary, it appears that renal structural changes in type 2 D are more complex than in type 1 D patients. Approximately one-third of the patients show atypical patterns of renal injury and these are related to greater body mass index and a paucity of advanced retinopathy findings. Type 2 D patients with MA or P may have minimal lesions. Further cross-sectional and longitudinal studies in type 2 D patients are required before the nature and natural history of these complexities can be better understood.

WHICH COMES FIRST: DIABETIC NEPHROPATHY LESIONS OR MICROALBUMINURIA?

To place this section of this chapter into perspective, it is necessary to consider the prognostic power of MA. The original articles suggested that about 80% or more of MA type 1 D patients would progress to overt P in the next 5–10 years, whereas this progression was rare among NA patients with long-standing type 1 D [24–26]. However, our recent review of this subject suggested that the risk of progression from MA to P over 5–10 years is about 35%, while a similar percent revert to NA [27]. Although only about 5–10% of NA long-standing type 1 D patients progress to P, perhaps 40% of the patients destined to develop P will be NA at initial evaluation [27]. Similar outcomes were seen for type 2 D patients, although the percent of MA type 2 D patients reverting to NA is not known [27]. Thus, the presence of MA confers about a 4–5-fold increase in risk of the development of P and measurement of AER remains the best available test for early risk prediction for both type 1 D and type 2 D patients [27] and should be regularly measured in these patients according to established guidelines and accepted definitions [28]. However, the imprecision of this test in predicting risk of or safety from progression to P requires investigation and explanation.

Given the above discussion, it is not surprising that many NA longstanding type 1 D patients have glomerular structural abnormalities while MA patients have a range of glomerular structure from normal to that bordering on the changes regularly associated with overt P [29]. Although glomerular DN lesions are, on average, greater in MA than in NA patients, considerable overlap exists [29]. As suggested above, even greater heterogeneity in glomerular structure and overlap in structural measurements among these AER categories is seen among type 2 D patients [21, 30]. It is clear, therefore, that diabetic glomerular lesions can precede the development of MA in both type 1 D [29] and type 2 D [21, 30, 31].

Furthermore, some NA D patients with more advanced DN lesions can have reduced GFR [32]. Our original studies demonstrated that this clinical picture of NA, reduced GFR and more advanced renal lesions, occurred among type 1 D women who tended to self select a reduced protein intake [32].

More recently, we have been studying a new cohort of 125 long-standing type 1 D patients (72 females) and have confirmed that, as a group, NA patients have glomerular lesions of DN, although some remain within the normal range for structure (Caramori L, Mauer M, unpublished data). We also confirmed that there is overlap in glomerular structure between the NA (N = 88), MA (N = 18) and P (N = 18) patient groups. Moreover, we identified a subset of 14 NA patients, all women, who had reduced GFR, thus confirming our earlier observations [32]. These patients had more advanced DN lesions, more retinopathy, more hypertension and, of course, were more frequently women, than the normal GFR group of NA patients. These two separate studies are important, not only in showing that DN lesions can preceed increases in AER, but in demonstrating that increased AER may not be the first clinical indicator of DN, but may be antedated by reduced GFR and hypertension. The relevance of these observations is suggested by the ultimate ≈10% risk of progression of longstanding type 1 and type 2 D NA patients to P. Also, our preliminary data suggests that more advanced DN lesions and lower GFR among NA patients are predictive of progression to P [33].

In summary, there are strong structural-functional relationships in DN. These relationships are directionally similar but less precise in type 2 D *vs.* type 1 D patients. Important DN lesions can develop before MA and may be associated with reduced GFR, hypertension and with increased risk of progression to P. However, the natural history of DN in type 1 D and type 2 D patients is far from completely described and continued longitudinal renal structural and functional observations are needed in order to further understand this complex and important disease.

ACKNOWLEDGMENTS

Dr M. Luiza A. Caramori is supported by a Juvenile Diabetes Foundation International (JDFI) Research Fellowship Award. Work presented here was supported by grants from JDFI, The National Institutes of Health (DK13003 and DK43605) and the National Center for Research Resources (MO1-KK00400) and the American Diabetes Association.

REFERENCES

1. Mauer SM. Nephrology Forum: Structural-functional correlations of diabetic nephropathy. Kidney Int 1994; 45: 612–622.
2. Østerby R. Early phases in the development of diabetic glomerulopathy: a quantitative microscopic study. Acta Med Scand 1974; 574 (Suppl): 1–82.
3. Brito P, Fioretto P, Drummond K, et al. Tubular basement membrane width in insulin-dependent diabetes mellitus (IDDM). Kidney Int 1997, in press.

4. Zhu D, Kim Y, Steffes MW, Groppoli TJ, Butkowski RJ, Mauer SM. Glomerular distribution of type IV collagen in diabetes by high resolution quantitative immunochemistry. Kidney Int 1994; 45: 425–433.
5. Moriya T, Groppoli TJ, Kim Y, Mauer M. Quantitative immunoelectron microscopy of type VI collagen in glomeruli in type I diabetic patients. Kidney Int 2001; 59: 317–323.
6. Vernier RL, Steffes MW, Sisson-Ross S, Mauer SM. Heparan sulfate proteoglycan in the glomerular basement membrane in type I diabetes mellitus. Kidney Int 1992; 41: 1070–1080.
7. Steffes MW, Bilous RW, Sutherland DER, Mauer SM. Cell and matrix components of the glomerular mesangium in type I diabetes. Diabetes 1992; 41: 679–684.
8. Mauer SM, Steffes MW, Ellis EN, Sutherland DER, Brown DM, Goetz FC. Structural-functional relationships in diabetic nephropathy. J Clin Invest 1984; 74: 1143–1155.
9. Ellis EN, Steffes MW, Goetz FC, Sutherland DER, Mauer SM. Glomerular filtration surface in type I diabetes mellitus. Kidney Int 1986; 29: 889–894.
10. Saito Y, Kida H, Takeda S, et al. Mesangiolysis in diabetic glomeruli: Its role in the formation of nodular lesions. Kidney Int 1988; 34: 389–396.
11. Lane PH, Steffes MW, Fioretto P, Mauer SM. Renal interstitial expansion in insulin-dependent diabetes mellitus. Kidney Int 1993; 43: 661–664.
12. Katz A, Kim Y, Sisson-Ross S, Groppoli T, Mauer M. An increase in the volume fraction of the cortical interstitium occupied by cells antedates interstitial fibrosis in type 1 diabetic patients. J Am Soc Nephrol 2000; 11: A0634 (abstract).
13. Falk RL, Scheinman JI, Mauer SM, Michael AF. Polyantigenic expansion of basement membrane constituents in diabetic nephropathy. Diabetes 1983; 32: 34–39.
14. Harris RD, Steffes MW, Bilous RW, Sutherland DER, Mauer SM. Global glomerular sclerosis and glomerular arteriolar hyalinosis in insulin-dependent diabetes. Kidney Int 1991; 40: 107–114.
15. Burkholder PM. Immunohistopathologic study of localized plasma proteins and fixation of guinea pig complement in renal lesions of diabetic glomerulosclerosis. Diabetes 1965; 14: 755–760.
16. Bohle A, Miller GA, Werkmann M, Mackensen-Haen S, Ziao JC. The pathogenesis of chronic renal failure in the primary glomerulopathy renal vasculopathies and chronic interstitial nephritis. Kidney Int 1996; 49 (Suppl 54): 2–9.
17. Fioretto P, Steffes MW, Sutherland DER, Mauer M. Sequential renal biopsies in insulin-dependent diabetic patients: structural factors associated with clinical progression. Kidney Int 1995; 48: 1929–1935.
18. Mauer SM, Sutherland DER, Steffes MW. Relationship of systemic blood pressure to nephropathy in insulin-dependent diabetes mellitus. Kidney Int 1992; 41: 736–740.
19. Hayashi H, Karasawa R, Inn H, et al. An electron microscopic study of glomeruli in Japanese patients with non-insulin dependent diabetes mellitus. Kidney Int 1992; 41: 749–757.
20. Østerby R, Gall M-A, Schmitz A, Nielsen FS, Nyberg G, Parving H-H. Glomerular structure and function in proteinuric type II (non-insulin-dependent) diabetic patients. Diabetologia 1993; 36: 1064–1070.
21. Pagtalunan ME, Miller PL, Jumping-Eagle S, et al. Podocyte loss and progressive glomerular injury in type II diabetes. J Clin Invest 1997; 99: 342–348.
22. Fioretto P, Mauer M, Brocco E, et al. Patterns of renal injury in NIDDM patients with microalbuminuria. Diabetologia 1996; 39: 1569–1576.
23. Nosadini R, Velussi M, Brocco E, Bruseghin M, Abaterusso C, Saller A, Dalla Vestra M, Carraro A, Bortoloso E, Sambataro M, Barzon I, Frigato F, Muollo B, Chiesura-Corona M, Pacini G, Baggio B, Piarulli F, Sfriso A, Fioretto P. Course of renal function in type 2 diabetic patients with abnormalities of albumin excretion rate. Diabetes 2000; 49: 476–484.
24. Viberti GC, Hill RD, Jarrett RJ, Argyropoulos A, Mahmud U, Keen H. Microalbuminuria as a predictor of clinical nephropathy in insulin-dependent diabetes mellitus. Lancet 1982; 1: 1430–1432.
25. Parving H-H, Oxenbøll B, Svendsen PA, Christiansen JS, Andersen AR. Early detection of patients at risk of developing diabetic nephropathy: a longitudinal study of urinary albumin excretion. Acta Endocrinol (Copenh) 1982; 100: 550–555.
26. Mogensen CE, Christensen CK. Predicting diabetic nephropathy in insulin-dependent patients. N Engl J Med 1984; 311: 89–93.
27. Caramori ML, Fioretto P, Mauer M. The need for early predictors of diabetic nephropathy risk: is albumin excretion rate sufficient? Diabetes 2000; 49: 1399–1408.
28. Mogensen CE, Chachati A, Christensen CK, Close CF, Deckert T, Hommel E, Kastrup J, Lefebvre P, Mathiesen ER, Feldt-Rasmussen B, Schmitz A, Viberti GC. Microalbuminuria: an early marker of renal involvement in diabetes. Uremia Invest 1986; 9: 85–95.

29. Fioretto P, Steffes MW, Mauer SM. Glomerular structure in non-proteinuric insulin-dependent diabetic patients with various levels of albuminuria. Diabetes 1994; 43: 1358–1364.
30. Fioretto P, Mauer M, Velussi M, Carraro A, Muollo B, Baggio B, Crepaldi G, Nosadini R. Ultrastructural measures of glomerular extracellular matrix accumulation in non-proteinuric type 2 diabetic patients. J Am Soc Nephrol 1996; 7: 1356–1357 (abstract).
31. Moriya T, Moryia R, Yajima Y, Steffes MW, Mauer M. Urinary albumin is a weeker predictor of diabetic nephropathy lesions in Japanese NIDDM patients than in Caucasians IDDM patients. J Am Soc Nephrol 1997; 8: 116A (abstract).
32. Lane PH, Steffes MW, Mauer M. Glomerular structure in IDDM women with low glomerular filtration rate and normal urinary albumin excretion. Diabetes 1992; 41: 581–586.
33. Caramori ML, Fioretto P, Mauer M. Long-term follow-up of normoalbuminuric longstanding type 1 diabetic patients: progression is associated with worse baseline glomerular lesions and lower glomerular filtration rate. J Am Soc Nephr 1999; 10: 126A, (abstract).

LUTHER T. CLARK AND OLUWOLE ABE

12. Prevention, Diagnosis and Management of Heart Disease in Patients with Renal Failure

Editor's Comment: Whether treated by hemodialysis, peritoneal dialysis, or a kidney transplant, renal failure in diabetes typically ends in a cardiac death. The remarkable advances in interventive cardiology that have improved the formerly bleak prognosis of coronary artery disease are just now being incorporated into uremia therapy. As observed by Clark and Abe, prior studies of the sensitivity and specificity of noninvasive diagnostic testing in patients with diabetes and end-stage renal disease have yielded highly variable findings. In those presenting for a kidney transplant, the prevalence of significant angiographically defined coronary artery disease (> 50% stenosis) is 25–50%. Significant coronary artery disease is usually associated with electrocardiographic (ECG) evidence of ischemia. Radionuclide scintigraphy is more sensitive and specific than routine treadmill exercise tests while dobutamine echocardiography, especially in patients with left ventricular hypertrophy, improves predictive accuracy. Perhaps the key point to be made in planning for long-term management of patients with kidney failure is that active and continuing collaboration with a cardiology service is essential to survival.

INTRODUCTION

Cardiovascular disease (CVD) is the leading cause of death in patients with end stage renal disease (ESRD), accounting for approximately 50% of all deaths in uremic patients [1–4]. According to the USRDS, approximately 16,000 patients on dialysis die each year from CVD [5]. This represents an overall CVD mortality of approximately 9% annually, which is about 30 times greater than that in the general population [5, 10]. The increased cardiovascular risk in patients with end-stage renal disease (ESRD) is due to a multiplicity of factors, including an excess burden of traditional risk factors as well as abnormalities associated with the chronic renal failure itself. In particular, an aging population of patients with ESRD, and high prevalences of diabetes and hypertension (the two most common causes of ESRD in the US [5, 11]), are powerful risk factors for cardiovascular morbidity and mortality. However, even after adjustment for these and other factors (gender, etc.), cardiovascular mortality rates remain considerably higher among patients with renal failure than in the general population. Importantly, the burden of CVD is increased in both those with ESRD and in those not yet requiring renal replacement therapy [6–9].

Prevention, early detection of cardiovascular disease, and optimal management of diagnosed cardiac disease in patients with renal failure are the cornerstones of effective therapy. In this review, we highlight the magnitude of burden of heart disease in patients with chronic renal failure as well as some of the special challenges faced by physicians and other health care providers who treat this very high risk population.

E.A. Friedman and F.A. L'Esperance, Jr. (eds.), Diabetic Renal-Retinal Syndrome, 129–145.

SPECTRUM OF CVD IN PATIENTS WITH CHRONIC RENAL FAILURE

The spectrum of cardiovascular disease in patients with chronic renal failure is similar to that in the general population and includes coronary disease (myocardial infarction, unstable angina, sudden cardiac death, stable angina), hypertension, left ventricular hypertrophy, heart failure, and calcific valvular disease. Coronary disease is by far the major cardiac cause of death.

New patients presenting for renal replacement therapy frequently already have co-existing cardiovascular disease, including hypertension (73.9%), diabetes (64.4%), congestive heart failure (33.4%) and ischemic heart disease (24.6%) [5]. In some studies as many as 50% of patients presenting for first dialysis already have both ischemic heart disease and cardiac failure, as many as 80% have left ventricular hypertrophy, dilatation, or systolic dysfunction [2].

Traditional cardiac risk factors such as hypertension, dyslipidemia, diabetes, smoking, and family history are major contributors to increased risk. In addition, abnormalities associated with chronic renal failure itself (anemia, extracellular volume overload, metabolic abnormalities, secondary hyperparathyroidism, hypotension and hypoxia during dialysis, and the arteriovenous fistula) may adversely affect myocardial oxygen supply:demand balance and contribute to increased cardiovascular risk.

Women with renal failure develop clinical coronary disease at a similar rate and severity as age-matched men with renal failure. The reasons for this have not been fully elucidated, but may be due to the high prevalence of diabetes, earlier onset of menopause, and abnormalities in the pituitary-gonadal axis in these women. Thus, both men and women with ESRD, chronic renal failure not yet requiring replacement therapy, and even those with only mild renal disease, should be targeted for intensive cardiovascular prevention and risk reduction efforts.

CORONARY HEART DISEASE

CHD Risk Factors

Patients with ESRD have an excess burden of cardiovascular risk factors. These include 1) traditional risk factors such as age (average age of US patients presenting for dialysis is 62.2 years), diabetes, hypertension, dyslipidemia, physical inactivity; 2) a higher level of thrombogenic factors, and homocysteine than in the general population; and 3) hemodynamic and metabolic factors (proteinuria, increased extracellular fluid volume, electrolyte imbalance, hypoalbuminemia, anemia).

Diabetes

Much of the excess cardiovascular mortality in patients with ESRD can be attributed to the high prevalence of diabetes mellitus, a common co-existing morbidity responsible for ESRD in 40–45% of patients [5]. Diabetes increases risk for cardiovascular events at least a two- to four-fold [12]. Also, diabetics experience greater mortality during the acute phase of myocardial infarction and during the post-infarction period, and more often have complications (arrhythmias, congestive heart failure, and recurrent ischemia and infarction) than nondiabetics.

Optimal diabetes management is one of the cornerstone of effective therapy for cardiovascular risk reduction in patients with ESRD. Although the achievement of strict glycemic control can be difficult and the risk of hypoglycemia increased, achieving optimal control is extremely important since the consequences of uncontrolled hyperglycemia are even more serious.

Hypertension

Approximately three-fourths of patients presenting for initial dialysis have hypertension and hypertension is the etiology of the ESRD 25–30% of the time [5]. Both systolic and diastolic hypertension are established risk factors for cardiovascular disease. Systolic blood pressure is a better predictor than diastolic blood pressure for overall mortality, and for the development of ESRD, CHD, stroke, and heart failure. Hypertension also predisposes to left ventricular hypertrophy, causes endothelial dysfunction and injury, and predisposes to accelerated development of atherosclerosis.

Optimal blood pressure control is an important goal of therapy in patients with ESRD. Except for diuretics, all classes of antihypertensives are effective. Calcium channel antagonists and ACE inhibitors (used with caution to avoid hypercalemia if the creatinine exceeds 3 mg/dL) are preferred agents. A key component of antihypertensive therapy is control of extracellular fluid volume and maintenance of 'dry weight'.

Left Ventricular Hypertrophy

Left ventricular hypertrophy (LVH) is highly predictive of ischemic heart disease morbidity and mortality, and is a stronger risk factor than hypertension, cigarette smoking, or hypercholesterolemia. In patients with ESRD on dialysis, more than two-thirds of deaths in patients with LVH are due to heart failure or sudden cardiac death [13].

Although echocardiography is an accurate and reproducible technique for assessing left ventricular mass and calculating left ventricular mass index, the latter is sensitive to changes in diastolic internal diameter of the LV (which may vary widely in dialysis patients). Measurement of LV parameters 15–20 hours following dialysis (when the extracellular fluid volume is more comparable with that in normal individuals) reduces this variability [13].

Control of hypertension and anemia may improve LV geometry, increase tolerance of myocardial ischemia, and improve cardiovascular morbidity and mortality. Over-judicious use of erythropoietin to correct anemia, however, may accelerate the risk of myocardial infarction or death [14]. Although few studies have evaluated LVH regression associated with antihypertensive drugs in dialysis patients, LV mass regression can be achieved provided the blood pressure is reduced decisively.

Dyslipidemia

Common lipid abnormalities in patients with renal failure include elevated LDL-cholesterol, elevated triglycerides, and decreased HDL-cholesterol. Elevated total or LDL-cholesterol are established independent risk factors for CHD and reductions in LDL-cholesterol have been shown to decrease risk for CHD events in several large clinical outcome trials. Low high-density lipoprotein (HDL)-cholesterol levels increase risk for development of CHD independent of LDL levels and other risk factors, whereas elevated HDL-cholesterol levels are protective. Several recent studies, including a meta-analysis [15] have reported an independent correlation of elevated triglycerides with CHD, suggesting that triglyceride-rich lipoproteins, such as VLDL remnants, are independently atherogenic. Patients with chronic renal failure have a cardiovascular mortality rate of almost 10% per year. Thus, these patients are in a very high risk category and should receive intensive lipid modifying therapy as a routine component of their risk reduction regimen. Although not specifically classified as a 'CHD risk equivalent', the excess risk (much greater than the 2% per year threshold defined by National Cholesterol Education Program) clearly puts them in the highest risk category and their goals of therapy should be an LDL < 100 mg/dL.

Hyperhomocysteinemia

Hyperhomcysteinemia is an independent risk factor for atherosclerotic cardiovascular disease in patients with and without renal failure [16, 17]. Elevated homocysteine levels are present in 85–100% of patients with chronic renal failure [18, 19] and may be an important contributor to the increased risk for premature cardiovascular disease in patients with uremia. The cause of the elevated homocysteine levels in patients with renal failure is unknown, and cannot be explained solely on the basis of abnormalities of homocysteine renal extraction and/or excretion. Vitamin B6, vitamin B12, and folic acid are effective therapies for decreasing homocysteine levels. In renal transplant recipients, homocysteine levels decrease, but usually remain higher than normal. In transplant patients, the combination of folic acid and vitamin B6 may provide effective reductions of elevated homocysteine levels.

Inflammation and Inflammatory Markers

Chronic inflammation is an important contributor to the pathogenesis of cardiovascular disease and atherosclerotic plaque instability. Acute-phase reactants such as C-reactive protein (CRP) have emerged as important markers of inflammation, as indicators of poor prognosis (short-term and long-term) in patients with coronary disease, and perhaps more specifically as a markers of 'unstable' atheromatous lesions. Increased levels of CRP and other markers of inflammation (interleukin 6) may be present in as many as 25% of patients with renal failure [21], even during the predialytic phase. Thus, patients with renal failure may have increased activation of inflammatory mediators that contribute to their enhanced cardiovascular morbidity and mortality.

Smoking

As in patients without renal disease, cigarette smoking is a powerful risk factor for CHD and other atherosclerotic diseases in patients with chronic renal disease. Smoking cessation counseling and nicotine replacement therapy as appropriate should be strongly recommended.

Thrombogenic factors

Patients with chronic renal disease have decreased platelet function and elevated procoagulant activity. Aspirin reduces risk of subsequent CVD events and should be considered/prescribed, although aspirin may further worsen platelet function in chronic renal disease.

CORONARY HEART DISEASE

It is well known that chronic heart disease patients have accelerated atherosclerosis [20]. Cardiac disease accounts for approximately 45% of all-cause mortality in patients with ESRD. Approximately 20% of the cardiac deaths are due to acute myocardial infarction. Among new dialysis patients, aged 45–64, more than one in five (21%) will already have CHD and 8% will have history of previous myocardial infarction [5]. Among new dialysis patients 65 years or older, more than one-third will have coronary disease and about one in eight will have a history of previous myocardial infarction [5].

Myocardial Infarction

Patients with renal failure (on dialysis and predialytic) are a particularly high risk group for acute myocardial infarction and poor outcomes following infarction. In an analysis of outcomes following first myocardial infarction in 34,189 patients on long-term dialysis [22], Herzog and colleagues found an overall mortality of 59% at one year, 73% at two years, and 90% at five years. The cardiac mortality was 41% at one year, 52% at two years, and 70% at five years. Older patients and those with diabetes had the highest mortality rates. Renal transplant recipients had somewhat better outcomes than dialysis patients with mortality rates at one, two, and five years of 41%, 52%, and 70% respectively [22]. The most powerful predictors of poor outcomes were older age (> 65 years) and diabetic nephropathy. However, even patients without diabetes or hypertension had poor outcomes with a two year mortality rates of 69 percent.

The reasons for the extremely poor outcomes in dialysis patients following myocardial infarction have not been fully elucidated, but appear to be related to 1) an excess of poor outcome indicators (increased age, diabetes mellitus, hypertension, LVH, etc.), 2) less use of thrombolytic and revascularization therapies [22, 23], and 3) increased predisposition to lethal arrhythmias [23, 24] and hypotension [25] during hemodialysis.

Thus, renal failure patients suffering acute myocardial infarction require particularly careful and intensive therapy during the acute and long term phases following myocardial infarction. This is especially true for older patients and those with diabetes mellitus. In addition to higher complication rates, including arrhythmias, congestive heart failure, and recurrent infarction, patients with diabetes and myocardial ischemia often present with non-chest pain symptoms including unexplained heart failure, malignant arrhythmias, or even with acutely uncontrolled blood sugars and diabetic ketoacidosis. New onset dyspnea and congestive heart failure in patients with diabetes and ESRD should always arouse suspicion of myocardial ischemia, as well as bilateral atherosclerotic renal artery stenosis [26].

Thrombolytic therapy and other forms of immediate revascularization should also not be denied to patients with ESRD. These patients are at increased risk from their infarction and there have been reported mortality risk reductions of almost 30% in dialysis patients who receive thrombolytic therapy [27].

Medical Management of Myocardial Ischemia

The medical management of myocardial ischemia in patients with uremia is similar to that in patients without kidney disease except that drug doses often have to be reduced. Nitrates, beta blockers, calcium channel blockers are effective and generally well tolerated in patients with uremia. However, these medications may have to be decreased or withheld prior to dialysis in order to avoid hypotension during the procedure. Maintenance of the hematocrit above 30 percent (Hemoglobin $\geq$ 10 gm/dl) may also decrease risk for myocardial ischemia. Hypertension, ECV overload and hyperparathyroidism should be controlled.

Cardiac Enzymes

The diagnosis of acute myocardial infarction in patients with ESRD can at times be very challenging since electrocardiographic abnormalities and elevation of cardiac enzymes may occur in the absence of myocardial ischemia or necrosis in dialysis patients. Creatinine phosphokinase (CK) levels may be elevated due to impaired renal clearance. Cardiac troponin I and troponin T have shown great promise as tools for diagnosis and risk stratification in patients with acute coronary syndromes. Although the usefulness of cardiac troponin I and troponin T to predict risk for subsequent adverse cardiac events in patients with renal failure and suspected acute coronary syndromes is reduced compared to patients without renal failure [28, 29], cardiac troponin I has higher specificity than cardiac troponin T in hemodialysis patients [30].

Silent Myocardial Ischemia

Patients with chronic renal failure have an increased prevalence of painless or atypical ischemic symptoms. This is due largely to the high prevalence of co-existing diabetes, and to the various nonischemic causes of chest pain in patients with ESRD (i.e. pericarditis). Diabetes is common in patients with chronic renal failure and diabetics frequently have silent or unrecognized myocardial ischemia. In the Framingham Heart Study, 32–42% of diabetic patients with myocardial infarction had atypical symptoms (dyspnea, fatigue, confusion, nausea, vomiting) compared to 6–15% of those patients without diabetes [31]. Silent infarction also occurs more often in patients with chronic hypertension. Atypical symptoms during myocardial ischemia are an important clinical problem since patients may not recognize the need for medical care and there is an increased risk of myocardial infarction, malignant arrhythmias, and sudden cardiac death [32]. The increased frequency of silent ischemia in diabetics may be due to the cardiac autonomic neuropathy that may be seen in these patients. Optimal glycemic control slows the development of cardiac autonomic neuropathy and may lessen propensity for silent ischemic episodes.

Congestive Heart Failure

Approximately 10% of patients with uremia are hospitalized for congestive heart failure annually [13]. The presence of congestive heart failure in patients with renal failure is a poor prognostic indicator with a mean patient survival of only 32 months [33]. One-third of patients with renal failure have congestive heart failure at the time of presentation for first dialysis [5], predisposing factors being increased age, presence of coronary disease, and left ventricular systolic dysfunction. In patients free of CHF at the onset of dialysis, age, severity of anemia, hypoalbuminemia, hypertension, and systolic dysfunction increase risk of developing heart failure [13]. Recognition that heart failure – and other powerful predictors of increased cardiovascular risk – are often present prior to the dialysis phase of renal failure emphasizes the importance of correction of these and predisposing abnormalities prior to the onset of overt ESRD.

Heart failure in patients at the beginning of dialysis is due to systolic dysfunction in about half of patients, and is associated with LV dilatation and diastolic dysfunction in the other half. The presence of diabetes increases the likelihood of development of heart failure from all causes. Congestive heart failure is twice as prevalent in diabetic men and five times as common in diabetic women as in their non-diabetic counterparts [34]. This is largely a consequence of the higher rates of coronary heart disease, hypertension, and diabetic cardiomyopathy.

The management of CHF in patients with chronic renal failure is similar to that in patients without renal disease. In addition to correcting reversible lesions (i.e. coronary disease) and improving contributory conditions (severe anemia, arrhythmias), preload and afterload should be optimized. Dietary sodium should be restricted. ACE inhibitors, digoxin, and vasodilators are often useful. Dialysis may be lengthened and additional fluid removed. Renal transplantation recipients usually have improvement in their left ventricular performance, hemodynamics, and symptoms.

Cardiomyopathy

The so-called 'uremic' cardiomyopathy is a heterogeneous (systolic and diastolic dysfunction) and multifactorial syndrome [13]. The cardiomopathy appears to be largely related to non-specific and potentially reversible factors such as anemia, hypertension, and overhydration. The role of uremic toxins in the development of cardiomyopathy is unclear and requires further study. In patients with diabetes, diabetes associated cardiomyopathy may be present. Diabetic cardiomyopathy appears to be a distinct pathologic entity that results in myocardial hypertrophy, increased interstitial connective tissue and microvascular pathology with systolic and diastolic ventricular dysfunction. The cardiomyopathy appears to be multifactorial in origin and may occur in the absence of coronary, valvular, or other known cardiac disease. In the Framingham Heart Study, approximately 15% of the individuals with diabetes and congestive heart failure were believed to have diabetic cardiomyopathy [34].

Hypertension and Ventricular Dysfunction

As noted previously in this review, hypertension is common in patients with renal failure and may result in left ventricular hypertrophy and heart failure. Patients with diabetic nephropathy are at particular risk since 75–85 percent of those with overt diabetic nephropathy will have hypertension [35, 36]. These patients may have diastolic dysfunction, systolic dysfunction, or both. In patients with normal left ventricular wall motion and ejection fraction, particularly in the presence of left ventricular hypertrophy, the predominant problem is usually diastolic dysfunction. Vigorous treatment of hypertension for optimal blood pressure control and regression of left ventricular hypertrophy are keys to improvement of symptoms and risk reduction.

Treatment and Prevention of Heart Failure

Primary prevention of heart failure involves vigorous control of hypertension, diabetes, and other risk factors for CHD. Secondary prevention following myocardial infarction involves control of myocardial ischemia and remodeling. In selecting pharmacologic therapy for patients with predominantly systolic dysfunction, a traditional stepped care approach to therapy is appropriate, including angiotensin converting enzyme (ACE) inhibition, digoxin, vasodilators (nitrates, hydralazine), and sometimes cautious use of certain beta blockers.

In patients whose heart failure is due predominantly to diastolic dysfunction, drug therapy can be problematic. Unlike with systolic dysfunction, clinical studies to direct patient care are inadequate. In addition to vigorous treatment of hypertension and other CHD risk factors, weight loss in obese patients is particularly important since obesity is a strong stimulus to ventricular hypertrophy. Angiotensin converting enzyme inhibitors, calcium channel blockers, selective alpha-1 blockers and beta-blockers may be beneficial. Digoxin should be avoided. Arterial vasodilators will at times be necessary, but can be problematic, and should be used cautiously to avoid hypotension. Control of myocardial ischemia is essential since muscle relaxation and ventricular function are worsened during ischemia. In patients with ESRD and low cardiac output, possible etiologies should be thoroughly investigated, including consideration of possible pericarditis with pericardial effusion and impending tamponade, especially if the patient has ben on chronic therapy with minoxidil.

VASCULAR AND TISSUE CALCIFICATIONS

Vascular and tissue calcifications occur commonly in patients with renal failure who are receiving long-term dialysis, and atherosclerotic lesions in dialysis patients have been found to have more extensive calcification and medial thickening that those from non-uremic patients [37–39]. Abnormalities of parathyroid function, calcium and phosphorus metabolism have been implicated as key factors in the development of soft tissue calcification in these patients [40].

Valvular Calcifications

Patients on chronic hemodialysis have an abnormally high incidence of premature calcification of cardiac valves – the mitral annulus and valve and the aortic valve being the preferential sites. Advanced age, male gender, duration of hemodialysis, and hyperphosphatemia associated with secondary hyperparathyroidism are predisposing factors. These abnormal calcifications have important clinical consequences since valvular calcifications may be associated with conduction abnormalities, arrhythmias, ventricular dysfunction and heart failure. In one study of 92 chronic hemodialysis patients compared to normal control subjects, Ribeiro found that 44.5% of dialysis patients had mitral and 52% had aortic valvular calcifications compared with 10% and 4.5% of control subjects, respectively [41]. Patient age and Ca × P product were the most predictive parameters for mitral calcification. In another study [42], 16 of 112 (14.3%) of hemodialysis patients developed aortic valvular calcification with aortic stenosis over a 3 year follow-up period. In addition to secondary hyperparathyroidism – which is almost universal in hemodialysis patients – other sources of excess calcium load in dialysis patients include dietary intake, calcium-containing dialysate solutions, intake of calcium-based phosphate binders, and vitamin D therapy. Long-term administration of phosphate-binding agents (e.g., calcium carbonate, calcium acetate) might improve abnormal calcium-phosphorous metabolism and decrease tissue and vascular calcific deposition.

PERICARDITIS

Although much less than in the predialysis era, pericarditis in dialysis patients remains an important problem. The pericarditis may be related to inadequate removal of uremic toxins, coincident diseases such as viral infections, tuberculosis, systemic lupus erythematosus, or to drugs such as minoxidil. Pericardial effusions frequently complicate pericarditis but cardiac tamponade is rare.

CARDIAC ARRHYTHMIAS

Cardiac arrhythmias occur frequently in patients with chronic renal failure, particularly during the dialysis and post-dialysis period. In one multicenter study, three-fourths (76%) of patients had ventricular arrhythmias during a 48 hour ambulatory monitoring period, 40% of whom had complex forms (ventricular couplets or nonsustained ventricular tachycardia) [43]. The arrhythmias occurred most frequently during the second hour of hemodialysis and lasted up to five hours following dialysis. Factors that contribute to increased risk of arrhythmias during dialysis include increased age (> 55 years old), left ventricular dysfunction, ischemic heart disease, conduction system calcification, pericarditis, hemodialysis-associated hypotension, dialysis associated electrolyte and acid-base disturbances, and hypoxemia. Dialysis patients receiving digoxin have an increased risk for both atrial and ventricular arrhythmias during dialysis because of the rapid shifts of potassium. Therefore digoxin, if required, must be used cautiously and in low doses.

The principles of management of cardiac arrhythmias are similar to those in patients without renal failure. If the arrhythmias are primarily related to dialysis, special attention should be directed to the patient's potassium and the potassium concentration of the dialysis bath. In patients with pericarditis, the arrhythmias usually respond to treatment of the pericarditis.

CHD RISK ASSESSMENT AND PRETRANSPLANT EVALUATION

Accurate diagnosis of coronary artery disease and risk assessment in patients with renal failure are important for prevention and risk reduction counselling, for interpretation of patient symptoms, and for patient evaluation prior to renal transplantation. In candidates for renal transplantation, detection and treatment of coronary disease are especially important since CHD is the leading cause of death following kidney transplantation [5, 44] . However, accurate risk assessment can be difficult since classic ischemic symptoms are often not present and the results of noninvasive studies may be inconclusive.

The prevalence of significant coronary artery disease in diabetic renal transplant candidates has been quite consistent in angiographic series; 25–40% of patients have one or more fixed coronary artery stenosis of at least 50% [45–48]. In the series where patients with significant disease did not receive pretransplant coronary revascularization, the two year mortality rate for diabetic ESRD patients was at least 50%, whereas outcome was improved in patients who had revascularization prior to surgery [48]. In the series by Koch and colleagues [45], routine coronary angiography was performed on 105 consecutive diabetic patients during the first six months of dialysis. Coronary artery disease was present in 47% of patients. Thirty six percent of patients had one or more significant stenoses and 10% had lesions that were less than 50%.

Because of the high prevalence of coronary artery disease and the limitations of clinical symptoms and noninvasive studies, some investigators have advocated routine coronary angiography prior to renal transplantation [45]. We do not find subjecting patients to the risks of routine coronary angiography an appealing strategy since less than half of patients will have lesions that require intervention.

Noninvasive Testing

Most studies of the sensitivity and specificity of noninvasive diagnostic testing in patients with renal failure have focused on patients with diabetes and end-stage renal disease with highly variable findings [49]. The reported prevalence of significant angiographically defined coronary artery disease (> 50% stenosis) presenting for renal transplant is 25–50%. In most patients with significant coronary artery disease, there will be electrocardiographic (ECG) evidence of ischemia upon adequate exercise cardiac stress testing. Stress testing with radionuclide scintigraphy is more sensitive and specific than routine treadmill exercise tests, especially in patients with baseline ECG abnormalities. Simultaneous evaluation of ventricular function (i.e. stress or dobutamine echocardiography), especially in patients with left ventricular hypertrophy improves predictive accuracy.

The sensitivity of radionuclide exercise stress testing in patients with renal failure is low in part because of poor exercise capacity in these patients. A frequently encountered problem is that patients are unable to exercise sufficiently to achieve an adequate stimulus for the desired heart rate because of coexisting morbidities such as obesity, diabetic foot ulcers, degenerative joint disease, claudication, amputations, or severe peripheral neuropathy. In these patients, pharmacologic stress testing with dipyridamole, dobutamine, or adenosine may provide a better evaluation of the coronary circulation. Our current screening test of choice is dobutamine echocardiography in patients with normal baseline left ventricular function and wall motion. In patients with possible previous myocardial infarction or abnormal baseline left ventricular systolic function and wall motion abnormalities, stress or pharmacologic testing with thallium scintigraphy have proven to be good alternatives.

Coronary Angiography

Coronary angiography is the gold standard for the diagnosis of coronary artery disease. It is indicated prior to transplantation in patients with evidence or symptomatic or asymptomatic myocardial ischemia or abnormal noninvasive studies. In very high risk patients, coronary angiography may be considered without prior noninvasive testing. Patients with severe coronary artery disease should get coronary revascularization prior to transplant surgery. Patients who have severe diffuse coronary disease who are not candidates for revascularization and those with severe left ventricular dysfunction will generally not be candidates for transplantation.

Contrast nephropathy

In patients with predialytic renal insufficiency, the risk of renal failure following radiocontrast dye rises substantially and contrast dye studies should be avoided. Noninvasive studies should be used to assess ventricular function and anatomy. In patients who do require contrast studies, appropriate precautionary measures should be taken including preprocedural hydration, postprocedural volume replacement, and minimizing the amount of contrast dye used.

CORONARY REVASCULARIZATION

Both percutaneous coronary interventions (angioplasty and stenting) and bypass graft surgery are effective revascularization techniques in patients with renal failure (with and without diabetes) and can be performed with acceptable risks. However, the complications are greater and the benefits of revascularization are considerably less than in patients without renal failure 50, 51].

Coronary Artery Bypass Surgery

The greatest benefits from coronary artery bypass surgery in terms of survival and improved quality of life have been seen in patients with left main coronary stenosis and in those with triple vessel disease and left ventricular dysfunction. Most recent studies have shown similar perioperative mortality between diabetic and nondiabetic patients. However, diabetes is associated with significantly greater perioperative morbidity because of wound infections, longer hospital stays, and reduced long-term survival compared to nondiabetics [52–54]. The poorer long-term prognosis in diabetics is due to more progressive disease in both the nonbypassed and bypassed native coronary arteries and a greater prevalence of associated risk factors [45]. The long-term benefits of bypass using internal mammary artery grafts is similar in patients with and without diabetes. However, surgeons generally avoid the use of bilateral internal mammary artery grafts in diabetic patients because of increased risk of sternal wound dehiscence.

Percutaneous Coronary Interventions (PTCA, Stenting)

Percutaneous coronary intervention (PCI) is a rapidly evolving technique and published outcome data rapidly become obsolescent. There have been a number of small and/or retrospective studies on PCI in dialysis patients, none of which suggested that dialysis patients do well with these procedures in terms of cardiac event-free survival [55–65]. Herzog and colleagues recently reported an analysis of comparative survival of dialysis patients in the US undergoing PCI (6,887 pts) or bypass surgery (7,419 pts) [66]. In this analysis, which covered a period from 1990–1995, they found that: 1) the in-hospital mortality was 12.5% in the bypass and 5.4% in the PCI (PTCA) group; 2) at two years the unadjusted survival was 57% in the bypass and 53% in the PTCA groups, respectively; 3) patients with diabetic nephropathy undergoing bypass vs PTCA had a 22% risk reduction in cardiac death, similar to findings in the BARI study [67] of non-ESRD patients. During the first three months there was a survival advantage with PTCA, due to the early post-operative mortality rate in the surgical cohort. The survival curves crossed at approximately 9 months for all-cause death and at six months for the cardiac death endpoint. The reasons for the poorer outcomes in the PTCA group long term have not been fully elucidated but are related at least in part to the high restenosis rates.

Coronary Stenting

The emergence of coronary stenting has increased the usefulness of PCI as an option in a number of high risk patient groups, including patients with renal failure and/or diabetes. Although there are currently no long-term outcome data on the benefits of coronary stenting in dialysis patients, preliminary data suggest that coronary stenting offers better outcomes than PTCA but not bypass surgery [68]. In a preliminary analysis of US dialysis patients undergoing coronary revascularization from 1995–1997, Herzog and colleagues reported on 4,142 PTCA patients, 2,246 patients receiving stents, and 5,929 patients having bypass surgery [68]. The unadjusted two year survival was 49.5% after PTCA, 52.2% after stenting, and 54.3% after bypass surgery [68]. The need for revascularization was lower in patients receiving stenting than PTCA but not those receiving bypass [69]. At 12 months the revascularization rate was 24.1%, 21.2% and 2.3% following PTCA, coronary stenting, and bypass surgery respectively. The risk of repeat revascularization was 19% lower after stenting and 90% lower after bypass compared to PTCA.

When compared to patients without ESRD, the benefits of coronary stenting are considerably less impressive [50, 51] in patients with ESRD. For example, Azar and colleagues [50] found that the rate of target vessel revascularization for ESRD was almost twice that of controls (34% vs 16.6%) at 9 months, following adjustment for diabetic status, lesion length, and reference vessel diameter.

Follow-up of Dialysis patients following PCI

The clinical follow-up of dialysis patients following coronary interventions presents special challenges. The presence of co-morbidities (diabetes, left ventricular hypertrophy, diastolic dysfunction) and the increased volume stress between and during dialysis sessions can make the accurate evaluation and etiologic assessment of chest pain and dyspnea very difficult. Furthermore, dialysis patients may have undetected critical coronary restenosis. One approach is to have dobutamine stress echocardiography performed at time intervals chosen to detect occult restenosis (i.e. 12–16 weeks after PCI) and periodically (i.e. annually) thereafter. A comparative clinical trial or large clinical tracking registry is needed to determine the optimal approach to revascularization, optimal ongoing surveillance and monitoring, and the relative benefits of various post-revascularization adjunctive therapies.

PREVENTION AND RISK REDUCTION

Cardiovascular disease, especially ischemic coronary disease, in patients with renal failure (predialytic, on dialyisis, and post-transplant) is highly prevalent. Although there have been a number of advances in revascularization techniques, the relative benefit of these therapies (i.e. angioplasty, stenting, and bypass surgery) are considerably less impressive than in patients without end-stage renal disease. Therefore, the cornerstone of treatment for cardiovascular risk reduction and improved survival is prevention of the malignant vasculopathy that is so widespread in renal failure patients. Since the poor outcome indicators (diabetes, hypertension, dyslipidemia, coronary disease) are frequently present prior to the development of end-stage renal disease, intensive medical intervention for risk factor modification should begin much earlier in the progression of renal disease. The high prevalence of modifiable risk factors provide great opportunities for prevention and risk reduction. Since approximately 70% of ESRD patients have diabetes and/or hypertension, optimal glycemic control and blood pressure control remain the cornerstones of an effective preventive strategy. Aggressive efforts to modify other risk factors are also essential as well as the development of appropriate algorithms for ongoing surveillance and monitoring for early disease detection. The development of better multidisciplinary, coordinated strategies for intensive risk modification and timely medical and/or invasive intervention for appropriate high risk individuals are essential if we are to achieve substantial reduction in the cardiovascular diseases that are so rampant in patients with renal failure.

REFERENCES

1. Baigent C, Burbury K, Wheeler D. Premature cardiovascular disease in chronic renal failure. Lancet 2000; 356: 147–152.
2. Levin A, Foley RN. Cardiovascular disease in chronic renal insufficiency. Am J Kidney Dis 2000; 36 (Suppl 3): S24–S30.
3. Vaitkus PT. Current status of prevention, diagnosis, and management of coronary artery disease in patients with kidney failure. Am Heart J 2000; 139: 1000–1008.
4. Herzog CA. Acute myocardial infarction in patients with end-stage renal disease. Kidney International 1999; 56 (Suppl 71): S130–S133.
5. USRDS US Renal Data System 2000. Annual Data Report. The National Institues of Health, National Institute of Diabetes and Digestive and Kidney Diseases. Bethesda, MD.
6. Shulman N, et al. Prognostic value of serum creatinine . . . HDFP. Hypertension 1989; 13: I80–I93.
7. Flack J, et al. MRFIT. Am J Kid Dis 1993; 21: 31–40.
8. Mann JFE, et al. Renal insufficiency as a predictor of cardiovascular outcomes . . . HOPE. Ann Intern Med 2001 (April); 134: 629–636.
9. Hall WD. Kidney CVD. Am J Med Sci 1999; 317: 176–182.
10. Levey AS, Beto JA, Coronado BE, et al. Controlling the epidemic of cardiovascular disease in chronic renal disease. Executive Summary. Report from the National Kidney Foundation Task Force on Cardiovascular Disease, October 1998. National Kidney Foundation document. Am J Kidney Dis 1998; 32 (Suppl 2): S5–S13.
11. Gomez-Farices MA, McClellan W, Soucie JM. A prospective comparison of methods for determining if cardiovascular disease is a predictor of mortality in dialysis patients. Am J Kidney Dis 1994; 23: 382.
12. Wingard DL, Barrett-Connor E. Heart Disease and Diabetes. In: Harris M (ed), Diabetes in America. 2nd ed. National Institute of Health, National Institute of Diabetes and Digestive and Kidney Diseases. NIH Publication No. 95-1468, 1995, pp 429–445.
13. Kunz K, Dimitrov Y, Muller S, Chantrel F, Hannedouche T. Uremic cardiomyopathy. Nephrol Dial Transplant 1998; 13(Suppl 4): 39–43.
14. Besarab A, et al. . . . NEJM 1998; 339: 584–590.
15. Austin MA, Hokanson JE, Edwards KL. Hypertriglyceridemia as a cardiovascular risk factor. Am J Cardiol 1998; 81: 7B–12B.
16. Bostom AG, Silbershatz H, Rosenberg IH, Selhub J, D'Agostino RB, Wolf PA, Jacques PF, Wilson PW. Nonfasting plasma total homocysteine levels and all-cause and cardiovascular disease mortality in elderly Framingham men and women. Arch Int Med 1999; 159: 1077–1080.
17. Bostom AG, Shemin D, Verhoef P, Nadeau Mr, Jacques PF, Selhub J, Dworkin L, rosengerg IH. Elevated fasting total plasma homocysteine levels and cardiovascular disease outcomes in maintenance dialysis patients. A prospective study. Arteriosclerosis, Thromb Vasc Biol 1997; 17: 2554–2558.

18. Perna AF, Castaldo P, Ingrosso D, De Santo NG. Homocysteine, a new cardiovascular risk factor, is also a powerful uremic toxin. J Nephrol 1999; 12: 230–240.
19. Bostom AG, Lathrop L. Hyperhomocysteinemia in end-stage renal disease: prevalence, etiology, and potential relationship to arteriosclerotic outcomes. Kidney Intern 1997; 52: 10–20.
20. Lindner A, et al. Accelerated athero – in prolonged maintenance hemodialysis. NEJM 1974; 290: 697–701.
21. Panichi V, Migliori M, De Pietro S, Taccola D, Bianchi AM, Norpoth M, Giovannini L, Palla R, Tetta C. C-reactive protein as a marker of chronic inflammation in uremic patients. Blood Purification 2000; 18: 183–190.
22. Hertzog CA, Ma JZ, Collins AJ. Poor long-term survival after acute myocardial infarction among patients on long-term dialysis. N Engl J Med 1998; 339: 799–805.
23. Herzog CA. Acute myocardial infarction in patients with end-stage renal disease. Kidney Intern 1999; 56 (Suppl 71): S130–S133.
24. Kong TQ, Dacanay S, Hsieh AA. Features of acute myocardial infarction in patients on chronic hemodialysis. Circulation 1993; 88 (Suppl I): I-49.
25. Ifudu O, Miles AM, Friedman EA. Hemodialysis immediately after acute myocardial infarction. Nephron 1996; 74:104–109.
26. Bloch MJ, et al. Prevention of recurrent pulmonary edema . . . Am J Hyperten 1999; 12: 1–7.
27. Herzog CA, Ma JZ, Collins AJ. Long-term survival of dialysis patients receiving thrombolytic therapy for acute myocardial infarction in the United States. Circulation 1999; 100 (Suppl 100): I-304.
28. Van Lente F, McErlean ES, DeLuca SA, Peacock WF, Rao JS, Nissen SE. Ability of troponins to predict adverse outcomes in patients with renal insufficiency and suspected acute coronary syndromes: a case matched study. J Am Coll Cardiol 1999; 33: 471–478.
29. Roppolo LP, Fitzgerald R, Dillow J, Ziegler T, Rice M, Maisel A. A comparison of troponin T and troponin I as predictors of cardiac events in patients undergoing chronic dialysis at a Veteran's Hospital: a pilot study. J Am Coll Cardiol 199; 34: 448–454.
30. Willging S, Keller F, Steinbach G. Specificity of cardiac troponins I and T in renal disease. Clin Chem Lab Med 1998; 36: 87–92.
31. Kannel WB. Lipids, diabetes and coronary heart disease: insights from the Framingham Study. Am Heart J 1985; 110: 1100–1107.
32. Tzivoni D, Benhorin J, Stern S. Significance and management of silent myocardial ischemia. Adv Cardiol 1990; 37: 312–319.
33. Harnett JD, Parfrey PS. Cardiac disease in uremia. Semin Nephrol 1994; 14: 245–252.
34. Kannel WB, Hjortland M, Castelli WP. Role of diabetes in congestive heart failure. The Framingham Study. Am J Cardiol 1974; 34: 29–34.
35. Sowers JR, Williams M, Epstein M, Bakris G. Hypertension in patients with diabetes. Strategies for drug therapy to reduce complications. Postgrad Med 2000; 107: 47–54, 60.
36. Guzman CB, Sowers JR. Special considerations in the therapy of diabetic hypertension. Progress Cardiovasc Dis 1999; 41: 461–470.
37. Hsu CH. Are we mismanaging calcium and phosphate metabolism in renal failure? Am J Kidney Dis 1995; 29: 241–649.
38. Schwartz U, Buzello M, Ritz E, Stein G, Raabe G, Wiest G, Mall G, Amann K. Morphology of coronary atherosclerotic lesions in patients with end-stage-renal failure. Nephrol Dial Transplant 2000; 15: 218–223.
39. Bommer J, Strohbeck E, Goerich J, Bahner M, Zuna I. Arteriosclerosis in dialysis patients. Int J Artif Organs 1996; 19: 638–644.
40. Raggi P. Detection and quantification of cardiovascular calcifications with electron beam tomography to estimate risk in hemodialysis patients. Clin Nephrol 2000; 54: 325–333.
41. Ribeiro S, Ramos A, Brandao A, Rebelo JR, Guerra A, Resina C, Vila-Lobos A, Carvalho F, Remedio F, Ribeiro F. Cardiac valve calcification in haemodialysis patients: role of calcium-phosphate metabolism. Nephrol Dial Transplant 1998; 13: 2037–2040.
42. Malergue MC, Urena P, Prieur P, Guedon-Rapoud C, Petrove M. Incidence and development of aortic stenosis in chronic hemodialysis. An ultrasonographic and biological study of 112 patients. Archives des Maladies du Coeur et des Vaisseaux 1997; 90: 1595–1601.
43. Gruppo Hemodialisi E Pathologie Cardiovasculari. Multicenter, cross-sectional study of ventricular arrhythmias in chronically hemodialyzed patients. Lancet 1988; 2: 305–309

44. Aker S, Ivens K, Guo Z, Grabensee B, Heering P. Cardiovascular complications after renal transplantation. Transpl Proc 1998; 30(5): 2039–2042.
45. Koch M, Gradaus F, Schoebel F, Leschke M, Grabensee B. Relevance of conventional cardiovascular risk factors for the prediction of coronary disease in diabetic patients on renal replacement therapy. Nephrol Dial Transplant 1997; 12: 1187–1191.
46. Braun WE, Philips DF, Vidt DG, Novic AC, Nakamoto S, Popowniak KL, Paganini E, Magnusson M, Pohl M, Steinmuller DR. Coronary artery disease in 100 diabetics with end stage renal failure. Transplant Proc 1984; 16(3): 603–607.
47. Lorber MI, Van Buren CT, Flechner SM, et al. Pretransplant coronary arteriography for diabetic renal transplant recipients. Transplant Proc 1987; 19: 1539–1541
48. Manske CL, Wang Y, Wilson RF. Coronary revascularization may improve short term survival in insulin-dependent diabetics considered for renal transplantation. Circulation 1991; 84 (Suppl II): II-516.
49. Herzog CA, Marwick TH, Pheley AM, White CW, Rao VK, Dick CD. Dobutamine stress echocardiography for the detection of significant coronary artery disease in renal transplant candidates. Am J Kidney Dis 1999; 33(6): 1080–1090.
50. Azar RR, Prpic R, Ho KKL, et al. Outcome of coronary stenting in patients with end-stage renal disease. Circulation 1999; 100 (Suppl I): I-365.
51. Gruberg L, Cury B, Duncan CC, et al. Does contemporary percutaneous coronary intervention improve the otherwise dismal prognosis of patients with end stage renal disease on dialysis? Circulation 1999; 100 (Suppl I): I-366.
52. Weintraub WS, Kosinski A, Culler S. Comparison of outcome after coronary angioplasty and coronary surgery for multivessel coronary artery disease in persons with diabetes. Am Heart J 1999; 138 (5 Pt 1): S394–S399.
53. King SB 3rd, Kosinski AS, Guyton RA, Lembo NJ, Weintraub WS. Eight-year mortality in the Emory Angioplasty versus Surgery Trial (EAST). J Am Coll Cardiol 2000; 35: 1116–1121.
54. Barsness GW, Peterson ED, Ohman EM, Nelson CL, DeLong ER, Reves JG, Smith PK, Anderson RD, Jones RH, Mark DB, Califf RM. Relationship between diabetes and long-term survival after coronary bypass and angioplasty. Circulation 1997; 96: 2551–2556.
55. Herzog CA. The optimal method of coronary revascularization in dialysis patients: choosing between a rock and a hard place. The Intern J Artifical Organs 2000; 23: 215–218.
56. Simsir SA, Kohlman-Trigoboff D, Flood R, Lindsay J, Smith BM. A comparison of coronary artery bypass grafting and percutaneous transluminal coronary angioplasty in patients on hemodialysis. Cardiovasc Surg 1998; 6: 500–505.
57. Marso SP, Gimple LW, Philbrick JT, DiMarco JP. Effectiveness of percutaneous coronary interventions to prevent recurrent coronary events in patients on chronic hemodialysis. Am J Cardiol 1998; 82: 378–380.
58. Schoebel FC, Gradaus F, Ivens K, et al. Restenosis after elective coronary balloon angioplasty in patients with end stage renal disease: a case-control study using quantitative coronary angiography. Heart 1997; 78: 337–342.
59. Schoebel FC, Gradaus F, Ivens K, Heering P, Jax TW, Grabensee B, Strauer BE, Leschke M. Restenosis after elective coronary balloon angioplasty in patients with end stage renal disease: a case-control study using quantitative coronary angiography. Heart 1997; 78: 337–342.
60. Koyanagi T, Nishida H, Kitamura M, et al. Comparison of clinical outcomes of coronary artery bypass grafting and percutaneous transluminal coronary angioplasty in renal dialysis patients. Ann Thorac Surg 1996; 61: 1793–1796.
61. Rinehart AL, Herzog CA, Collins AJ, Flack JM, Ma JZ, Opsahl JA. A comparison of coronary angioplasty and coronary artery bypass grafting outcomes in chronic dialysis patients. Am J Kidney Dis 1995; 25: 281–290.
62. Ahmed WH, Shubrooks SJ, Gibson M, Baim DS, Bittl JA. Complications and long-term outcome and after percutaneous coronary angioplasty in chronic hemodialysis patients. Am Heart J 1994; 128: 252–255.
63. Reusser LM, Osborn LA, White HJ, Sexson R, Crawford MH. Increased morbidity and coronary angioplasty in patients on chronic hemodialysis. Am J Cardiol 1994; 73: 965–967.
64. Takeshita S, Isshiki T, Tagawa H, Yamaguchi T. Percutaneous transluminal coronary angioplasty for chronic dialysis. J Invasive Cardiol 1993; 5: 345–350.

65. Kahn JK, Rutherford BD, McConahay DR, Johnson WL, Giorgi LV, Hartzler GO. Short- and long-term outcome of percutaneous transluminal coronary angioplasty in chronic dialysis patients. Am Heart J 1990; 119: 484–489.
66. Herzog CA, Ma JZ, Collins AJ. Long-term outcome of dialysis patients in the United States with coronary revascularization procedures. Kidney Int 1999; 56: 324–332.
67. The Bypass Angioplasty Revascularization Investigation (BARI) Investigators. Comparison of coronary bypass surgery with angioplasty in patients with multivessel disease. N Engl J Med 1996; 335: 217–225.
68. Herzog CA, Ma JZ, Collins AJ. Long-term survival of dialysis patients in the United States after coronary artery bypass surgery, coronary angioplasty and coronary stenting. Circulation 1999; 100 (Suppl I): I-365.
69. Ma JZ, Collin AJ, Herzog CA. The likelihood of repeated coronary revascularization procedures (CRP) after first CRP in dialysis patients. J Am Soc Nephrol 1999; 10: 249A.

Yalemzedwd Worerdekal

13. Reducing Mortality for Diabetic Patients on Hemodialysis

Editor's Comment: Truly simple and common sense measures translate into enhanced patient survival as Worerdekal illustrates in what amounts to a patient care manual for hemodialysis patients with diabetes. Reading as if extracted from a nonexistent American Diabetes Association Dialysis Practice Guide, the lessons learned from supervising a large kidney program ring through as sound and appropriate. It took more than a decade of maintenance hemodialysis care before the question of adequate dose of dialysis was properly addressed. Today, there are Medicare standards for minimal dosage. The check points listed by Worerdekal: dose, blood pressure control, blood glucose regulation, correction of anemia, reduction of hyperlipidemia, and provision of sufficient nutrition all appear reasonable but may be lacking in an individual patient unless a routine surveillance of progress is made. What needs to be added for completeness is a check list for periodic consultations with cardiology, ophthalmology, podiatry, and other services as indicated by the patient's individual course.

Diabetic nephropathy is the leading cause of end stage renal disease (ESRD) in the United States, Japan, and most industrialized countries in Europe and Asia. The number of new diabetic patients accepted for renal replacement therapy (RRT) has been increasing steadily for the last decade. The United States Renal Data System (USRDS) reported in 1987 that the percentage of diabetics newly started on renal replacement therapy was 27%, but this number had increased to 44.5% by 1997 [1]. The great majority of diabetic patients starting RRT suffer from type 2 diabetes mellitus, and the increasing number of diabetic patients, particularly type 2, is explained by the increasing prevalence of type 2 diabetes in the general population, and partly because diabetic patients with nephropathy are now living long enough to develop end stage renal disease.

Although survival of diabetic patients on RRT has significantly improved over the last decade, it has remained shorter when compared to non-diabetics. USRDS 2000 reports higher mortality in diabetic patients than non-diabetics undergoing peritoneal dialysis (PD), hemodialysis (HD), or after kidney transplantation [1].

FACTORS AFFECTING SURVIVAL OF DIABETIC ESRD PATIENTS

1. Age

Age is the strongest factor associated with increased mortality. Since the majority of diabetics with ESRD are type 2 patients, the age of new diabetic patients accepted for RRT has been increasing steadily. Currently, the mean age of diabetic patients starting

E.A. Friedman and F.A. L'Esperance, Jr. (eds.), Diabetic Renal-Retinal Syndrome, 147–154.

RRT in the United States is 62.5 years [1]. As the age of the patients' increase the number and severity of co-morbid conditions also increase [2]. Foucan et al. did survival analysis of 784 patients (22% diabetics) who began chronic dialysis between 1978–1997. Median survival was significantly lower in diabetic than in non-diabetic patients (3.5 years verses 6.9). The RR for death (95% CI) were 1.90 (1.10–3.22) and 3.43 (2.00–5.87) for diabetics began dialysis at age 55–64 years and 65–83 years of age respectively [3].

2. *Co-morbid Conditions (Table 2)*

The presence of preexisting diabetic complications such as cardiovascular disease, peripheral vascular disease, neuropathy, and retinopathy have significant impact on the survival of diabetics on hemodialysis [4, 5]. A study by Chantiel et al. evaluated 84 consecutive type 2 diabetic patients starting dialysis between 1995–1996. Cardiovascular disease was highly prevalent at initiation of dialysis, with a history of myocardial infarction in 26%, angina in 36%, and acute left ventricuar dysfunction in 67%; 27 of 87 patients (32%) died after a mean follow-up of 211 days, mostly from cardiovascular disease [6]. Adding to the difficulty in management of diabetic nephropathy complicated by cardiac disease is the reality that extensive coronary artery disease is often asymptomatic in diabetic patients. Koch et al., evaluated all of their diabetic patients (71 type 1, 28 type 2) for coronary artery disease during their first six months of dialysis treatment. Coronary angiography was performed in all regardless of clinical symptoms of coronary artery disease, in 38 of whom (36%) coronary artery disease was documented (17 patients with single vessel, 6 patients with two vessel, and 15 patients with three vessels disease). Angina pectoris was present only in 9 (24%) of these 38 patients. In 11 patients cardiac intervention was felt to be indicated and was performed (3 patients underwent coronary artery bypass surgery, and 8 had angioplasty). In this study risk factors such as hypertension, smoking, and cholesterol and lipoprotein (a) level were not significantly different in patients with and without coronary artery disease [7]. Vascular disease is not limited to the coronary arteries in diabetics starting dialysis. According to a recent survey of 25,037 patients with diabetic nephropathy, 18% of them had amputation (above ankle) at the time of starting RRT [8].

3. *Adequacy of dialysis*

Numerous studies have demonstrated a correlation between adequacy of hemodialysis and patient survival [9–11]. Although the prescribed dose of dialysis is the same for both diabetic and non-diabetic patients, the delivered dose of dialysis is usually less for diabetic patients because of either insufficient blood flow secondary to malfunction of their vascular access or interruption of treatment due to intradialytic hypotension. Palevsky et al. studied 29 dialysis facilities whose average urea reduction rate was < 67%, and found that reduced treatment time and use of a catheter for angioaccess to be the two barriers to the delivery of adequate dialysis [12]. Both of which are more frequently seen in diabetic patients.

4. *Malnutrition*

Protein energy malnutrition is common in diabetic patients at the start of RRT [13]. Hypoalbuminemia and malnutrition are strong predictors of morbidity and mortality in dialysis patients [14, 15]. A recent study by Obrador et al. using data from the US Renal Data Systems looked at 110,843 incident chronic dialysis patients in the United States between April 1995 and June 1997. At initiation of dialysis the median serum albumin was 3.3 gm/dl, and sixty percent of patients had a serum albumin level below the normal limit (3.5 gm/dl). In multivariate analysis diabetes was one of the associated factor with low serum albumin levels [16]. Causes of malnutrition in diabetic azotemic patients are poor intake, gastroparesis, and urinary protein loss during the pre-dialysis period.

STRATEGIES TO REDUCE MORTALITY IN DIABETIC ESRD PATIENTS

Most complications of diabetes mellitus including those that may evolve into life threatening conditions develop before initiation of renal replacement therapy and are not abolished by dialysis or renal transplant. Cardiovascular disease is the number one cause of death in diabetic ESRD patients. Recently, more focus has been placed on treating diabetic patients early in order to prevent major organ damage. In the following we will discuss the possible ways to reduce mortality and morbidity in diabetic patients, as, most start early in the course of diabetic nephropathy (pre-ESRD), greater attention has to be paid to their development.

PRE-ESRD CARE

5. *Glycemic Control*

Many studies have shown that the development of diabetic complications depends to a large extent, on the quality of glycemic control throughout the duration of diabetes [17–20]. The Diabetes Control and Complication Trial demonstrated that meticulous control of glucose in type 1 diabetic patients reduce the development of retinopathy, nephropathy, and neuropathy [17]. Similarly, the United Kingdom Prospective Diabetes Study (UKPKS) has shown that tight glycemic control (hemoglobin A1c < 7%) in newly diagnosed type 2 diabetic patients correlates with improvement of all diabetic end points by 12%, microvascular end points by 25%, and myocardial infarction by 16% [18]. In addition Wu et al. showed that in diabetic patients on hemodialysis, glycemic control before the start of dialysis had a significant impact on patient survival [19]. Among 137 type 2 diabetic patients starting hemodialysis in a single university hospital, patients with good glycemic control 6 months before the start of treatment had significantly higher 1 year and 5 year survival than patients with poor control (1 year, 95.5% vs. 80%; 5 year, 75.8% vs. 21.8%). Although attempting to achieve normoglycemia is certainly justified, advising diabetic patients with renal insufficiency has to be tempered by consideration of the higher incidence of hypoglycemia in these patients.

6. Adequate Blood Pressure Control

The prevalence of hypertension increases linearly as the glomerular filtration rate falls. At the start of RRT almost 100% of patients have hypertension [21], in addition blood pressure control during pre-ESRD period is often suboptimal [22]. Hypertension is a major risk factor for left ventricular hypertrophy (LVH) and congestive heart failure (CHF), which are strong predictors of death in dialysis patients [23, 24]. There are many studies showing the high prevalence of significant LVH in patients with chronic renal insufficiency, long before they reached end stage renal disease, and its correlation with mean 24-hour systolic blood pressure [25–28]. A marked reduction in LVH has been shown to occur with adequate blood pressure control [29, 30]. Adequately control of blood pressure during the early conservative phase of treatment for diabetic nephropathy may prevent or slow down the progression of LVH, and decrease the cardiovascular burden when these patients reach end stage renal disease.

7. Correction of Anemia

Chronic anemia is commonly seen in patients with chronic renal insufficiency. The degree of anemia correlates with the degree of renal impairment. Anemia is also considered one of the most important risk factor for the development of left ventricular dilation and hypertrophy [31, 32]. Studies have demonstrated that correction of anemia with recombinant human erythropoietin (rHuEPO) results in partial regression of the left ventricular mass in LVH, both among dialysis and pre-ESRD patients [33, 34]. Currently in the United States, the use of rHuEPO for correction of anemia during the predialysis period is suboptimal [16]. When Obrador et al. looked at over 110,000 chronic dialysis patients between 1995 and 1997 at the initiation of dialysis, (data from the Health Care Financing Administration (HCFA)) the median hematocrit was 28%, and only 23% had received rHuEPO before the start of dialysis. Hematocrit level should be monitored closely during pre-ESRD period, and after non-renal causes of anemia are ruled out, oral iron supplementation and correction of nutritional deficiencies are the appropriate steps. Although the hematocrit level at which rHuEPO therapy should be started in pre-ESRD patients have not yet been established, it is justifiable to start rHuEPO treatment when hematocret level is below 30%. Therefore, along with strict adherence to glycemic and blood pressure control, correction of renal anemia is highly recommended.

8. Early Detection and Treatment of Cardiovascular Disease

Cardiovascular disease is the major cause of death in diabetic patients with end-stage renal disease receiving chronic dialytic therapy: the disease accounts for over 60% of all cause mortality [1]. Studies have shown a high prevalence of LVH, CHF and ischemic heart disease in diabetic patients at the initiation of dialytic therapy [6, 7, 35]. Cardiovascular disease begins early in the course of diabetic nephropathy. The impact of multiple risk factors for the development of cardiovascular disease in diabetic azotemic patients, include the traditional risk factors of hypertension, hyperglycemia, dyslipidemia, smoking, physical inactivity, and the uremia associated risk factors: anemia, arteriovenous fistula, hypervolemia, hyperparathyroidism, malnutrition, and thrombogenicity. Recognition of ischemic heart disease in diabetic patients is usually difficult because of the asymptomatic nature of the disease in this group of patients [36, 37]. Early identification of cardiovascular disease and aggressive risk factor modification during the pre-ESRD period is highly recommended to reduce morbidity and mortality associated with cardiovascular disease during dialysis treatment. In our renal clinic, all new diabetic patients with nephropathy are evaluated by a cardiologist; if the resting EKG shows LVH, they will undergo Dobutamin echocardiography for evaluation of cardiac function and, if this test is positive, patients will undergo coronary angiography and others will be followed by a cardiologist as needed. Symptomatic patients will go directly to cardiac catheterization. The National Kidney Foundation Task Force on Cardiovascular Disease recently published recommendations regarding cardiovascular risk intervention in patients with chronic renal insufficiency or ESRD [38]. Similar recommendations are applicable to diabetic azotemic patients as well. They are summarized below.

Table 1. Primary and Secondary prevention of cardiovascular disease in Diabetic patients with nephropathy.

Risk factors	Intervention	Goal
Glycemic control	Diet Oral hypoglycemic agents Insulin	$HgA1_c < 7\%$
Hypertention	Low-salt diet Anti-hypertensive drugs	< 125/75
Hyperlipedemia	Diet HMO-COA reductase	LDL < 100
Smoking	Counseling Nicotime patch	Cessation of smoking
Physical activity	Regular exercise program	Moderate physical activity most days of the week

9. Placement of Vascular Access

Creation of vascular access in diabetics is the most challenging problem in delivering adequate hemodialysis. It is now clear that a well functioning vascular access at the start of hemodialysis saves the cost of hospitalization and avoids placement of temporary central venous access and its complications. In a prospective nationwide study of all patients starting dialytic treatment in Scotland, Metcalfe et al. found that those patients who had planned far enough in advance to start dialysis with a matured access were 3.6 times more likely to survive beyond three months than those with no access [39]. A native arteriovenous fistula (AVF) is a preferred type of permantnt vascular access. Patients with AVFs have fewer complications related to infection and thrombosis rates compared to patients with arteriovenous grafts (AVGs) [40–42]. Pre-dialysis care should include patient education about preserving veins in the forearm and antecubital region by avoiding vein puncture and intravenous catheter placement in non-dominant arm, early when serum creatinine is around 3 mg/dl. The DOQI guidelines recommend placement of an AVF placement of an AVF when the serum creatinine exceeds 4 mg/dl, creatinine clearance falls below 25 ml/min, or ESRD is anticipated is anticipated within one year [41]. Diabetic patients should be referred to a well-experienced vascular surgeon at least 6 months prior to the expected start of hemodialysis, as such patients need long time for maturation of their access. Whenever possible, the first priority should be an attempt at creation of arteriovenous fistula. This requires careful preoperative investigations for selection of the best location for placement of the initial AV fistula. In our hospital since surgeons started construction of AV fistulas in the elbow region using large diameter aarteries and veins, we have observed more successful AVFs in our diabetic patients..

MEDICAL CARE DURING UNDERGOING HEMODIALYSIS

Although little or no hard data exist to support the contention, there is widespread consensus among nephrologist that diabetics develop uremic symptoms at higher levels of renal function and that dialysis should be initiated earlier than in non-diabetics. As renal failure progresses, the diabetic kidney seems to become particularly ineffective at the regulation of volume and sodium balance and, as a result, hypertension and fluid overload may become refractory at a relatively high GFR. Concomitant gastroparesis and nephrotic proteinuria may also exacerbate protein-calorie malnutrition during pre-ESRD period. Kjellstrand, in 1972, observed that retinopathy in diabetics progresses rapidly during the one to two year period prior to the initiation of dialysis [43]. Early initiation of dialysis may prevent malnutrition and also eliminate the adverse effects of hypertension and volume overload on retinopathy and heart disease.

Many diabetic patients present with fluid overload at the start of dialysis. Even on dialysis excessive intradialytic weight gain is a problem in diabetic patients. Ifudu et al. compared intradialytic weight gain between 33 diabetic and 27 non-diabetic patients undergoing hemodialysis at an ambulatory dialysis unit, findings that interdialytic weigh gain was greater in diabetic patients than non-diabetics. This correlated with the glycemic control in the diabetic patients [44]. At times removal of fluid during dialysis is difficult in diabetic patients because of interdialytic hypotension resulted from autonomic neuropathy and diffuse arteriosclerosis leading to a delayed response to changevolume. It has been suggested that long, slow dialysis sessions may facilitate successful fluid removal in diabetic patients with poor cardiac function and severe autonomic neuropathy. Kjellstrand et al. reported a much better quality of life in non-diabetic patients with various cardiac problems who underwent dialysis as frequent as 6 days per week. In his group of patients, fluid removal was better tolerated and inter-dialytic weight gain was reduced [45]. This approach may be useful in diabetic ESRD patients with poor cardiac function for whom excessive fluid gain is a problem.

REFERENCES

1. United States Renal Data System USRDS 2000. Annual Data Report. The National Institutes Diabetes and Digestive and kidney Diseases. Bethesda, MD, 2000.
2. Toriyama T, Yokoya M, Nishida Y, et al. Increased incidence of coronary artery disease and cardiac death in elderly diabetic nephropathy patients undergoing chronic hemodialysis therapy. J Cardiol 2000; 36(3): 165–171.
3. Foucan L, Merault H, Deloumeaux J, et al. Survival analysis of diabetic patients on dialysis in Guadeloupe. Diabetes Metab 2000; 26(4): 307–313.
4. Collins AJ, Hanson G, Umen A, et al. Changing risk factor demographics in end-stage renal disease patients entering hemodialysis and the impact on long-term mortality. Am J Kidney Dis 1990; 15(5): 422–432.
5. Bruno RM, Gross JL. Prognostic factors in Brazilian diabetic patients starting dialysis: a 3.6-year follow-up study. J Diabetes Complications 2000; 14(5): 266–271.
6. Chantreal F, Enache I, Bouiller M, et al. Abysmal prognosis of patients with Type 2 diabetics entering dialysis. Nephrol Dial Transplant 1999; 14(1): 129–156.
7. Koch M, Gradaus F, Schocbet FC, et al. Relevance of conventional cardiovascular risk factors for the prediction of coronary artery disease in diabetic patients on renal replacement therapy. Nephrol Dial Transplant 1997; 12(6): 1187–1191.
8. Eggers PW, Gohdes D, Pugh J. Nontraumatic lower extremity amputation in the Medicare end-stage renal disease population. Kidney Int 1999; 56: 1524–1533.
9. Charra B, Calemanrd E, Laurent G. Importance of treatment time and blood pressure control in achieving long-term survival on dialysis. Am J Nephrol 1996; 16: 35–44.
10. Keshaviah P, Ma J, Thorpe K, et al. Comparison of 2-year survival on hemodialysis and peritoneal dialysis with dose of dialysis matched using the peak concentration hypothesis. J Am Soc Nephrol 1995; 6: 540–544.
11. Stefanovic V, Stojanovic M, Djordjevic V. Effect of adequacy of dialysis and nutrition on morbidity and working rehabilitation of patients treated by maintenance hemodialysis. Int J Artif Organ 2000; 23(2): 83–89.
12. Palevsky PM, Washington MS, Stevenson JA, et al. Barriers to the delivery of adequate hemodialysis in ESRD Network 4. Adv Ren Replace Ther 2000; 7 (4 Suppl 1): S11–S20.
13. Espinosa A, Cueto-Manzano AM, Velasquez-Alva C, et al. Prevalence of malnutrition in Mexican CAPD diabetic and non-diabetic patient patients. Adv Perit Dial 1996; 12: 302–306.
14. Port F. Morbidity and mortality in dialysis patients. Kidney Int 1994; 46: 1728–1737.
15. US Renal Data System. Co-morbid conditions and correlations with mortality risk among 3,399 incident hemodialysis patients. Am J Kidney Dis 1992; 19: 214–234.
16. Obrador G, Ruthazer R, Arora P, et al. Prevalence of and factors associated with suboptimal care before initiation of dialysis in the United States. J Am Soc Nephrol 1999; 10: 1793–1800.
17. The Diabetes Control and Complications Trial (DCCT) Research Group. The effect of intensive treatment of diabetes on the development and progression of long-term complications in insulin-dependent diabetes mellitus. N Engl J Med 1993; 329: 977–986.
18. United Kingdom Prospective Diabetes Study (UKPDS) Group. Intensive blood-glucose control with sulphonylureas or insulin compared with conventional treatment and risks of complications in patients with type 2 diabetes. Lancet 1998; 352: 837–.
19. Wu MS, Yu CC, Yang CW, et al. Poor pre-dialysis glycemic control is a predictor of mortality in type 2 diabetic patients on maintenance hemodialysis. Nephrol Dial Transplant 1997; 12: 2105–2110.
20. Ohkubo Y, Kishikawa H, Araki E, et al. Intensive insulin therapy prevents the progression of diabetic microvascular complications in Japanese patients with non-insulin dependent diabetes mellitus: a randomized prospective 6-year study. Diabetes Res Clin Pract 1995; 28: 103–117.
21. Vendemia F, Forniasieri A, Velis O, et al. Different prevalence rates of hypertension in various reno-parenchymal diseases. In: Balufox MD, Bian-Chi C (eds), Secondary Forms of Hypertension. Current Diagnosis and Management. Baven, New York, 1990, pp 1583–1599.
22. Buckalew VM, Berg RL, Wang S-R, et al. Prevalence of hypertension in 1,795 subjects with chronic renal disease: The Modification of Diet in Renal Disease Studycohort. Am J Kidney Dis 1996; 28: 811–821.

23. Foley RN, Parfrey PS, Harnett JD, et al. Impact of hypertension on cardiomyopathy, morbidity and mortality in end-stage renal disease. Kidney Int 1996; 49: 1379–1385.
24. Trocha A, Schmidtke C, Didjurgeit U, et al. Effects of intensified antihypertensive treatment in diabetic nephropathy: mortality and morbidity results of a prospective controlled 10-uear study. J Hypertension 1999; 17(10): 149–1503.
25. Fucker B, Fabbian F, Giles M, et al. Left ventricular hypertrophy and ambulatory blood pressure monitoring in chronic renal failure. Nephrol Dial Transplant 1997; 12(4): 724–728.
26. Ha SK, Park HS, Kim SJ, et al. Prevalence and patterns of left ventricular hypertrophy in patients with predialysis chronic renal failure. J Korean Med Sci 1998; 13(5): 488–494.
27. Washio M, Okuda S, Mixoue T, et al. Risk factors for left ventricular hypertrophy in chronic hemodialysis patients. Clin Nephrol 1997; 47(6): 362–366.
28. ACE inhibitors captopril and enalpril induce regression of left ventricular hypertrophy in hypertensive patients with chronic renal failure. Nephrol Dial Transplant 1997; 12(5): 945–951.
29. Cuspidi C, Lonati L, Sampieri L, et al. Impact of blood pressure control on prevalence of left ventricular hypertrophy in treated hypertensive patients. Cardiology 2000; 93(3): 143–154.
30. London GM, Marchais SJ, Guerin AP, et al. Cardiovascular function in hemodialysis patients. Adv Nephrol Necker Hosp 1991; 20: 249–273.
31. Madore F, Lowerie E, Brugnara C, et al. Anemia in hemodialysis patients: variables affecting this outcome predictor. J Am Soc Nephrol 1997; 8: 1921–1929.
32. Silverberg D, Wexler D, Blum M, et al. The use of subcutaneous erythropoietin and intravenous iron for the treatment of the anemia of severe, resistant congestive heart failure improves cardiac and renal function and functional cardiac class, and markedly reduces hospitalizations. J Am Coll Cardiol 2000; 35: 1737–1744.
33. Besarab A, Bolton W, Browne J, et al. The effect of normal as compared with low hematocrit value in patients with cardiac disease receiving hemodialyssis and epoetin. N Engl J Med 1998; 339: 584–590.
34. Hayashi T, Suzuki A, Shoji T, et al. Cardiovascular effect of normalizing the hematocrit level during erythropoietin therapy in predialysis patients with chronic renal failure. Am J Kidney Dis 2000; 35(2): 250–256.
35. Foley RN, Culleton BF, Parfrey PS, et al. Cardiac disease in diabetic end-stage renal disease. Diabetologia 1997; 40: 1307–1312.
36. Weinraueb LA, D'Elia JA, Healy RW, et al. Asymptomatic coronary artery disease: angiography in a diabetic patients before renal transplantation. Relation of findings to postoperative survival. Ann Intern Med 1978; 88: 346–348.
37. Manske Cl, Wilson RF, Wang Y, et al. Prevalence of, and risk factors, for angiographically determined coronary artery disease in type 1 diabetic patients with nephropathy. Arch Intern Med 1992; 152: 2450–2455.
38. National Kidney Foundation Task Force on Cardiovascular Disease: Controlling the epidemic of cardiovascular disease in chronic renal disease: What do we know? What do we nee to learn? Where do we go from here? Am J Kidney Dis 1998; 32 (Suppl): S1–S199.
39. Metcalfe W, Khan IH, Prescott GJ, et al. Can we improve early mortality in patients receiving renal replacement therapy? Kidney Int 2000; 57(6): 2539–2545.
40. Kalman PG, Pope M, Bhola C, et al. A practical approach to vascular access for hemodialysis and predictors of success. J Vasc Surg 1999; 30: 727–733.
41. NKF-DoQI. Clinical Practice Guidelines of Vascular Access. Am J Kidney Dis 1997; 30 (Suppl 3): S150–S191.
42. Schwab SJ. Vascular access for hemodialysis. Kidney Int 1999; 55: 2078–2090.
43. Kfellstrand CM, Simmons RI, Goetz FC, et al. Mortality and morbidity in diabetic patients accepted for renal transplantation. Proc Eur Dial Transplant Assoc 1972; 9: 345–368.
44. Ifudu O, Dulin AL, Frieman EA. Interdialytic weight gain correlates with glycosylated hemoglobin in diabetic hemodialysis patients. Am J Kidney Dis 1994; 23(5): 686–691.
45. Kjellstrand CM, Ting G. Daily hemodialysis: Dialysis for next century. Adv Renal Replace Ther 1998; 5: 267–274.

PLOUMIS PASSADAKIS AND DIMITRIOS OREOPOULOS

14. Long-Term Perspective on Utility of Peritoneal Dialysis in Diabetes

Editor's Comment: Ever since its introduction as an alternative to hemodialysis, peritoneal dialysis has been viewed as either an exploitable desirable resource or an inferior, though nevertheless enthusiastically promoted modality in uremia therapy. According to the 2001 Report of the United States Renal Data System, diabetic end-stage kidney failure patients have less than a 4% chance of surviving a decade no matter what form of dialysis is employed to sustain their lives. In this chapter, emanating from a kidney program renowned for the quality of its innovative peritoneal dialysis, the long-term problems of treating diabetic patients are analyzed. In the authors' own program, survival in 224 diabetic patients after 12, 24, 36, 48 and 60 months was 93%, 81%, 70%, 50% and 33% respectively, with more type 1 than type 2 diabetic patients alive after 1, 2, 3, and 4 years on CAPD, (95%, 84%, 76%, 56% versus 92%, 79%, 66%, 47%). While the relative merits of types of peritoneal dialysis, or the advantages of peritoneal dialysis versus hemodialysis may be debated, it seems fair to generalize that any regimen that loses two-thirds of its patients in five years is not likely to permit rehabilitation to home, employment, or educational responsibilities. As the threats to survival during dialytic therapy are unmasked, including the toxicity of advanced glycosylated end-products, and aggressive cardiac intervention is practices, the still grim loss of diabetic dialysis patients can be expected to improve. For the present, we are indebted to Passadakis, and Oreopoulos for so diligently observing and reporting their unique experience thereby providing a reference standard and stimulus for all who accept the challenge of dialyzing those with diabetes.

Abstract: Diabetic patients are vulnerable to considerable morbidity, which has been linked to diabetic nephropathy neuropathy and retinopathy whereas, diabetic nephropathy is the most common cause of end-stage renal disease (ESRD). Early initiation of dialysis, technological improvements in renal replacement therapy, and advances in the management of coronary-artery disease and critical-care medicine, have resulted in a dramatic improvement of survival of diabetics on dialysis.

However, long-term mortality rates in diabetic patients are still twice as high as in non-diabetic dialysis patients and only a small number of diabetic dialysis patients have been followed up for more than 5 years. This is largely due to coexisting, far-advanced, target-organ damage at the initiation of dialysis and its progression during the course of dialysis, as well as to the presence of various comorbid conditions at the start of dialysis.

Survival and technique success rates in diabetic patients undergoing dialysis therapy are lower than their nondiabetic counterparts patients of comparable age, whereas survival of diabetic ESRD patients undergoing either peritoneal dialysis (PD) or hemodialysis (HD) probably is similar.

This paper was designed to describe the long-term perspective utility of peritoneal dialysis in patients with end stage renal disease patients due to diabetes.

Key words: diabetes mellitus, ESRD, hemodialysis, long-term survival, peritoneal dialysis

E.A. Friedman and F.A. L'Esperance, Jr. (eds.), Diabetic Renal-Retinal Syndrome, 155–172.

Diabetes mellitus, the fastest growing cause of end stage renal disease (ESRD), has become the leading cause of ESRD worldwide. In the USA between 1984 and 1998, the proportion of new patients starting renal replacement therapy (RRT) whose ESRD was due to diabetes increased from 27 to 45.4% [1]. This may be related to the advancing obesity, carbohydrate intolerance, and insulin resistance that suggested by recent trends of increased weight and body mass index at initiation of therapy [1]. Besides the increasing prevalence of diabetic nephropathy is due primarily to the greater number of patients with type 2 diabetes [2].

Early initiation of dialysis, technological improvements in renal replacement therapy, and advances in the management of coronary-artery disease and critical-care medicine, have resulted in a dramatic improvement of survival of diabetics on dialysis. The first-year mortality on PD decreased from 40.4 per 100 patient years in 1986 to < 20 per 100 patients-years in 1997 [1]. Similarly the life expectancy in our diabetic continuous ambulatory peritoneal dialysis (CAPD) population in Toronto was clearly longer in 1996 with 2- and 4-year survival rates for type 1 and type 2 diabetic patients of 84%, 56% and 79% and 47% respectively compared with the a 2-year survival of 78% and 47% in 1986 [3]. These findings were more important as ESRD diabetics on CAPD were 10 years older than those who dialyzed in 1986.

However, long-term mortality rates in diabetic patients are still twice as high as in nondiabetic dialysis patients and only a small number of diabetic dialysis patients have been followed up for more than 5 years. This is largely due to the coexistence of far-advanced target-organ damage at dialysis initiation and its progression during the course of dialysis, with the presence of several comorbid conditions.

This article was designed to describe the perspective utility of peritoneal dialysis in patients with end stage renal disease patients due to diabetes.

COMORBIDITY IN DIABETIC PATIENTS AT INITIATION OF PD

Diabetic patients are vulnerable to considerable morbidity, which has been linked to diabetic nephropathy, neuropathy and retinopathy. The development and progression of these complications of diabetes mellitus depend on the quality of glycemic control throughout the duration of long-lasting diabetes, which presuppose intensive medical care, careful management, educational intervention as well as patient's cooperation and compliance.

Except for diabetic complication, the presence of a variety of comorbid conditions at the initiation of peritoneal dialysis such as cardiac disease, vascular peripheral and cerebrovascular disease, hypoalbuminemia and hyperparathyroidism (Table 1), may determine the clinical course as well as morbidity and mortality of diabetic patients on PD. The incidence of these factors is higher in diabetic than nondiabetic patients as well as and in older than younger patients in both groups (Figure 1) [4].

Advanced age at initiation of PD, a history or presence of cardiac disease and a systolic blood pressure of above 160 mmHg represent important predictors of survival in CAPD patients with diabetic nephropathy [5–7].

As regard the cardiovascular disease the risk factors in diabetic patients do not differ from those of in non-diabetic, including a positive family history for heart disease, hypertension, hyperlipidemia, lack of exercise, cigarette smoking, low-HDL-C,

Table 1. Comorbidity in diabetic patients at initiation of PD.

Advanced Age
Cardiovascular disease
Peripheral vascular disease (PVD) – Cerebrovascular disease (CVD)
Hypertension
Protein-energy malnutrition – Hypoalbuminemia
Disorders of lipid metabolism – Atherosclerosis
Hyperparathyroidism
Infection
Malignancy
Cigarette smoking
Advanced glycosylation end products (AGEs)

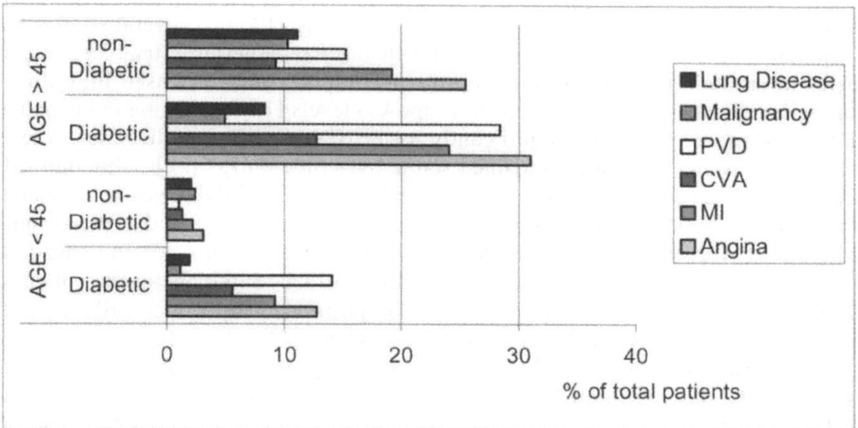

Figure 1. Comorbidity among registered patients by age and diabetes status, Canada 1988–1994, [3]. [PVD = peripheral vascular disease, CVA = cerebrovascular accident, MI = myocardial infarction).

increased platelet adhesiveness, elevated fibrinogen and probably the level of glucose control [8].

However among ESRD patients, diabetic patients have 2- to 3-fold higher death rates than nondiabetic patients from myocardial infarction and other cardiac causes of death. This could be explained by the high prevalence of risk factors for cardiac disease or pre-existing heart disease at the start of dialysis. Diabetic patients have a significantly higher frequency of hypertension, cardiac disease, and peripheral vascular disease than do nondiabetic patients. Therefore investigation for cardiac disease is warranted in diabetic patients, even though the patient may be asymptomatic, because of the high incidence of cardiovascular disease and the increased chance of silent myocardial infarction.

In addition protein-energy malnutrition, a strong predictor of morbidity and mortality, is common among diabetic dialysis patients [9, 10], and many diabetics already are malnourished at the initiation of PD [11]. This is a result of decreased intake, increased protein losses and altered metabolism, whereas coexisting conditions such as uremia, intercurrent illness, gastroparesis, socioeconomic condition and depression also may participate in the development of malnutrition in these patients. Furthermore during CAPD diabetic patients often develop protein-energy malnutrition more rapidly than do nondiabetic patients and this is a common cause of CAPD technique failure on long-term dialysis that contributes to increased mortality [12–15]. Early nutritional assessment by anthropometric measurements, diet history, biochemical indices and subjective global assessment (SGA) should be used to identify and address malnutrition, because CAPD can aggravate this state by additional peritoneal protein losses, which are greater in diabetics than in nondiabetic patients [16].

Lipid disorders may also play an important role in the pathogenesis of atherosclerotic cardiovascular disease in PD diabetic patients. Except for atherogenic lipid profiles that have been reported in diabetic nephropathy [17, 18], several studies have shown that CAPD patients have increased levels of total cholesterol, triglycerides, low density lipoprotein (LDL) and apolipoprotein B (apoB) and decreased high density lipoprotein (HDL) and apolipoprotein A-1 (apoA1) levels [19–21]. Diabetic PD patients in a cross-sectional study, had significantly lower HDL-C than non-diabetics, whereas apolipoproteins and risk ratios remained stable over time on PD [22]. Although Little et al. [23] suggested that the worsening dyslipidemia during CAPD was not associated with aspects of treatment such as glucose load or protein losses and that the strongest predictors of worsening lipid profiles were weight gain and pre-existing cardiovascular comorbidity, lipid profiles have to carefully evaluated in diabetic PD patients.

Furthermore increased serum levels of lipoprotein a (Lp(a)) contribute in the development of atherosclerotic complications, whereas levels > 30 mg/dl has been shown to be an independent risk factor for the early development of coronary and cerebrovascular atherosclerotic disease [24–26]. Thus, whether or not Lp(a) levels are affected by diabetes [27, 28], the increased levels seen in CAPD patients [29, 30] and the possible underlying effect of hypoalbuminemia [31] prompt for effective management of lipid disorders in CAPD diabetic patients.

The presence of advanced glycosylated end products (AGEs), which normally are excreted by the kidney but accumulate in patients with renal failure, may also affect outcome of PD patients. Plasma levels are elevated in uremic patients, while AGE's in serum are significantly higher in diabetic than non-diabetic uremic patients [32, 33]. It has been suggested that the excessive accumulation of AGEs in tissues exerts its main pathological effect through abnormal permeability of blood vessels; at the same time, diabetes accelerates the synthesis and tissue deposition of AGEs [34].

LONG-TERM OUTCOMES OF DIABETIC PATIENTS UNDERGOING PD

Diabetic patients are at a higher risk of developing concurrent illnesses than the general population, and among ESRD patients, comorbidity is more common in diabetic than in nondiabetic patients. Insensitive feet at risk of ulceration, deformation, and amputation, are the clinical sequel of peripheral sensorimotor neuropathy, while autonomic neuropathy is expressed as cardiovascular disorders and abnormal visceral function. These devastating complications that increase with age and duration of diabetes, may continue during the renal replacement therapy, increasing ESRD patients' morbidity and hospitalization requirements. However, there was no evidence that retinopathy or other severe diabetic complications might be accelerated by dialytic therapy.

Among other parameters efficacy of RRT therapy can be gauged by use of the hospitalization rates, as the length and frequency of hospitalizations are closely linked to morbidity. Recent USRDS data showed that diabetics have higher admission rates and hospital days than nondiabetic patients in all modalities of RRT, whereas female diabetics tend to have higher rates of hospitalization than their male counterparts [1]. However since 1994 the numbers of hospital days per admission and the hospital days per patient year at risk have decreased for patients undergoing PD, which may be due to the increased dialysis therapy and hematocrit levels [1].

In addition patients undergoing PD tend to have higher overall infection rates, complicated by high rates of peritonitis, and hospital days for infections are higher among these patients. Amputation rates are similar across both RRT modalities as well as hospital days for cardiovascular procedures whereas amputations are similar for both hemodialysis and peritoneal dialysis patients [1].

a. Technique Survival

The reduction in peritonitis rates since the late 1980s had reduced the 'dropout' rates from CAPD and transfers from CAPD to HD declined primarily because of the use of the Y-set. However frequent severe peritonitis episodes have contributed to alterations in peritoneal membrane permeability and ultrafiltration (UF) loss, which in association with the higher incidence of comorbid factors may shorten CAPD longevity.

Peritonitis remains the major cause of 'drop out' (discontinuation of CAPD) among all CAPD patients but there is no evidence that diabetic patients are at increased risk of peritonitis and catheter-related infection. Besides intraperitoneal insulin administration has not been shown to increase the risk of peritonitis [35].

Ultrafiltration failure usually is due to rapid peritoneal transport of glucose, and hence rapid dissipation of the osmotic gradient. Diabetic patients with frequent hyperglycemia may develop fluid retention because of reduced transperitoneal osmotic gradient and of increased fluid intake, stimulated by thirst. Nevertheless no differences in the frequency of ultrafiltration failure have been observed between diabetics and non-diabetics, and UF failure is not a frequent cause of discontinuation of CAPD (technique failure) in diabetics [12].

As regard transfer between modalities, Serkes et al. [36] initially reported that diabetics on CAPD have a higher risk of changing modes of dialysis than nondiabetic patients, although Gentil et al. [37] observed the opposite trend. In a multicenter study comparing CAPD diabetics with non-diabetics, Viglino et al. [12] found that technique survival and the relative risk of dropout did not differ significantly between the two groups. Also the probability and relative risk of dropout due to peritonitis, as well as of the time to first peritonitis episode, did not differ significantly between the two groups, or between diabetics using (or not using) intraperitoneal insulin.

In our study of 224 diabetic CAPD patients [3] technique survival rates for the 1st,

Table 2. Long-term survival results from studies including varying proportions of diabetic patients undergoing CAPD.

Source/year	Patients [n] (%CAPD)	Mean age (years)	Comparison	Patients survival 3-year	4-year	5-year	Study type/ duration
Mc Millan et al. 1990 [39]	82 D (50%) Type 1/Type 2	40/58[1]	Diabetics	72%	72%	63%	SC, R/
	82 non-D (50%)		Non-diabetics	95%	92%	87%	12–132 months
Catalano et al. 1990 [40]	42 D (65%)	43	Diabetics	40%	17%	0	SC, R/
	42 non-D	43	Non-diabetics	80%	60%	60%	42 months (2–103)
Rotellar et al. 1991 [41]	44 D (26%)	47	Diabetics	50%	32%	32%	SC, R/
	127 non-D		Non-diabetics	75%	67%	60%	10 year study
Viglino et al. 1994 [12]	301 D (15.1%)	58.9	Diabetics	37%	20.6%		Multicenter/
	1689 non-D	57.8	Non-diabetics	65%	55%		10-year survey
Zimmerman et al. 1996 [5]	118 D (41.7%)	48.1	Diabetics < 55 years	80%	68%	68%[2]	SC, R/
	165 non-D	52.5	Non-diabetics < 55 years	85%	80%	80%	10-year survey
			Diabetics > 55 years	32%	11%	11%	
			Non-diabetics > 55 years	48%	32%	27%	
Marcelli et al. 1996 [38]	347 D on CAPD (38.8%)	62.2	Diabetics on CAPD	46%		20%	Multicenter/
	545 D on HD (61.2%)	58.1	Diabetics on HD	54%		37%	11 year survey
			Adjusted results	52%		34%	
Passadakis et al. 1998 [3]	224 D on CAPD	59.8	CAPD Diabetics	70%	50%	33%	SC, R/
	Type 1: 137	43.3	Type 1	76%	56%	33%	7-year study
	Type 2: 87	65.6	Type 2	66%	47%	30%	

Abreviations: SC = Single center; R = retrospective; D = diabetic patients; non-D = nondiabetic patients.
[1] Mean age for type 1 and type 2 diabetic patients; the mean ages of control patients were 41 and 57 years, respectively.
[2] Type 1, 69 patients, 5-year survival 57%; type 2, 49 patients, 5-year survival 5%.

3rd and 5th year of CAPD were 93%, 72% and 44% respectively. Also Marcelli et al. [38] found that the PD technique survival in diabetic patients, unadjusted for pretreatment prognostic differences, was 91% at one year, 73% at three years, and 61% at five years, while the HD technique survival was 94%, 80%, and 75% respectively.

b. Long-Term Survival of Diabetic and Nondiabetic Patients on CAPD

Several studies compared outcome between diabetic and nondiabetic patients undergoing CAPD; Table 2 summarizes their findings regarding the study and patients' characteristics, and the probability of surviving beyond three years. All are retrospective studies published after 1990, either from a single center or multicenter studies and compared long-term outcomes between diabetic and nondiabetic patients undergoing CAPD. Patient survival data were obtained mainly from retrospective multicenter or single center studies, for varying follow up periods. Most of these studies compared the outcomes of diabetic and nondiabetic patients undergoing CAPD. Some of these data are estimates from the graphical presentations of actuarial patients survival of the published studies [12, 39–42]. However caution is needed in interpreting studies of survival in ESRD diabetic patients, because they involve dissimilar treatment groups, with regards to age, race, diabetes type, factors of comorbidity and their severity, various complications, dialysis dose and degree of metabolic control.

A multicenter analysis [12] of 301 diabetic and 1,689 nondiabetic patients undergoing CAPD showed that diabetic patients' survival was significantly lower than that of nondiabetic patients; 20.6% of diabetics were alive at 4 years compared to 55.6% of non-diabetics, and the relative risk of death was 2.13 times higher for diabetic patients. Among the diabetic patients, cardiovascular diseases and cachexia were nearly twice as frequent, and infections other than peritonitis, were more than 3 times as frequent as in nondiabetic patients. Thus, the higher probability of death for diabetic patients could be correlated with the presence of baseline disease and comorbidity.

On the contrary Zimmerman et al. [5], who analyzed the long-term outcome of 118 diabetic patients and compared them to 165 nondiabetic patients receiving PD over a 10-year period, found no significant difference in survival between diabetic and nondiabetic patients under the age of 55 years. For nondiabetic patients, 1- and 5-year survival rates were of 98% and 79% respectively, whereas survival for diabetic patients less than 55 years was 95% and 68% respectively. However diabetic patients 55 years or older had a significantly decreased survival compared to nondiabetic patients, whereas this tendency was more significant in patients over 60 years of age. The authors concluded that advanced age, low serum albumin levels (after 3 months on treatment) were significant risk factors in both diabetic and non-diabetic patients, whereas previous cardiac disease conferred increased risk only in diabetic patients. The cardiac-related death rate, which was the most common cause of death, did not differ between diabetic and nondiabetic patients. Also peritonitis rate, and diabetes of type 2 were the most significant risk factors for technique failure.

According to the published data [3, 5–7, 12, 38–46] for diabetic patients undergoing PD short-term survival rates for the 1st year range from 74 to 95% (average 85%), for the 2nd year 52 to 84% (average 71%), and 37 to 72% for the 3rd year (average 51%); beyond the third year, the 4-year survival of diabetics on CAPD varies from 17%

to 72% (average 39%) and the 5-year survival rates from 19% to 63% (average 35%) (Figure 2).

In our study [3] survival for all 224 diabetic patients after 12, 24, 36, 48 and 60 months was 93%, 81%, 70%, 50% and 33% respectively (Figure 3); this was significantly better than the previously reported 2-year survival rates from this center of 78% and 47% for type 1 and type 2 diabetic patients respectively [47]. As with the previous study more type 1 than type 2 diabetic patients survived after 1, 2, 3, and 4 years on CAPD, (95%, 84%, 76%, 56% versus 92%, 79%, 66%, 47%), whereas these differences in survival may be due to that type 2 patients are older and usually have severe atherosclerotic heart disease and other comorbid diseases.

Six diabetic patients (3 males and 3 females) in our center, 2 patients with type 1 diabetes and 4 patients with type 2 diabetes, survived longer than 5 years. The mean age of these patients was 59.8 ± 4.9 years, and the mean duration at the end of follow up was 64.3 ± 5.7 months. On peritoneal equilibration test (PET), 3 patients were high (H) transporters and 3 patients showed high average (HA) peritoneal solute transport.

To identify the characteristics of long-term peritoneal dialysis survivors, Rahman et al. [48] compared data of 20 patients, who survived on PD for more than 100 months, with data of 103 patients who died or were switched to hemodialysis (HD) after less than 100 months. Seven patients with type 1 diabetes were among the long-term survivors but no patients with type 2 diabetes mellitus survived longer than 100 months on CAPD. Long-term survivors were significantly younger, weighed less, had fewer episodes of peritonitis had fewer hospital days, and were prescribed more dialysis clear-

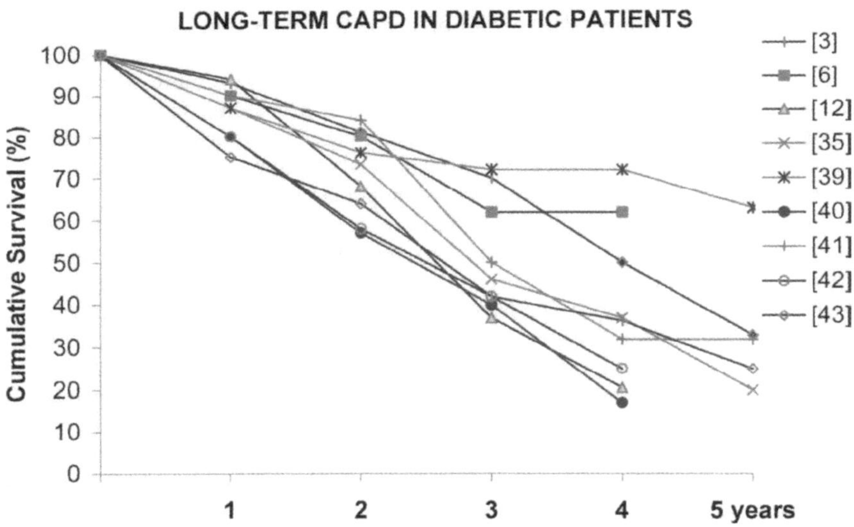

Figure 2. Actuarial patients survival. Long-term survival of CAPD patients with diabetes mellitus. On the right top corner the references, from which the data were obtained are presented. Some of the values were estimated from the graphical presentation in the references. Mean values for 1 to 5 years were 85%, 71%, 51%, 39% and 35% respectively.

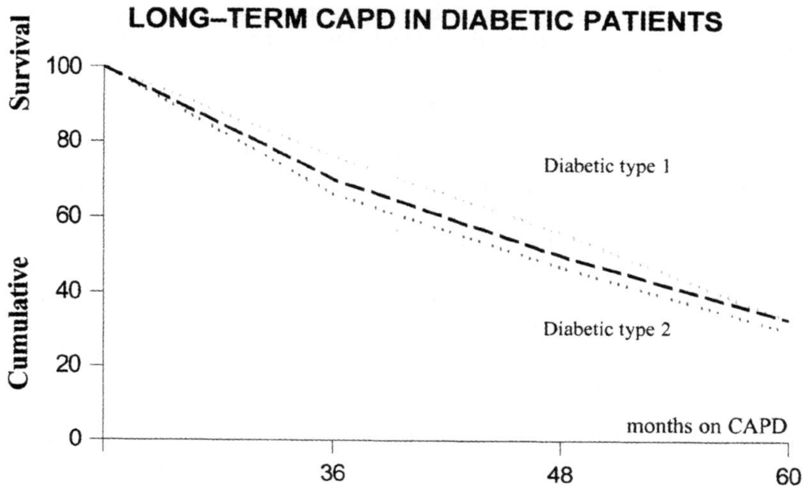

Figure 3. Actuarial patients survival of 224 CAPD patients with diabetes mellitus. More type 1 than type 2 diabetic patients survived after 1, 2, 3, and 4 years on CAPD, (95%, 84%, 76%, 56% versus 92%, 79%, 66%, 47%).

ance per kg body weight, than those who died or were switched to HD before 100 months. Patients' age, predialysis comorbidity, prolonged residual renal function (RRF), maintenance of adequate nutrition, low solute transport rates and low rates of peritonitis were the most important characteristics of patients surviving on long-term peritoneal dialysis, in several studies [49–51] (Table 3).

c. *Comparison of survival of diabetic ESRD patients on PD and HD*

Although there are many technical differences between these two RRT, several studies of dialysis populations, including a varying proportion of diabetic patients, have attempted to compare long-term clinical outcomes in PD and HD. Because population characteristics may influence the overall mortality and morbidity many studies attempted

Table 3. Important characteristics of patients surviving on long-term PD.

Younger age
Lower body weight
Few predialysis comorbid conditions
Prolonged maintenance of RRF
Maintenance of good nutrition
Low solute transport rates
Low hospitalization rate
Low peritonitis rates

to control patients' heterogeneity, by using groups matched for sex, age at onset, year of starting dialysis, or by specific statistical techniques such as Cox proportional hazards model. Also, several adjustments for the presence of pretreatment prognostic factors have been used because such factors may adversely affect long-term survival [7, 10, 13, 14].

Despite the disparity of results, most medium and long-term studies have concluded that overall survival rates of patients undergoing peritoneal dialysis and hemodialysis are not statistically significantly different [36–38, 52, 53], while other studies favor HD [54–56] and others favor PD [57–59] (Table 4). Whenever a study made an adjustment for comorbid factors, the comparison between modalities showed no statistically significant difference or they favored CAPD.

The annual 1991 USRDS data reported that 1-year survival for the diabetic patients undergoing HD was better than for those undergoing CAPD/Continuous cycling peritoneal dialsysis (CCPD) group (69.6% vs. 65.7%) [54]. Bloembergen et al. [55] showed that diabetic patients undergoing CAPD had a 38% higher mortality (RR = 1.38, $P < 0.001$) than diabetic patients undergoing HD. However, they found that mortality risk was insignificant ($P > 0.05$) or small for age < 55 years and became increasingly large and significant ($P < 0.05$) for age > 55 years.

Similarly, after adjustment for patient characteristics including the presence of comorbid conditions at onset of ESRD, Held et al. [56] compared patient survival and found a higher adjusted mortality RR for diabetic patients undergoing CAPD (RR = 1.26 for CAPD vs. HD, $P = 0.03$) (Table 5). This adjusted relative risk translated to an annual absolute mortality of approximately 27% for diabetic HD patients and 34% for diabetic patients undergoing CAPD. Compared with that in HD, the mortality risk for CAPD was higher in older diabetic patients; thus, diabetic patients, who were 58 years of age or older at onset of ESRD showed elevated mortality in CAPD (RR = 1.25 for CAPD vs. HD, $P < 0.05$), whereas they found no statistically significant difference in mortality between these two modality groups for diabetic patients less than 58 years of age at onset of ESRD ($P > 0.05$). Among diabetic patients, survival was similar for CAPD and HD at 1 year (83.3% and 85.4%, for CAPD and HD respectively), but was reduced for CAPD compared to HD at 2 years (54.0% vs. 64.6%) and through the remainder of the follow-up period.

On the contrary, using an 'intent-to-treat' analysis, Nelson et al. [57] found that diabetic patients undergoing CAPD experienced lower mortality rates than their HD-treated counterparts. Among diabetic patients, this difference in mortality relative risk (RR) ranged from 0.40 to 0.70, being lowest for younger diabetics and statistically significant ($P \leq 0.05$) for ages 20 to 52 years (Table 5). Moreover, evaluation of mortality trends showed a significant ($P < 0.01$) decrease in diabetic CAPD mortality rates during the decade, whereas diabetic HD mortality rates increased ($P = 0.06$). They found no differences in mortality by sex and race among non-diabetic patients, but mortality did increase significantly with age in all groups.

Using Poisson regression analysis, Fenton et al. [58] compared mortality rates on hemodialysis and CAPD/CCPD in 11,970 ESRD patients undergoing dialysis, which had started treatment between 1990 and 1994 and were followed-up for a maximum of 5 years. They found that the mortality rate ratio for CAPD/CCPD relative to HD was 0.73 (0.62–0.87). This decrease in mortality for PD relative to HD was less pronounced among patients over 65 years; in these patients, the difference was not statistically significant. They also found that the increased mortality on hemodialysis compared with CAPD/CCPD was concentrated in the first 2 years of follow-up.

After adjusting for statistical differences, Vonesh and Moran [53] found that, among diabetic patients, the PD-HD death rate ratio varied significantly according to gender and age. The average male diabetic patient had little or no difference in risk between PD and HD from 1989–1993, whereas diabetic patients, under the age of 50 years, treated with PD had a significantly lower risk of death than those treated with HD (1989–1993: 0.84 < or = RR < or = 0.89, $P < 0.005$). Over the same period, female diabetic patients treated with PD and all diabetic patients older than 50 years had a higher risk than those undergoing HD. Unlike the study of Bloembergen et al. [55], which was restricted to prevalent-only patients, this national study of both prevalent and incident patients found little or no difference in overall mortality between PD and HD.

Recently Collins et al. [59], in an attempt to reconcile the above differences, evaluated incident Medicare patients treated between 1994 and 1996 by using Poisson regression and compared death rates, adjusting for age, gender, race and primary renal diagnosis. They showed that CAPD/CCPD patients have outcomes comparable with or significantly better than hemodialysis, although results varied with time on dialysis. Cox regression analysis showed a lower PD-HD mortality in diabetic CAPD/CCPD patients younger than 55 years. In contrast, the risk of all-cause death for female diabetics 55 years of age and older was 1.21 for CAPD/CCPD, while the risk of death was lower (1.03) in men over 55 years. (Table 5).

A recent study Schaubel and Fenton [60] studied the trends in mortality among the 17,900 patients receiving PD in Canada during the period 1981–1997 used Poisson regression to adjust for age, race, gender, primary renal diagnosis, follow-up time, and type of PD (CAPD/CCPD *versus* intermittent PD). These workers found a significant reduction in adjusted mortality rates by calendar period when compared with the years 1981–1985, that served as the reference period (RR = 1, fixed; RR = 0.81, for 1986 through 1989; RR = 0.73, for 1990 through 1993; RR = 0.63, for 1994 through 1997). This improvement in mortality was fairly consistent across patient subpopulations, except for diabetic patients > 65 years of age from 1986-1989 (RR in this interval 1.01). The authors believe that more extensive data on practice patterns would empower future studies to elucidate the cause/effect relationship between PD practice patterns and patient survival.

A more recent reanalysis by Krishnan [61] of the HD versus PD mortality in Canada showed that the previously reported results were holding true over the period from 1990 to 1998 with 7,581 patients on PD and 18,031 on HD. The overall adjusted mortality rate ratio (RR) for PD relative to HD was 0.85 (95% CI 0.81–0.89), whereas for the diabetic patients the differences were significant only for younger (< 64 years) males (RR 0.80 (0.71–0.91)). Diabetic females on PD had a significantly higher mortality rate than males in both age groups, whereas in non-diabetic females on PD the significantly higher mortality rate than males was seen in the ≥ 65 years age group.

Regarding comparative mortality of hemodialysis and peritoneal dialysis in Canada, Murphy et al. [62] recently have shown that the apparent survival advantage of PD is due to lower comorbidity and a lower burden of acute-onset ESRD at the inception of dialysis therapy. Hemodialysis and peritoneal dialysis, as practiced in Canada in the 1990s, are associated with similar survival rates.

Table 4. Comparison of long-term survival between CAPD and HD from studies including varying proportions of CAPD-treated diabetic patients.

Comparison/ Source	Data control	CAPD patients, [n] (%)	HD patients, [n] (%)	Type of study	Study duration
No differences between					
Serkes et al. 1990 [36]	Adjusted the mortality by age, sex, race, hypertensive nephrosclerosis, diabetes, ASHD and PVD, modality and time on RRT	325 total, 119 diabetics (37%)	342 total, 104 diabetics (30%)	Multicenter	1982–1985
Gentil et al. 1991 [37]	Account for the heterogeneity of the patients	272 total, 78 diabetics (28.7%)	842 total, 19 diabetics (2.3%)	Multicenter	1984–1988
Marcelli et al. 1996 [38]	Results adjusted for comorbid factors	347 diabetics (38.8%)	545 diabetics (61.2%)	Multicenter	11-years survey
Maiorca et al. 1996 [52]	Results adjusted for age and comorbid factors	297 total, 63 diabetics (21%)	HD 281 total, 32 diabetics (11%)	Multicenter	10 years
Vonesh et al. 1999 [53]	Results adjusted for statistical differences	Patients treated with PD versus HD in the United States continue to hold true over the period 1987–1993		USRDS annual reports for the cohort periods: 1987–1989, 1988–1990, 1989–1991, 1990–1992, and 1991–1993	

HD better					
USRDS 1991 [54]		41,317 incidence patients 13,595 diabetics (32.9%)		USRDS data	Annual report
Bloembergen et al. 1995 [55]	Adjusted the mortality by age, sex, race, modality and time on RRT	Patients prevalent on January 1 of the years 1987–1989, each with 1 year of follow-up		USRDS data	4 years
Held et al. 1994 [56]	Adjustment for patient characteristics and the presence of comorbid conditions	681 total, (36.4% diabetic)	3,376 total, (28.1% diabetic)	USRDS data	2.25–4.25 years
PD better					
Nelson et al. 1992 [57]	Results adjusted for patient characteristics and comorbid conditions	4,288 Michigan residents start in 1980s (20–59 years) 613 diabetics, 20–40% on CAPD		Michigan residents	4–113 months
Fenton et al. 1997 [58]	Results adjusted for age, primary renal diagnosis, center size, and predialysis comorbid conditions	11,970 all dialysis (31.9% diabetics)		CORR	5 years
Collins et al. 1999 [59]	Poisson regression to compare death rates, adjusting for age, sex, race and primary renal diagnosis	Incident Medicare patients (18,110 on CAPD/CCPD, 99,048 on HD)		USRDS data	1994–1996

Abbreviations: ASHD = arteriosclerotic heart disease; PVD = peripheral vascular disease; CORR = Canadian Organ Replacement Register.

Table 5. Mortality relative risk (RR) for CAPD/CCPD versus HD for diabetic patients as reported in some comparative studies.

Source	HD	Diabetic patients		Non-diabetic patients	
		CAPD	*P* – value	CAPD	*P* – value
Bloembergen et al. [55]					
	1.00	1.38	$P < 0.001$	1.11[a]	$P < 0.001$
Age < 55 years	1.00	1.00	$P > 0.05$	1.00	$P > 0.05$
Age > 55 years	1.00	RR > 1.00	$P < 0.05$	RR > 1.00	$P < 0.05$
Held et al. [56]					
	1.00	1.26	$P = 0.03$	0.84	$P = 0.25$
Age < 58 years[b]	1.00	1.11	$P > 0.05$		
Age ≥ 58 years	1.00	1.25	$P < 0.05$		
Age ≥ 63 years	1.00	1.34	$P < 0.01$		
Vonesh et al. [53]					
Average male:					
1989–91	1.00	1.02	$P > 0.05$	1.05	$P > 0.05$
1990–92	1.00	1.05	$P > 0.05$	1.04	$P > 0.05$
1991–93	1.00	1.08	$P < 0.01$	1.07	$P < 0.01$
< 50 years: 1989–93	1.00	0.84–0.89	$P < 0.05$		
> 50 years: 1989–93	1.00	1.28–1.30	$P < 0.001$		
Female: 1989–93	1.00	1.18–1.19	$P < 0.001$		
Serkes et al. [36]					
	1.00	0.90	$P > 0.05$	0.62	$P = 0.08$
Collins et al. [59][c]					
Women < 55	1.00	0.88		0.61	
Women > 55	1.00	1.21		0.87	
Men < 55	1.00	0.86		0.72	
Men > 55	1.00	1.03		0.87	
Fenton et al. [58][d]					
Age 0 – 64 years	1.00	0.73 (0.62–0.87)	$P < 0.01$	0.54	$P < 0.01$
Age over 65 years	1.00	0.88	$P > 0.05$	0.76	$P < 0.01$
Nelson et al. [57]					
Age 20–52 years	1.00	0.40–0.70[e]			

[a] RR was 1.19 for all prevalent CAPD patients ($P < 0.001$).
[b] Age at onset of ESRD.
[c] These values of RR reported to be significantly lower.
[d] RR controlled for age, predialysis comorbid conditions and center size.
[e] This relative mortality risk was lowest for younger diabetics and statistically significant ($P \leq 0.05$) for ages 20 to 52 years (0.40, 0.48, 0.58, 0.70 for 20–29, 30–39, 40–48 and 50–59 years respectively); also among diabetics, mortality rates rose significantly faster ($P = 0.03$) in CAPD patients (RR = 1.37 per ten years of age) rather than HD patients (RR = 1.14 per ten years).

However the increasing clinical application of the newer hemodialysis techniques, such as high-flux and high efficiency hemodialysis and/or the use of biocompatible membranes, may improve long-term survival on HD. Thus in newer studies, we should reevaluate the mortality comparisons between HD and PD in patients with ESRD due to diabetes.

SURVIVAL OF DIABETIC ESRD PATIENTS ON VARIOUS PD MODALITIES

Regarding survival of diabetics on different types of PD (continuous ambulatory PD (CAPD), continuous cyclic PD (CCPD), nocturnal intermittent PD (NIPD), daytime ambulatory PD (DAPD), and automated PD (APD)) there is no evidence for differences in long-term results. Most comparative studies of mortality between PD and HD have treated CAPD/CCPD patients as a single PD population [58–60]. Strauss et al. [63] compared high transporters with inadequate ultrafiltration, treated with NIPD and DAPD, with patients on CAPD and CCPD and found similar dialysate protein losses, adequacy, and nutrition indices. However diabetic patients demonstrated lower levels of serum albumin and normalized protein catabolic rate (nPCR) than nondiabetic patients while maintaining equivalent Kt/V urea indices. Also Saade et al. [64] analyzing the clinical results of CAPD/CCPD population including 41% patients with diabetic nephropathy, in relation to treatment modality systems, compliance, rehabilitation characteristics, complications, and survivals reported good patient survival rates at 1, 2, and 3 years of 87.9%, 76.6%, and 67%, respectively.

In diabetic patients the main indications for using various automated PD techniques (APD-CCPD) include: patient preference, young diabetic patients waiting for renal transplantation, and diabetics requiring partner support – older, blind and partner – dependent patients. Recently the use of various combinations of APD and CAPD (use of cycler at night and at least one manual exchange per day) has been increasing. Such techniques can deliver an adequate dialysis dose if the peritoneal permeability is not too low and if the prescription is adapted to each patient. Also APD may be used with success in patients in whom continuation of CAPD or HD therapy is very difficult due to its complications or comorbid conditions [65]. These factors and the use of icodextrin may further increase PD patients' survival.

CONCLUSION

Diabetic patients frequently initiate their chronic renal replacement therapy with several devastating complications due to long-lasting diabetes mellitus that has been linked to diabetic nephropathy, neuropathy and retinopathy. Besides they are at a higher risk of developing concurrent illnesses than the general population, and among ESRD patients, comorbidity is more common in diabetic than in nondiabetic patients.

Although several clinical studies have evaluated the factors affecting survival rates and compared patient survival rates between diabetic patients and nondiabetic patients undergoing the two major dialysis modalities, CAPD and HD, most of them were conducted at a time when the dose of peritoneal dialysis was inadequate, the importance of residual renal function was poorly understood, and the presence of comorbid conditions was underestimated. Inadequate adjustments for all patients' characteristics as well as for the presence of serious comorbid factors, renal residual failure, and the delivered dose of dialysis make comparisons very complicated.

Despite survival improvement, survival and technique success rates in diabetic patients undergoing CAPD are lower than their nondiabetic counterparts, whereas worse prognosis observed in older and female diabetic patients over 65 years.

Early dialysis initiation, adequate dialysis dose, intensive medical care to prevent and manage of serious complications in concert with patient's compliance and cooperation, may ensure the higher long-term survival in ESRD patients with diabetes mellitus undergoing PD.

REFERENCES

1. USRDS (United States Renal Data System) USRDS 2000 Annual Data Report. Atlas of end – stage renal disease in the United States. National Institute of Health. NIDDK/DKUHD Am J Kidney Dis 2000; 36 (Suppl 2): S38–S176.
2. Piccoli GB, Quarello F, Bonello F, et al. Diabetic patients on dialysis: a changing picture. Kidney Int 1993; 41 (Suppl): S14.
3. Passadakis P, Oreopoulos DG. Continuous ambulatory peritoneal dialysis in 224 diabetics with end stage renal disease: evidence of improved survival over the past 10 years. In: Friedman EA, L'Esperance FA (eds), Diabetic Renal-Retinal Syndrome, 21st Century Management Now. Kluwer Academic, London, England, 1998, pp 89–115.
4. Schaubel D, Fenton SSA. Diabetes and Renal Failure. Section 6. Canadian Organ Replacement Register. Annual Report 1996. Canadian Institute of Health Information, Ottawa, Ontario, 1996.
5. Zimmerman SW, Oxton LL, Bidwell D, Wakeen M. Long-term outcome of diabetic patients receiving peritoneal dialysis. Perit Dial Int 1996; 16: 63–68.
6. Coronel F, Hortal L, Naranjo P, Pozo C, Torrente J, Prats D, Barrientos A. Analysis of factors in the prognosis of diabetics on continuous ambulatory peritoneal dialysis (CAPD): long-term experience. Perit Dial Int 1989; 9: 121–125.
7. Kemperman FA, van Leusen R, van Liebergen FJ, Oosting J, Boeschoten EW, Struijk DG, Krediet RT, Arisz L. Continuous ambulatory peritoneal dialysis (CAPD) in patients with diabetic nephropathy. Neth J Med 1991 (Jun); 38: 236–245.
8. Zimmerman S. The diabetic ESRD patient: specific needs. Perit Dial Int 1997; 17 (Suppl 3): S9–S11.
9. Tzamaloukas AH, Yuan ZY, Balaskas E, Oreopoulos DG. CAPD in end stage patients with renal disease due to diabetes mellitus – an update. In: Khanna R, Nolph KD, Prowant BF, Twardowski ZJ, Oreopoulos DG (eds), Advances in Peritoneal Dialysis, Vol 8. Peritoneal Dialysis Bulletin Inc, Toronto, 1992, pp 185–191.
10. Espinosa A, Cueto-Manzano AM, Velazquez-Alva C, Hernandez A, Cruz N, Zamora B, et al. Prevalence of malnutrition in Mexican CAPD diabetic and nondiabetic patients. In: Advances in Peritoneal Dialysis, Vol 12. Peritoneal Dialysis Publications, Toronto, 1996, pp 302–306, .
11. Miller DG, Levine S, Bistrian B, D'Elia JA. Diagnosis of protein calorie malnutrition in diabetic patients on hemodialysis and peritoneal dialysis. Nephron 1983; 33: 127–132.
12. Viglino G, Cancarini GC, Catizone L, Cocchi R, De Vecchi A, Lupo A, Safomone M, Segoloni GP, Giangrande A. Ten years experience of CAPD in diabetics: comparison of results with non-diabetics. Nephrol Dial Transpalnt 1994; 9: 1443–1448.
13. Maiorca R, Vonesh E, Cancarini GC, Cantaluppi A, Manili L, Brunori G, Camerini C, Feller P, Strada A. A six-year comparison of patient and technique survivals in CAPD and HD. Kidney Int 1988; 34: 518–524.
14. Avram MM, Fein PA, Bonomini L, Mittman N, Loutoby R, Avram DK, Chattopadhyay J. Predictors of survival in continuous ambulatory peritoneal dialysis patients: A five-year prospective study. Perit Dial Int 1996; 16 (Suppl 1): S190–S194.
15. Scanziani R, Dozio B, Bonforte G, Surian M. Nutritional parameters in diabetic patients on CAPD. Adv Perit Dial 1996; 12: 280–283.
16. Kriedet RT, Zuyderhoudt FMJ, Boeschoten EW, Arisz L. Peritoneal permeability to proteins in diabetic and non-diabetic continuous ambulatory peritoneal dialysis patients. Nephron 1986; 42: 133–140.
17. Dullaart RP, Dikkeschei LD, Doorenbos H. Alterations in serum lipids and apolipoproteins in male type 1 (insulin-dependent) diabetic patients with microalbumi-nuria. Diabetologia 1989; 32: 685–689.

18. Watts GF, Naumova R, Slavvin BM, et al. Serum lipids and lipoproteins in insulin-dependent diabetic patients with persistent microalbuminuria. Diabetic Med 1989; 6: 25–30.
19. Siamopoulos KC, Elisaf MS, Bairaktari HT, Pappas MB, Sferopoulos GD, Nikolakakis NG. Lipid parameters including lipoprotein (a) in patients undergoing CAPD and hemodialysis. Perit Dial Int 1995; 15: 342–347.
20. Avram MM, Goldwasser P, Burrell DE, Antignani A, Fein PA, Mittman N. The uremic dyslipidemia: a cross-sectional and longitudinal study. Am J Kidney Dis 1992; 20: 324–335.
21. Lindholm B, Norbeck HE. Serum lipids and lipoproteins during continuous ambulatory peritoneal dialysis. Acta Med Scand 1986; 220: 143–151.
22. Burrell D, Antignani A, Fein P, Goldwasser P, Mittman N, Avram MM. Longitudinal survey of apolipoproteins and atherogenic risk in hemodialysis and continuous ambulatory peritoneal dialysis patients. ASAIO Trans 1990; 36: M331–M335.
23. Little J, Phillips L, Russell L, Griffiths A, Russell GI, Davies SJ. Longitudinal lipid profiles on CAPD: their relationship to weight gain, comorbidity, and dialysis factors. J Am Soc Nephrol 1998; 9: 1931–1939.
24. AssmannG, Schulte H, von Eckardstein A. Hypertriglyceridemia and elevated lipoprotein (a) are risks factors for major coronary events in middle-aged men. Am J Cardiol 1996; 77: 1179–1184.
25. Dahlen GH. Lp(a) lipoprotein in cardiovascular diseases. Atherosclerosis 1994; 108: 111–126.
26. Raitakari OT, Adams MR, Celermajer DS. Effect of Lp(a) on the early functional and structural changes of atherosclerosis. Arterioscler Thromb Vasc Biol 1999; 19: 900–905.
27. Purnell JQ, Marcovina SM, Hokanson JE, et al. Levels of lipoprotein (a), apolipoprotein B, and lipoprotein cholesterol distribution in IDDM. Diabetes 1995; 44: 1218–1226.
28. Babazono T, Miyamae M, Tomonaga O, Omori Y. Cardiovascular risk factors in diabetic patients undergoing continuous ambulatory peritoneal dialysis. Advances in Peritoneal Dialysis 1996; 12: 120–125.
29. Misra M, Webb AT, Reaveley DA, Doherty E, O'Donnell M, Seed M, Brown EA. The effect of change of renal replacement therapy on serum lipoprotein (a) concentration. Adv Perit Dial 1997; 13: 168–173.
30. Avram MM, Sreedhara R, Patel N, Chattopadhyay J, Thu T, Fein P. Is an elevated level of serum lipoprotein (a) a risk factor for cardiovascular disease in CAPD patients? Advances in Peritoneal Dialysis 1996; 12: 266–271.
31. Yang WS, Min WK, Park JS, Kim SB. Effect of increasing serum albumin on serum lipoprotein (a) concentration in patients receiving CAPD. Am J Kidney Dis 1997 (Oct); 30: 507–513.
32. Korbet SM, Makita Z, Firanek CA, Vlassara H. Advanced glycation end products in continuous ambulatory ρ peritoneal dialysis patients. Am J Kidney Dis 1993; 22: 588–594.
33. Monnier VM, Sell DR, Nagarají RN, Miyata S, Grandhee S, Odetti P, Maillard. Reaction-mediated molecular damage to extracellular matirx and other tissue proteins in diabetes, aging, and uremia. Diabetes 1992; 41: 36–41.
34. Brownlee M, Cerami A, Vlassara H. Advanced glycosylation end products in tissue and the biochemical basis of diabetic complication. N Engl J Med 1988; 18: 1315–1321.
35. Lindblad AS, Nolph KD, Novak JW, Friedman EA. A survey of the NIH CAPD Registry population with end-stage renal disease attributed to diabetic nephropathy. J Diabetic Compl 1988; 2: 227–232.
36. Serkes KD, Blagg CR, Nolph KD, Vonesh EF, Shapiro F. Comparison of patient and technique survival in continuous ambulatory peritoneal dialysis (CAPD) and hemodialysis: a multicenter study. Perit Dia Int 1990; 10: 15–19.
37. Gentil MA, Carriazo A, Pavon MI, et al. Comparison of survival in continuous ambulatory peritoneal dialysis and hospital haemodialysis: a multicentric study. Nephrol Dial Transplant 1991; 6: 444–451.
38. Marcelli D, Spotti D, Conte F, et al. Survival oI diabetic patients on peritoneal dialysis or hemodialysis. Perit Dial Int 1996; 16: S283–S287.
39. McMillan MA, Briggs JD, Junor BJ. Outcome of renal replacement treatment in patients with diabetes mellitus. BMJ 1990; 301: 540–544.
40. Catalano C, Goodship TH, Tapson JS, Venning MK, Taylor RM, Proud G, Tunbridge WM, Elliot RW, Ward MK, Alberti KG, et al. Renal replacement treatment for diabetic patients in Newcastle upon Tyne and the Northern region, 1964–88. BMJ 1990; 301: 535–540.
41. Rotellar C, Black J, Winchester JF, Rakowski TA, Mosher WF, Mazzoni MJ, Amiranzavi M, Garagusi V, Alijani MR, Argy WP. Ten years' experience with continuous ambulatory peritoneal dialysis. Am J Kidney Dis 1991; 17: 158–164.
42. Lupo A, Tarchini R, Cancarini G, Catizone L, Cocchi R, de Vecchi A, Viglino G, Salomone M, Segoloni G, Giangrande A. Long-term outcome in continuous ambulatory peritoneal dialysis: a 10-year survey by the Italian Cooperative Peritoneal Dialysis Study Group. Am J Kidney Dis 1994; 24: 826–837.

43. Rodriguez JA, Cleries M, Vela E. Diabetic patients on renal replacement therapy: analysis of Catalan Registry data. Renal Registry Committee. Nephrol Dial Transplant 1997 (Dec); 12: 2501–2509.
44. Zimmerman SW, Glass N, Sollinger H, Miller D, Belzer F. Treatment of end-stage diabetic nephropathy: over a decade of experience at one institution. Medicine (Baltimore) 1984; 63: 311–317.
45. Rottembourg J, Issad B, Allouache M, Ruotolo C, Deray G, Baumelou A, Kahn JF, Jacobs C. Clinical aspects of continuous ambulatory peritoneal dialysis in diabetics Nephrologie 1988; 9: 227–232.
46. Miles AM, Friedman EA. Dialytic therapy for diabetic patients with terminal renal failure. Curr Opin Nephrol Hypertens 1993; 2: 868–875.
47. Khanna R, Wu G, Prowant B, Jastrzebska J, Nolph KD, Oreopoulos DG. Continuous ambulatory peritoneal dialysis in diabetics with end-stage renal disease. A combined experience of two North American Centers. In: Friedman EA, L'Esperance FA (eds), Diabetic Retinal Syndrome, Vol 3. Grune and Stratton, New York, NY, 1986, pp 363–381.
48. Rahman EA, Wakeen M, Zimmerman SW. Characteristics of long-term peritoneal dialysis survivors: 18 years experience in one center. Perit Dial Int 1997; 17: 151–156.
49. Maiorca R, Cancarini GC, Brunori G, Zubani R, Camerini C, Manili L, Movilli E. Comparison of long-term survival between hemodialysis and peritoneal dialysis. Adv Perit Dial 1996; 12: 79–88.
50. Davies SJ, Phillips L, Griffiths AM, Russell LH, Naish PF, Russell GI. What really happens to people on long-term peritoneal dialysis? Kidney Int 1998 (Dec); 54: 2207–2217.
51. De Vecchi AF, Maccario M, Scalamogna A, Castelnovo C, Ponticelli C. Nine patients treated for more than 10 years with continuous ambulatory peritoneal dialysis. Am J Nephrol 1996; 16: 455–461.
52. Maiorca R, Cancarini GC, Zubani R, Camerini C, Manili L, Brunori G, Movilli E. CAPD viability: a long-term comparison with hemodialysis. Perit Dial Int 1996; 16: 276–287.
53. Vonesh EF, Moran J. Mortality in end-stage renal disease: a reassessment of differences between patients treated with hemodialysis and peritoneal dialysis. J Am Soc Nephr 1999; 10: 354–365.
54. US Renal Data System. USRDS 1991 Annual Data Report, the National Institute of Diabetes and Digestive and Kidney Diseases. Am J Kidney Dis 1991; 18 (Suppl 2): 49-60.
55. Bloembergen WE, Port FK, Mauger EA, Wolfe RA. A comparison of mortality between patients treated with hemodialysis and peritoneal dialysis. J Am Soc Nephrol 1995; 6: 177–183.
56. Held PJ, Port FK, Turenne N, Gaylin DS, Hamburger RJ, Wolfe RA. Continuous ambulatory peritoneal dialysis and hemodialysis: comparison of patient mortality with adjustment for comorbid conditions. Kidney Int 1994; 45: 1163–1169.
57. Nelson, CB, Port, FK, Wolfe, RA, Guire, KE. Comparison of continuous ambulatory peritoneal dialysis and hemodialysis patient survival with evaluation of trends during the 1980s. J Am Soc Nephrol 1992; 3: 1147–1155.
58. Fenton SS, Schaubel DE, Desmeules M, Morrison HI, Mao Y, Copleston P, Jeffery JR, Kjellstrand CM. Hemodialysis versus peritoneal dialysis: a comparison of adjusted mortality rates. Am J Kidney Dis 1997; 30: 334–342.
59. Collins AJ, Hao W, Xia H, et al. Mortality risks of peritoneal dialysis and hemodialysis. Am J Kidney Dis 1999; 34: 1065–1074.
60. Schaubel D, Fenton SSA. Trends in Mortality on Peritoneal Dialysis: Canada, 1981–1997. Journal of the American Society of Nephrology 2000; 1: 126–133.
61. Krishnan M, Schaubel D, Bargman MJ, Oreopoulos DG, Fenton SSA. HD versus PD in Canada Revised. Abstract, Meeting of the International Society of Peritoneal Dialysis, Montreal, June 26–28, 2001.
62. Murphy WS, Foley NR, Barreti JB, et al. Comparative mortality of hemodialysis and peritoneal dialysis in Canada. Kidney International 2000; 57: 1720–1726.
63. Strauss FG, Holmes DL, Dennis RL. Dialysis adequacy indices in high membrane transporters treated with short-dwell peritoneal dialysis. Adv Perit Dial 1995; 11: 110–113.
64. Saade M, Joglar F. Chronic peritoneal dialysis: seven-year experience in a large Hispanic program. Perit Dial Int 1995; 1: 37–41.
65. Liberek T, Renke M, Lichodziejewska-Niemierko M, Rutkowski B. The role of Automated Peritoneal Dialysis in peritoneal dialysis programme: one centre experience. Int J Artif Organs 1999; 11: 734–738.

MARIANA S. MARKELL

15. Complications of Post-transplant Diabetes

Editor's Comment: Markell provides a keen perspective on the syndrome of diabetes encountered in kidney and other solid organ transplant recipients. Both cyclosporine and tacrolimus, in combination with prednisone, are associated with what appears to be type 2 diabetes with onset after the start of an immunosuppressive regimen. Whether these calcineurin inhibitors precipitate a diabetic state destined by genetic predisposition to appear later in the patient's life or are in some poorly understood way able to induce a unique form of insulin resistance is unknown. Like gestational diabetes, however, the clinical course emulates that of typical type 2 diabetes and has been associated with recurrent nodular and diffuse intercapillary glomerulosclerosis after seven or more years. For the present, a rational management strategy, correctly termed 'prudent' by Markell, is to minimize the dose of the calcineurin inhibitor while optimizing metabolic control, utilizing all of the educational and motivational techniques applicable to the individual with 'ordinary' diabetes,

Abstract: Post-transplant diabetes mellitus (PTDM) remains a common complication of immunosuppressive therapy, especially among minority populations, with an incidence that ranges from 10–30%. Postulated etiologies include defects in peripheral insulin action (insulin resistance) and direct toxicity to the beta-cell, resulting in decreased insulin secretion. Although excess mortality has not been demonstrated, PTDM increases the subsequent risk of graft loss. De novo development of diabetic nephropathy has been documented, and probably accounts for many allograft failures that are diagnosed as 'chronic rejection'. The true incidence of diabetic complications, including retinopathy and neuropathy is not known. It is imperative that patients with PTDM be appropriately educated and monitored and optimal glucose control maintained. The extent to which PTDM remits following loss of the allograft or whether it represents acceleration of the disease course in a patient population 'primed' to develop Type 2 diabetes remains to be studied.

BACKGROUND

One of the earliest recognized side effects of the immunosuppressants used for transplantation was an increase in the incident rate of diabetes/glucose intolerance. In the era when high dose corticosteroids and azathioprine were used, post-transplant diabetes was called 'steroid-induced' diabetes due to the relationship between steroid dose and development of diabetes [1–3]. The true incidence of PTDM during this time period is unclear, as the definition of diabetes in the literature ranged from 'fasting glucose of > 200 mg/dl' [5] to 'requirement for insulin' [6], however, prevalence rates were quoted at 6 to 46% (4).

In the early 1980's, the first of the calcineurin inhibitors, cyclosporine, was introduced, followed in the 1990's by tacrolimus. As calcineurin inhibitors are potent immunosuppressants, allowing for the use of lower doses of corticosteroids, it was hoped

E.A. Friedman and F.A. L'Esperance, Jr. (eds.), Diabetic Renal-Retinal Syndrome, 173–181.

that the rate of new cases of PTDM would decrease. Review of the literature reveals, however, that prevalence rates did not decrease, leading to the conclusion that the calcineurin inhibitors themselves have adverse effects on glucose tolerance. Further evidence that CSA and tacrolimus have independent effects on glucose tolerance was provided by the observation that although steroid withdrawal protocols were associated with a lower incidence of diabetes [7, 8] the incidence was not zero. Unfortunately, steroid withdrawal protocols were associated with an increased risk for acute rejection, which required high dose steroid exposure, offsetting the mild improvement in glucose tolerance resulting from the absence of steroids. Thus, the problem of new-onset diabetes continues to haunt the post-transplant population, and leads to myriad complications as detailed below.

CYCLOSPORINE THERAPY

Pathophysiology

Studies of PTDM in patients treated with CSA report prevalence rates ranging from 11–20% [5, 9–12]. CSA is believed to affect glucose tolerance through several mechanisms. Alteration of peripheral insulin sensitivity [13, 14] and decreased β-cell function secondary to direct toxicity, including decreased synthesis and secretion of insulin [15, 16] have been noted in animal models. Studies in human kidney transplant recipients [17] suggest that the effect is minor when cyclosporine is used as monotherapy, but that it may be additive when used in conjunction with corticosteroids [9, 18].

High dose cyclosporine (15–20 mg/kg) causes decreased glucagon-stmulated insulin secretion in a dog model which is reversible upon discontinuation of the treatment [19]. In addition, rat models administered extremely high doses (40–50 mg/kg) of cyclosporine exhibit signs of pancreatic toxicity, including cytoplasmic vacuolization of islet cells, dilatation of the endoplasmic reticulum, and decreased mRNA content [20, 21], suggesting that CSA is an islet toxin, at least at supra-therapeutic doses.

CSA may affect insulin secretion, independent of synthesis. Pancreatic islet cells in culture, incubated with CSA at clinically relevant levels (100 ng/ml) demonstrate increased insulin content over control, however, upon exposure to glucose, insulin release is significantly diminished [22]. In models of isolated perfused pancreata, CSA exposure suppresses both the first and second phase insulin response to glucose challenge, and as impairs the response to arginine stimulation [23, 24]. The molecular mechanisms by which these effects occur are unknown at the present time.

Finally, CSA may affect the peripheral action of insulin. In a rat model, glucose area-under-the-curve rises compared with controls, within 2 weeks of initiation of CSA therapy, whereas insulin secretion rates and pancreatic insulin content do not fall until 4 weeks of therapy, two weeks later than the observed decrease in glucose tolerance [25].

Risk factors for the development of post-transplant diabetes (Table 1) are similar to factors which increase risk for Type 2 diabetes in the general population [11]. In a study of 337 renal allograft recipients whose grafts survived more than 1-year, increased risk was not associated with type of induction or immunosuppressant therapy, incidence of

Table 1. Reported risk factors for the development of post-transplant diabetes.

Older age
Obesity
Elevated trough levels of cyclosporine or tacrolimus
Family history of diabetes mellitus
Afro-Caribbean or Hispanic descent

Adapted from Markell [54].

rejection, total steroid dose or cyclosporine dose, percentage of body weight gain, serum creatinine concentration or patient sex [26]. Increased risk was associated with race [11], advanced age [9, 11, 26], being the recipient of a cadaveric kidney [26, 27], and presence of HLA-A 30 and Bw42 antigens [26]. Additional risk factors have been described as weight over 70 kg and elevated cyclosporine trough level [9].

More recently, a survey of over 2000 cyclosporine-treated patients, transplanted at a single center since 1983, revealed that the prevalence of diabetes has increased from 5.9% to 10.5% at 1 year, with the largest increase occurring after 1995. The authors note that the patient population has become heavier and older, and that the more bioavailable cyclosporine, Neoral was introduced around the time of the increase in prevalence rate [27].

Tacrolimus therapy

Tacrolimus, a more potent calcineurin inhibitor than CSA, was introduced in the mid-1990's. It was hoped that by allowing lower does of maintenance steroids, that diabetes incidence would decrease. Early multi-center trials, however, reported incidence rates for PTDM ranging from 18–37% [28–32], with the higher rates occurring in patients of Afro/Carribean-American or Hispanic descent [28], and of more concern were the reports of PTDM in pediatric age renal transplant, which had not been observed with CSA therapy [33].

The tacrolimus effect on glucose tolerance may be dose-related. In a study of 395 patients, of whom 18% initially required insulin, 40% had their insulin discontinued (although some required oral hypoglycemic agents) after tacrolimus and steroid doses were tapered [32]. Additionally, in a study of 15 pediatric age renal transplant patients, both abnormal first phase insulin response, and decreased glucose constant decay following intravenous glucose tolerance test were inversely correlated with tacrolimus trough level, and both findings improved following dose reduction of tacrolimus, although they did not normalize [34].

There are several postulated mechanisms by which tacrolimus could alter glucose tolerance, including direct effects on pancreatic beta-cells, and alteration of peripheral insulin sensitivity. Suggesting direct effects are the observations that rats develop vacuolization of the islets, decreased insulin secretion and hyperglycemia after exposure to pharmacologic doses of tacrolimus (10 mg/kg), abnormal findings that return to normal following discontinuation of the drug [35]. Also, exposure of cultured pancreatic beta-cells to tacrolimus results in alteration in the rate of insulin gene transcription [36, 37], possibly through inhibition of the serine-threonine phosphatase, calcineurin. However, a study of normal dogs following administration of tacrolimus at 1 mg/kg, demonstrated decreased glucose disposal but no decrease in insulin secretion initially, suggesting an effect on peripheral mechanisms of insulin action. In this study, after 4 weeks of treatment, insulin secretion did decrease and, of concern, did not return to baseline after discontinuation of the drug [38].

Although a study of patients who were being treated with tacrolimus for autoimmune diseases did not find an alteration of glucose tolerance, as assessed by hyperglycemic clamp technique [39], it is possible that sensitizing genetic factors exist are present in patients who develop kidney failure, or that the synergy of other agents (corticosteroids) may plays a significant role, or both. Supporting the former hypothesis is a study of 97 cardiac transplant recipients, which found that family history of diabetes and need for insulin after the first 24 hours were factors which predicted long-term PTDM [40] and an interesting case report that described a Japanese patient who had an HLA-type believed to confer increased risk for Type I diabetes in people of Japanese ancestry, who developed insulin-requiring diabetes, associated with anti-glutamic acid decarboxylase antibody following tacrolimus treatment for a kidney transplant [41]. The diabetes remitted after conversion to cyclosporine therapy and the authors hypothesize that exposure of beta-cell antigen due to damage from tacrolimus treatment, resulted in antibody generation and subsequent diabetes. In addition, there is a report of three pediatric age renal transplant recipients who developed PTDM while being treated with tacrolimus, who also experienced remission of diabetes upon conversion to cyclosporine [42]. Anti-islet and glutamic acid antibodies were not reported in the latter paper.

COMPLICATIONS

Patient Survival

Most studies of the effects on diabetes on post-transplant course do not differentiate between post-transplant diabetes and that which occurred prior to transplantation. All studies of long-term transplant survival document inferior survival in patients with diabetes [43, 44], predominantly due to cardiovascular death [43].

There are only 3 studies in which survival of patients with post-transplant diabetes is compared to controls (Table 2). The first, by Sumrani, compared 39 patients with PTDM to 298 cyclosporine-treated controls, reported 5 year actuarial survival rates

Table 2. Graft and patient survival in patients with PTDM vs. controls.

Study author	Graft survival (PTDM vs. control)	Patient survival (PTDM vs. control)
Sumrani et al., 1991 [11]	70% vs. 90% (P = NS)	87% vs. 93% (P = NS)
Vesco et al., 1996 [45]	67% vs. 93% (P = NS)	86% vs. 93% (P = NS)
Miles et al., 1998 [46]	48% vs 70% ($P < 0.05$)	73% vs 79% (P = NS)

See text for details.

which were similar for the two groups: 87% for patients with PTDM vs 93% for controls [11]. The second, a study by Vesco et al. of 33 patients and 33 'paired control' recipients, again suggested that 6-year actuarial survival was similar (PTDM 86% vs. controls 93%) [45]. The third, a prospective analysis of 78 cyclosporine-treated patients [46], was reported by Miles in 1998, and has the longest follow-up to date. In the latter study, presence of PTDM did not assign increased risk of patient death; 73% of patients with PTDM were alive at 9-year follow-up vs 79% of controls. It is difficult to interpret these data however, as, except for the study by Miles, the actual follow-up periods are not long. Also, patients with PTDM lose their allografts at a higher rate than controls (Table 2), thus sicker patients might be censored earlier because of allograft loss and return to dialysis.

Allograft Survival and Recurrent Disease

In all of the above quoted studies, graft survival was inferior in patients with PTDM (Table 2), but the results were not always statistically significant. In the first study, actuarial 5-year graft survival was 70% in patients with PTDM and 90% in controls (P = NS) [11]. In the second quoted study, 6-year actuarial graft survival was 67% vs. 93% in controls [45]. The third study demonstrated greatly decreased 12 year allograft survival, with survival rates of 48% for patients with PTDM vs. 70% in control patients ($P = 0.04$) [46]. By logistic regression, relative risk of graft loss for patients with PTDM was 3.72 times that of control.

The cause of graft loss in the above quoted papers varies, with chronic rejection most frequently reported. There was no difference in the rate of graft loss from chronic rejection in patients with PTDM and controls in the paper by Miles, and *de novo* diabetic nephropathy was reported as a cause of allograft loss in only 1 of 17 patients who lost their grafts [46]. The authors state, however, that changes consistent with diabetic nephropathy were observed in 2 of the 10 patients with PTDM who had allograft biopsies.

There are fewer than 15 cases of de novo diabetic nephropathy in patients with PTDM reported in the literature [47–49]. It is likely, however, that many cases of 'chronic rejection' are in fact diabetic nephropathy, as biopsies are not commonly performed in patients who present with proteinuria and a slow deterioration of renal function, clinical hallmarks of both diseases.

Diabetic Complications

It is known that pre-existent diabetes increases risk for both cardiovascular events [50, 51] and vascular complications [52] following renal transplantation. In a study of 427 patients prior to and following transplantation, presence of diabetes afforded a relative risk of 4.3 for the occurrence of a cardiovascular event following transplantation, higher than smoking (RR 2.50), LDL cholesterol > 180 (RR 2.3) and obesity (RR 2.6) [50].

Rao et al. compared the impact of vascular disease on post-transplant morbidity in a population of 283 non-diabetic and 99 (pre-existent) diabetic patients. Of note, diabetic patients had a greatly increased prevalence of vascular disease prior to transplantation

(33% vs. 13%). Of patients with pre-existent disease, the recurrence rate was higher in patients with diabetes (67% vs. 40%) as was the development of disease in patients without pre-existent vascular disease (33% vs. 13%). In addition, the amputation rate was significantly higher in diabetic patients (18% vs. 0.4%) [52].

An unanswered question in patients with PTDM is whether they had undiagnosed pre-transplant glucose intolerance, as has been suggested by some studies [53], which could predispose them to silent vascular disease development at an accelerated rate. At the present time, however, there are no systematic reviews of either pre-transplant glucose tolerance or cardiovascular complications in populations of patients with PTDM. The previously quoted paper by Miles et al., reported no difference in cardiovascular mortality between the patients with PTDM and controls, although 5 patients with PTDM suffered myocardial infarctions (13%) vs. 2 control patients (5%) [46].

There are no reports of *de novo* development of diabetic retinopathy in patients with PTDM, however, after reviewing the literature, it does not appear that this complication has been studied. The paper by Miles et al. Reports 2 cases of 'neuropathy' in the patients with PTDM, but does not specify the type, and whether the finding was believed to be secondary to diabetes [46].

PREVENTION STRATEGIES

The optimal therapeutic regimen for PTDM has not been established. Most centers do not report their treatment regimens, or only report patients treated with insulin. In a large study of 395 renal transplant recipients, although 18% of the patients were insulin-treated, the number of patients who received oral hypoglycemic agents was not reported [33]. In our own patient population, 55% of 30 cyclosporine-treated patients with PTDM could be managed with oral hypoglycemic agents as their initial choice of therapy, and 42% of 20 tacrolimus-treated patients (unpublished data).

We have recently published our approach to the patient with newly diagnosed post-transplant diabetes [54]. Patients are referred to an ophthalmologist and podiatrist, and close follow-up through glycosylated hemoglobin levels is attempted in order to achieve optimal blood glucose control. The Diabetes Complications and Control Trial (DCCT) suggests that optimizing blood glucose control may delay or prevent the development of diabetic complications in patients with non-transplant related diabetes, and in all likelihood, patients with PTDM as well [55].

FUTURE DIRECTIONS AND RESEARCH AREAS

Until such time as we develop methods for specifically targeting which patients are sensitive to the diabetogenic effects of the immunosuppressants, such that we can tailor immunsuppression accordingly, the prudent course is to maintain each patient on the lowest possible dose of medication. Patients with known risk factors (Table 1) should be intensively monitored for the development of weight gain and glucose intolerance, especially during the first post-transplant year. Intensive teaching of self-blood glucose monitoring must be commenced immediately after PTDM is detected, in the hopes of preventing diabetic complications.

Careful population-based studies are needed, in order to elucidate the risks of cardiovascular disease, retinopathy and neuropathy in the patient with PTDM, and follow-up studies of patients once they return to dialysis are imperative. At the present time, it is not known to what extent patients with 'post-transplant' diabetes will remain glucose intolerant after withdrawal of immunosuppression. It is likely that a subset of the population exists, who have Type 2 diabetes, the course of which is accelerated by exposure to the pancreato-toxic calcineurin inhibitors, and who will remain diabetic upon return to dialysis. It is also possible, that the period of glucose intolerance experienced while transplanted will increase cardiac risk even after withdrawal of medications. These important issues remain to be evaluated.

REFERENCES

1. Siegel RR, Luke RG, Hellebusch AA. Reduction of toxicity of corticosteroid therapy after renal transplantation. Am J Med 1972; 53: 159–169.
2. Gunnarsson R, Lundgren G, Magnusson G, Ost L, Groth CG. Steroid diabetes-a sign of overtreatment with steroids in the renal allograft recipient? Scand J Urol Nephrol 1980; 54: 135–138.
3. Hricik DE, Almawi W, Strom TB. Trends in the use of glucocorticoids in renal transplantation. Transplantation 1994; 57: 979–989.
4. Weir MR, Fink JC. Risk for post-transplant diabetes mellitus with current immunosuppressive medications. Am J Kid Dis 1999; 34(1): 1–13.
5. Jindal RM. Post-transplant diabetes mellitus – a review. Transplantation 1994; 58: 1289–1298.
6. Yoshimura N, Nakai I, Ohmori Y, Aikawa I, Fukuda M, Yasumura T, Matsui S, Hamashima T, Oka T. Effect of cyclosporine on the endocrine and exocrine pancreas in kidney transplant recipients. Am J Kid Dis 1988; 12: 11–17.
7. Hricik DE, O'Toole M, Shulak JA, Herson J. Steroid-free, cyclosporine-based immunosuppression after renal transplantation: A meta-analysis of controlled trials. J Am Soc Nephrol 1993; 4: 1300–1305.
8. Hricik DE, Bartucci MR, Moir EJ, Mayes JT, Schulak JA. Effects of steroid withdrawal on post-transplant diabetes mellitus in cyclosporine-treated renal transplant recipients. Transplantation 1991; 51: 374–377.
9. Boudreux JP, McHugh L, Canafax DM, Ascher N, Sutherland DE, Payne W, Simmons RL, Najarian JS, Fryd DS. The impact of cyclosporine and combination immunosuppression on the incidence of post-transplant diabetes in renal allograft recipients. Transplantation 1987; 44(3): 376–381.
10. Rao M, Jacob CK, Shastry JCM. Post-renal transplant diabetes mellitus – a retrospective study. Nephrol Dialysis Transplant 7:1039-1-42, 1992
11. Sumrani NB, Delaney V, Ding Z, Davis R, Daskalakis P, Friedman EA, Butt KH. Diabetes mellitus after renal transplantation in the cyclosporine era-an analysis of risk factors. Transplantation 1991; 51: 343–347.
12. Yoshimura N, Nakai I, Ohmori Y, Aikawa I, Fukuda M, Yasumura T, Matsui S, Hamashima T, Oka T. Effect of cyclosporine on the endocrine and exocrine pancreas in kidney transplant recipients. Am J Kidney Dis 1988; 12: 11–17.
13. Dresner LS, Anderson DK, Khang KU, Munshi IA, Wait RB. Effects of cyclosporine on glucose metabolism. Surgery 1989; 106(2): 163–169.
14. Ost L, Tyden G, Fehrman I. Impaired glucose tolerance in cyclosporine treated renal graft recipients. Transplantation 1988; 46(3): 370–372.
15. Fehmann HC, Haverich R, Stockmann F, Creutzfeldt W. Cyclosporine A in low doses induces functional and morphologic changes in rat pancreatic B cells. Trans Proc 1987; 19(5): 4015–4016.
16. Garvin PJ, Niehoff M, Staggenborg J. Cyclosporine's effect on canine pancreatic endocrine function. Transplantation 1988; 45(6): 1027–1031.
17. Esmatjes E, Ricart MJ, Ferrer JP, Oppenhaimer F, Vilardell J, Casamitjana R. Cyclosporine's effect on insulin secretion in patients with kidney transplants. Transplantation 1991; 52(3): 500–503.
18. Roth D, Milgrom M, Esquenazi V, Fuller L, Burke G, Miller J. Posttransplant hyperglycemia. Increased incidence in cyclosporine-treated renal allograft recipients. Transplantation 1989; 47(2): 278–281.

19. Wahlstrom HE, Akimoto R, Endres D, Kolterman O, Moosa AR. Recovery and hypersecretion of insulin and reversal of insulin resistance on withdrawal of short-term cyclosporine treatment. Transplantation 1992; 53: 1190–1195.
20. Eun HM, Pak CY, Kim CJ, McArthur RG, Yoon JW. Role of cyclosporine A in macro-molecular synthesis of beta-cells. Diabetes 1987; 36: 952–958.
21. Andersson A, Borg M, Mallberg A, Mellerstrom C, Sandler S, Schnell A. Long-term effects of cyclosporine A on cultured mouse pancreatic islets. Diabetologia 1984; 17: 66–69.
22. Neilson JH, Mandrup-Poulson T, Nerup J. Direct effects of cyclosporine A on human pancreatic beta-cells. Diabetes 1986; 35: 1049–1052.
23. Gillison SG, Bartlett ST, Curry DL. Synthesis-secretion coupling of insulin: effects of cyclosporine. Diabetes 1989; 38: 465–470.
24. Gillison SG, Bartlett ST, Curry DL. Inhibition by cyclosporine of insulin secretion – a beta cell specific alteration of islet tissue function. Transplantation 1991; 52: 890–895.
25. Menegazzo LA, Ursich MJ, Fukui RT, Rocha DM, Silva ME, Ianhex LE, Sabbaga E, Wajchenberg BL. Mechansim of the diabetogenic action of cyclosporine A. Horm Metab Res 1998; 30(11): 663–667.
26. Mejia G, Arbelaez M, Henao JE, et al. Cyclosporine-induced diabetes mellitus in renal transplants. Clin Transplant 1989, 3(5): 260–265.
27. Cosio FG, Pesavento TE, Osei K, Henry ML, Ferguson RM. Post-transplant Diabetes Mellitus: increasing incidence in renal transplant recipients transplanted in recent years. Kidney Int 2001; 59(2): 732–737.
28. Roth D, Milgrom M, Esquenazi V, Fuller L, Burke G, Miller J. Posttransplant hyperglycemia. Increased incidence in cyclosporine-treated renal allograft recipients. Transplantation 1989; 47(2): 278–281.
29. Pirsch JD, Miller J, Deierhoi MH, Vincenti F, Filo RS. A comparison of tacrolimus (FK506) and cyclosporine for immunosuppression after cadaveric renal transplantation. Transplantation 1997; 63: 977–983.
30. Vincenti F, Laskow DA, Neylan JF, Mendez R, Matas AJ. One-year follow-up of an open label trial of FK506 for primary kidney transplantation: A report of the US Multicenter FK506 Kidney Transplant Group. Transplantation 1996; 61: 1576–1581 .
31. Todo S, Fung JJ, Starzl TE, Tzakis A, Demetris AJ, Kormos R, Jain A, Alessani M, Takaya S, Shapiro R. Liver, kidney, and thoracic organ transplantation under FK506. Ann Surg 1990; 212: 295–307.
32. Scantlebury V, Shapiro R, Fung J, Tzakis A, McCauley J, Jordan M, Jensen C, Hakala T, Simmons R, Starzl TE. New onset of diabetes mellitus in FK506 vs. cyclosporine-treated kidney transplant recipients. Transplant Proc 1991; 23: 3169–3170.
33. Shapiro R, Jordan M, Scatlebury V, Fung J, Jensen C, Tzakis A, McCauley J, Carroll P, Ricordi C, Demetris AJ, Mitchell S, Jain A, Iwaki Y, Kobayashi M, Reyes J, Todo S, Hakala TR, Simmons RL, Starzl TE. FK506 in clinical kidney transplantation. Trans Proc 1995; 27: 814–817.
34. Filler G, Neuschulz I, Vollmer I, Amendt P, Hocher B. Tacrolimus reversibly reduces insulin secretion in pediatric renal transplant recipients. Nephrol Dialysis Transplant 2000; 15(6): 867–871.
35. Hirano Y, Fujihira S, Ohara K, Katsui S, Noguchi H. Morphological and functional changes of islets of Langerhans in FK506-treated rats. Transplantation 1992; 53: 889–894.
36. Tamura K, Fujimura T, Tsutsumi T, Nakamura K, Ogawa T, Atumaru C, Hirano Y, Ohara K, Ohtsuka K, Shimomura K, Kobayashi M. Transcriptional inhibition of insulin by FK506 and possible involvement of FK-binding protein 12 in pancreatic beta-cell. Transplantation 1995; 59: 1606–1613.
37. Redmon JB, Olson LK, Armstrong MB, Greene, MJ, Robertson RP. Effects of Tacrolimus (FK506) on human insulin gene expression, insulin mRNA levels, and insulin secretion in HIT-T15 cells. J Clin Invest 1996; 98: 2786–2793.
38. Strasser S, Alejandro R, Shapiro ET, Ricordi C, Todo S, Mintz DH. Effect of Fk506 on insulin secretion in normal dogs. Metabolism 1992; 41(1): 64–67.
39. Strumph P, Kirsch D, Gooding W, Carroll P. The effect of FK506 on glycemic response as assessed by hyperglycemic clamp technique. Transplantation 1995; 60(2): 147–151.
40. Depczynski B, Daly B, Campbell LV, Chisholm DJ, Keogh A. Predicting the occurrance of diabetes mellitus in heart transplant recipients. Diabet Med 2000; 17(1): 15–19.
41. Yoshioka K, Sato T, Okada N, Ishii T, Imanishi M, Tanaka S, Kim T, Sugimoto T, Fujii S. Post-transplant diabetes with anti-glutamic acid decarboxylase antibody during tacrolimus therapy. Diabetes Res Clin Pract 1998; 42(2): 85–89.
42. Butani L, Makker SP. Conversion from tacrolimus to neural for postrenal transplant diabetes. Pediatr Nephrol 2000; 15(3–4): 176–178.

43. Ojo AO, Hanseon JA, Wolfe RA, Leichtman AB, Agodoa LY, Port FK. Long-term survival in renal transplant recipients with graft function. Kidney Int 2000; 57(1): 307–313.
44. Cecka M. Outcome statistics of renal transplants with an emphasis on long-term survival. Clin Transplant 1994; 8 (3 part 2): 324–327.
45. Vesco L, Busson M, Bedrossian J, Bitker MO, Hiesse C, Lang P. Diabetes mellitus after renal transplantation: characteristics, outome, and risk factors. Transplantation 1996; 61(10): 1475–1478.
46. Miles AMV, Sumrani N, Horowitz R, Homel P, Maursky V, Markell MS, Distant DA, Hong JH, Sommer BG, Friedman EA. Diabates Mellitus after renal transplantation: as deleterious as non-transplant associated diabetes? Transplantation 1998; 65(3): 380–384.
47. Giminez LF, Watson AJ, Burrow CR, Olson JL, Klassen DK, Cooke CR. De novo diabetic nephropathy with functional impairment in a renal allograft. Am J Nephrol 1986; 6: 378–381.
48. Schwarz A, Krause PH, Offermann G, Keller F. Recurrent and de novo renal disease after kidney transplantation with and without cyclosporine A. Am J Kidney Dis 1991; 17: 524–531.
49. Sharma UK, Jha V, Gupta KL, Joshi K, Sakhuja V. Am J Kidney Dis 1994; 23(4): 597–599.
50. Aker S, Ivens K, Grabensee B, Heering P. Cardiovascular risk factors and diseases after renal transplantation. Int Urol Nephrol 1998; 30(6): 777–778.
51. Lindholm A, Albrechtsen D, Frodin L, Tufveson G, Persson NH, Lundgren G. Ischemic heart disease – major cause of death and graft loss after renal transplantation in Scandinavia. Transplantation 1995; 60: 451–457.
52. Rao KV, Andersen RC. The impact of diabetes on vascular complications following cadaveric renal transplantation. Transplantation 1987; 43(2): 193–197.
53. Tanabe K, Koga S, Takahashi K, Sonda K, Tokumoto T, Babazono T, Yagisawa T, Toma H, Kawai T, Fuchinoue S, Teraoka S, Ota K. Diabetes mellitus after renal transplantation under FK506 (tacrolimus) as primary immunosuppression. Trans Proc 1996; 28(3): 1304–1205.
54. Markell MS. Post-transplant diabetes: Incidence, relationship to choice of immunosuppressive drugs and treatment protocol. Adv Renal Replacement Ther 2001; 8(1): 64–69.
55. Report of the expert committee on the diagnosis and classification of Diabetes Mellitus. Diabetes Care 2000; 23 (supp 1): S4–S19.

AMY L. FRIEDMAN AND ELI A. FRIEDMAN

16. Prevalent US Transplant Center Policies Towards Pancreas Transplantation for Patients with Type 2 Diabetes Mellitus

Editor's Comment: If type 2 diabetes is a disorder of insulin resistance in which circulating insulin levels are high, why should a pancreas transplant be of any benefit? That is the dilemma addressed by this essay. Following the initial reports of successful pancreas transplants in type 2 diabetic recipients, the question of just what the approach of US transplant centers might be to extending this procedure to the overall population of type 2 diabetic patients. A key finding of the authors' survey is that at present, so termed 'mainstream centers' practice highly restrictive candidate selection for pancreas transplantation with significantly more liberal policies applicable for those seeking renal transplantation alone. Type 2 diabetic patients are methodically identified during pre-transplant evaluation. but in practice, rarely are selected as pancreas recipients. Until some relief is in hand for the stress of pancreas shortage in the face of a growing waiting list, it is unlikely that the benefits of this procedure in type 2 diabetic recipients will be widely assessed.

Pancreas transplantation is firmly established as an excellent means of achieving euglycemia for selected individuals with type 1 diabetes mellitus [1]. In those patients restoration of autonomous glucose regulation through implantation of a vascularized pancreas abrogates a pre-existing insulinopenia ascribed to auto-immune destruction of native beta-cells. Data on application of this approach to individuals with type 2 diabetes mellitus whose hyperglycemia is caused by relative, not absolute, insulin deficiency is scarce but encouraging. Although 129 type 2 recipients are reported to the International Transplant Registry, they represent only 4% of all registered pancreas transplant recipients [2], while it is estimated that approximately 90% of all American diabetics [3], and 78% of diabetics with endstage renal disease are type 2 [4]. Both graft and patient survival (the standard indicators of successful transplantation) among this small cohort of type 2 pancreas transplant recipients match the registry results of pancreatic allografts in type 1 diabetics [2]. To ascertain whether this limited initial success has altered prior reluctance to proffer pancreas transplantation for type 2 diabetes we conducted a mail survey of American pancreas transplant programs.

METHODS

US centers with active pancreas and/or islet transplant programs on September 15, 2000 were identified through the United Network for Organ Sharing (UNOS). Surveys con-

E.A. Friedman and F.A. L'Esperance, Jr. (eds.), Diabetic Renal-Retinal Syndrome, 183–190.

taining 58 queries were mailed to the program directors of 145 programs with a response requested by FAX or mail within 2 weeks.

RESULTS

A total of 44 pancreas transplant programs (30.3%) returned survey responses. The cumulative experience reported by these programs was 6,014 transplants, while the largest cumulative experience in pancreas transplantation is reported from the UNOS Registry and includes 12,939 cases from international sources (approximately 9,000 of these are reported by US programs). Collectively these centers performed 872 pancreas transplants in 1999.

Classification of Diabetes

The large majority (86%, 38/44) of responding programs routinely attempt to establish a pancreas transplant candidate's type of diabetes. Table 1 demonstrates the method(s) utilized by these centers for classification. Only 13.6% (6/44) considered the classification made by the referring physician adequate for the purpose of transplantation. Most (88.6%, 39/44) incorporated an independent review of the candidate's medical history, including age of disease onset, requirement for insulin and body habitus in their assessment. C-peptide values were usually (88.6%, 39/44) included in the classification. The presence (or absence) of anti-islet antibodies was rarely needed for classification at these programs.

Among the recipients of pancreas transplants in 1999, diabetes mellitus was classified as type 2 in 16/872 (1.8%). Selection of those recipients with type 2 diabetes mellitus was intentional in 14/16 (87.5%) and inadvertent in 2/16 (12.5%).

Fewer programs (28/44, 63.6%) routinely classify the type of diabetes among candidates for renal transplantation. Those centers utilize the medical history (11/44, 25%), C-peptide values (21/44, 47.7%), or presence of anti-islet antibodies (1/44, 2.3%).

Candidate Age

Program directors were queried regarding their consideration of age as a criterion for kidney or pancreas transplantation in a series of 12 questions. Advanced age was viewed as a relative contraindication to renal transplantation in 63.6% (28/44). Only 3/44 (6.8%)

Table 1. Methods of classification of diabetes type employed by US pancreas transplant programs.

C-Peptide	88.6%
Medical history	100%
HLA type	2.3%
Islet antibodies	6.8%
Endocrine Stimulation Test	6.8%

responded that advanced age was an absolute contraindication. When asked to cite the specific age cut-off used for patient exclusion only 21 centers responded; for them the definitive age was 95 years (1/21), 90 years (1/21), 75 years (8/21), 70 years (7/21), 65 years (3/21) or 60 years (1/21). In contrast, 14/44 (31.8%) of centers utilized advanced age as an absolute reason to decline a candidate for pancreas transplantation while 35/44 (79.5%) viewed it as a relative contraindication. The specific age cut-off for pancreas candidate exclusion was cited by 33 centers; with an upper age of 60 years (8/33), 55 years (17/33), 50 years (3/33), 45 years (4/33), or 40 years (1/33). Figure 1 demonstrates the prevalent relative rates of declination of hypothetical kidney and pancreas candidates based solely on age among these responding centers asked the questions 'would an age of X years preclude pancreas (kidney) transplantation for an otherwise suitable patient?' At all ages there is a consistently lower rate of acceptance for pancreas versus renal candidates indicating a more restrictive approach to the older patient desiring pancreatic transplantation than kidney transplantation.

Candidate Obesity

The impact of a candidate's obesity on a center's willingness to proceed with transplantation was explored through a series of 16 questions. Obesity is an absolute contraindication to pancreas transplantation at 8/44 (18.2%) and a relative contraindication at 39/44 (88.6%) programs. The degree of obesity is routinely assessed through subjective physical examination at 23/44 (52.2%), body mass index (BMI) at 22/44 (50%), or weight alone at 10/44 (22.7%) of centers. When asked to identify the upper limit of obesity considered acceptable for a pancreas transplant candidate, 28 responses were made. Among these 28 centers the specific limit cited was BMI > 35 kg/m^2 (n = 4), BMI > 31–34 kg/m^2 (n = 4), BMI > 28–30 kg/m^2 (n = 9), BMI > 28 kg/m^2 for females

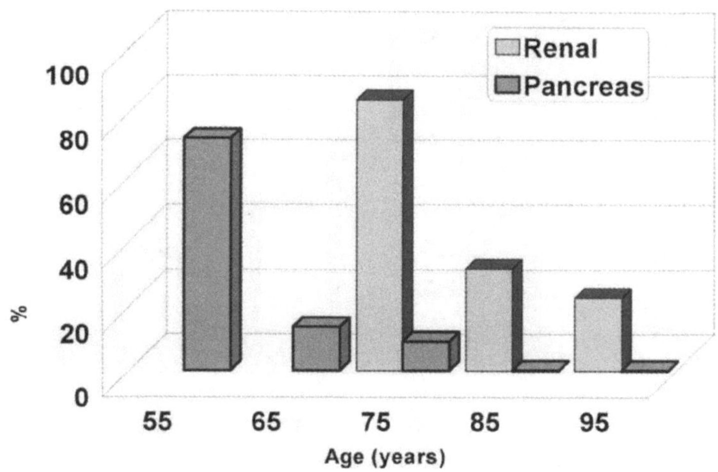

Figure 1. Prevalent age restrictions for US kidney and pancreas transplant candidates.

but BMI > 30 kg/m^2 for males (n = 1), weight of 250 lbs (n = 1), weight of 220 lbs (n = 2), 50 lbs above ideal body weight (n = 1), 75% above ideal body weight (n = 1), 50% above ideal body weight (n = 2), 25% above ideal body weight (n = 2) or moderate obesity (n = 1).

At 9/44 (22.7%) centers obesity is considered an absolute contraindication to renal transplantation while at 32/44 (72.7%) it is viewed as a relative contraindication. The upper limit of obesity used to exclude renal transplantation was specifically cited by 25/44 (56.8%) programs as BMI > 40 kg/m^2 (n = 3), BMI > 35 kg/m^2 (n = 10), BMI > 30–34 kg/m^2 (n = 4), weight > 300 lbs (n = 1), weight > 220 lbs (n = 1), 100 lbs above ideal body weight (n = 1), 75% above ideal body weight (n = 1), or 25–30% above ideal body weight (n = 4).

To clarify the differential impact of candidate obesity on exclusion from transplantation responses to specific patient descriptions were requested through the questions 'would obesity preclude PANCREAS (KIDNEY) transplantation for an otherwise suitable patient with a height of 5 feet 5 inches and a weight of X pounds?' Figure 2 demonstrates that except for the leanest hypothetical patient with a BMI of 31 kg/m^2 who would be considered acceptable by all programs for either pancreas or kidney transplantation, there is a consistently lower rate of acceptance for candidates with the same body habitus for pancreas versus kidney transplantation.

DISCUSSION

Transplantation of the pancreas is fundamentally different from the replacement of other organs. (Table 2). Implanted into the diabetic patient in a 'life enriching' procedure,

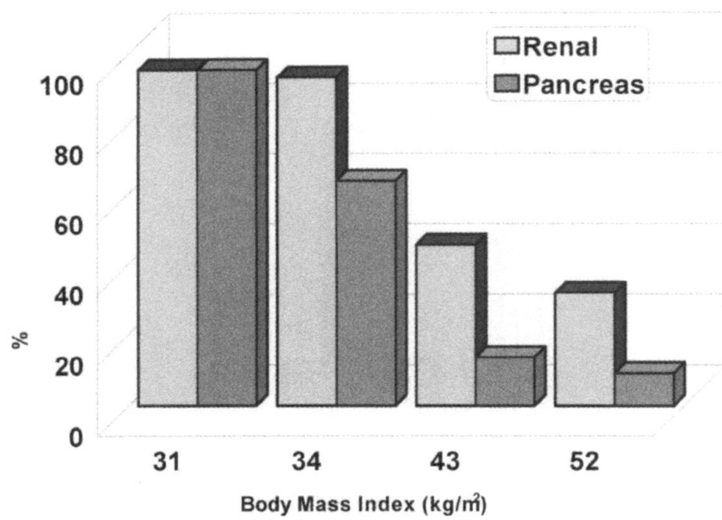

Figure 2. Prevalent obesity restrictions for US kidney and pancreas transplant candidates.

Table 2. Comprehensive comparison of kidney and pancreas transplantation.

	Kidney	Pancreas
Technical difficulty	5/10	7/10
Candidate selection	Most suitable	Restricted
Organ selection	Most suitable	Very restricted
Immunosuppression	Moderate	Heavy

the pancreatic allograft does not replace an indispensable failed organ. Since hyperglycemia is manageable with the use of insulin and/or oral hypoglycemic agents and the control of dietary and exercise behavior, this procedure cannot be considered a life saving procedure over either the immediate or the short term. With insulin readily available on the shelf and the substantial morbidity and mortality still associated with pancreas transplantation [5, 6], it behooves the transplant team to be highly selective in applying this procedure.

Consistent with this conservative approach is evidence that the resource upon which transplantation depends, pancreatic grafts from cadaver donors, is indeed used in a very discriminating way. Although strong advocates suggest the use of less than ideal donors still results in successful transplantation [7], in practice, the majority of available organs are discarded, having been rejected for clinical use. In the year 1999 1,287 pancreas transplants were reported to the United Network for Organ Sharing, while 5,851 cadaver donors were available [8]. In comparison, 8,007 cadaveric kidneys were transplanted from the same donor pool. Few hard data describing the specific donor attributes that either encourage or dissuade pancreatic procurement are available. Yet, it is certain that the ill defined group of 'marginal' quality organs are generally turned down. With few prime organs available transplant centers are compelled to offer pancreas transplantation to only those candidates for whom the very tangible risks are indeed justified by intolerable life quality. The patient's medical condition must not pose undue jeopardy and patients must be expected to tolerate the procedure and its potential complications as well as the generally higher level of immunosuppression used in pancreatic compared to kidney transplantation.

Age of the patient is an obvious consideration in making all significant decisions about medical care. Tissue strength and the ability to heal must be expected to decline [9] and the incidence and severity of comorbid conditions increase with aging [10]. The inexorable physical progress of growing older makes an undertaking as significant as organ transplantation increasingly more challenging in elderly patients. In the earliest days of the era of organ replacement technology neither dialysis nor transplantation were judged feasible for individuals older than 30 to 40 years. Today, physiologic rather than chronologic age is usually assessed and candidates older than 60 reported to prosper with renal [11], liver [12] and cardiac transplants [13].

With the median age of incident ESRD patients now 63 years [4] it is not surprising that transplant centers are asked to evaluate increasingly older candidates. Respondents to our survey indicated clearly that advanced age is usually considered a relative but rarely an absolute contraindication to transplantation. Indeed, there is generalized willingness to offer renal transplantation to surprisingly elderly patients if they are otherwise good candidates. Most (39/44, 88.6%) would accept candidates as old as 75 years and 10/44 (22.7%) even profess to accept 95 year olds.

Consistent with the conservative theoretical approach to pancreas transplantation is the finding that advanced age is an absolute contraindication at 31.8% of programs. A 55 year old diabetic candidate searching for a pancreas transplant center would be accepted at only 32/44 (72.7%) of centers while a 65 year old would have great difficulty, finding that only 13.6% (6/44) of centers would agree to placement on the waiting list.

Obesity generally increases the risks of anesthesia [14] and poor surgical wound healing [15]. Although rarely discussed frankly, the presence of significant quantities of adipose tissue obviously complicates the technical conduct of surgery. Identification and manipulation of relevant anatomic structures is more difficult both because of increased wound depth and the additional layers of tissue that tend to obscure them. Operative and anesthetic times are likely to be prolonged as a consequence. Limited data suggests that renal transplantation in the moderately obese is also associated with a high incidence of wound infections but has not been definitively linked to diminished survival of either the patient or the allograft [16, 17]. No data is currently available regarding the outcome of transplantation in morbidly obese populations (BMI > 35 kg/m^2).

More centers consider obesity a strong relative contraindication to pancreas than for kidney transplantation, consistent with the generally greater conservatism for this procedure. Class I obesity (BMI 3–35 kg/m^2) would not prevent the majority of centers from accepting candidates for either renal or pancreas transplantation although there is again a more restrictive approach to the latter. Class III obesity (BMI > 40 kg/m^2) is a strong deterrent to pancreas transplantation at all but a small minority of programs.

Type 1 and type 2 diabetes mellitus are etiologically distinct, if still incompletely clarified. Clinical patient classification remains challenging. It has previously been generally assumed that pancreas transplantation was performed only for treatment of type 1 diabetes. The only prior published reference to this approach for type 2 diabetes comes from the International Pancreas Transplant Registry and includes 129 patients [2]. Respondents to this survey uniformly attempt to classify the pancreas candidate's type of diabetes, usually utilizing the specific medical history combined with c-peptide levels. Categorization using HLA type and specific islet antibodies was rare. Our survey confirms that transplantation of the pancreas for type 2 diabetes is performed in a highly limited group of patients, usually by specific intent. Advanced age and obesity, two features generally characteristic of type 2 versus type 1 disease are each independent deterrents to pancreas transplantation as noted above. Though patients with type 2 disease represent 78% of incident ESRD diabetics in the United States, they are older (median age 65 versus 54 years), less likely to undergo even renal transplantation (1.5% versus 9.3%) and more likely to die (22.5% versus 18.0%) within the first year of renal replacement therapy than the cohort with type 1 disease [4]. The rare selection of type 2 diabetics for pancreas transplantation is consistent with the cautious overall approach to patient selection and likely accounts for the excellent outcomes reported thus far.

Respondents to this survey represent centers that performed the majority of American pancreas transplants in the year 1999, with 872 transplants reported to us, equaling 70.5% of the 1,237 cadaveric transplants reported to UNOS during the same time period. These mainstream centers report highly restrictive candidate selection for pancreas transplantation with significantly more liberal policies applicable for those seeking renal transplantation alone. Type 2 diabetic patients are methodically identified during pre-transplant evaluation. While they may be intentionally accepted as pancreas transplant candidates in practice they rarely undergo the procedure. Without dramatic changes in the availability of high quality organs the rate of pancreas transplantation for those individuals with type 2 diabetes seems unlikely to change substantially despite the excellent results reported by the UNOS registry.

REFERENCES

1. Friedman AL, Friedman EA, Sommer BG. Appropriateness and timing of kidney and/or pancreas transplants in type 1 and type 2 diabetes. Advances in Renal Replacement Therapy 2001.
2. Gruessner AC, Sutherland DER. Analyses of pancreas transpland outcomes for United States cases reported to the United Network for Organ Sharing (UNOS) and non-US cases reported to the international pancreas transplant registry (IPTR). In: Cecka JM, Terasaki PI (eds), Clinical Transplants 1999. UCLA Immunogenetics Center, Los Angeles, 2000.
3. American Diabetes Association. Physician's guide to insulin-dependent (type 1) diabetes: diagnosis and treatment, Second Edition, 1988.
4. United States Renal Data System, USRDS 2000 annual Data Report, The National Institutes of Health, National Institute of Diabetes and Digestive and Kidney Diseases. Bethesda, MD, August 2000.
5. Humar A, Sutherland DE, Ramcharan T, Gruessner RW, Gruessner AC, Kandaswamy R. Optimal timing for a pancreas transplant after a successful kidney transplant. Transplantation 2000; 70(8): 1247–1250.
6. Odorico JS, Leverson GE, Becker YT, Pirsch JD, Knechtle SJ, D'Alessandro AM, Sollinger HW. Pancreas transplantation at the University of Wisconsin. Clinical Transplants 1999: 199–210.
7. Odorico JS, Heisey DM, Voss BJ, Steiner DS, Knechtle SJ, D'Alessandro AM, Hoffmann RM, Sollinger HW. Donor factors affecting outcome afterpancreas transplantation. Transplantation Proceedings 1998; 30(2): 276–277.
8. 2000 Annual report of the US scientific registry for transplant recipients and the organ procurement and transplantation network-transplant data: 1988–1999. UNOS, Richmond, VA, and the Division of Transplantation, Office of Special Programs, Health Resources and Services Administration, US Department of Health and Human Services, Rockville, MD.
9. Hurley BF, Roth SM. Strength training in the elderly: effects on risk factors for age-related diseases. Sports Medicine 2000; 30(4): 249–268.
10. Ishida K, Sato Y, Katayama K, Miyamura M. Initial ventilatory and circulatory responses to dynamic exercise are slowed in the elderly. Journal of Applied Physiology: Respiratory, Environmental & Exercise Physiology 2000; 89(5): 1771–1777.
11. Doyle SE, Matas AJ, Gillingham K, Rosenberg ME. Predicting clinical outcome in the elderly renal transplant recipient. Kidney International 2000; 57(5): 2144–2150.
12. Collins BH, Pirsch JD, Becker YT, Hanaway MJ, Van der Werf WJ, D'Alessandro AM, Knechtle SJ, Odorico JS, Leverson G, Musat A, Armbrust M, Becker BN, Sollinger HW, Kalayoglu M. Long-term results of liver transplantation in older patients 60 years of age and older. Transplantation 2000; 70(5): 780–783.
13. Pederzolli C, Martinelli L, Grande AM, Goggi C, Minizioni G, Castiglione N, Gavazzi A, Campana C, Vigano M. Heart transplant over 55 years [Italian]. Giornale Italiano di Cardiologia 1997; 27(3): 263–269.
14. Adams JP, Murphy PG. Obesity in anaesthesia and intensive care. British Journal of Anaesthesia 2000; 85(1): 91–108.
15. Lamm P, Godje OL, Lange T, Reichart B. Reduction of wound healing problems after median sternotomy by use of retention. Annals of Thoracic Surgery 1998; 66(6): 2125–2126.
16. Drafts HH, Anjum MR, Wynn JJ, Mulloy LL, Bowley JN, Humphries AL. The impact of pre-transplant obesity on renal transplant outcomes. Clinical Transplantation 1997; 11 (5 Pt 2): 493–496.
17. Modlin CS, Flechner SM, Goormastic M, Goldfarb DA, Papajcik D, Mastroianni B, Novick AC. Should obese patients lose weight before receiving a kidney transplant? Transplantation 1997; 64(4): 599–604.

JOHN S. NAJARIAN

17. Pancreas Transplantation: Does it Have a Future?

Editor's Comment: Najarian individually championed introduction in 1966 and subsequent expansion of pancreas transplantation thoughout his tenure as Chief of Surgery at the University of Minnesota. Now, looking back on the multiple cohorts of recipients given a pancreas alone or a pancreas plus kidney, Najarian observes that though more expensive than exogenous insulin treatment, pancreas transplantation is so effective that a new American Diabetes Association position statement proposing that a pancreas transplant should be routine in type 1 diabetic kidney transplants recipients and is appropriate for nonuremic though labile diabetic individuals. Beyond its value as a means of forestalling diabetic complications, Najarian advocates pancreas transplants to preempt secondary complications and/or for adult patients who would rather manage immunosuppression and its risks than diabetes and its risks. Thus, until a simpler and more reliable means of 'curing' type 1 diabetes is in hand, utilization of whole organ pancreas transplants will likely continue for the indefinite future.

Clinical pancreas transplantation, began at Minnesota in 1966, now encompasses a third of a century. Transplantation of endocrine tissue (pancreatic islet beta cells) is the only treatment that can induce insulin independence for Type I diabetic patients. To date, transplantation of islets within an immediately vascularized graft (pancreas) has been much more successful than as a free cellular graft.

The promise that islet transplantation can replace pancreas transplantation has been propagated over nearly 3 decades. Recent successes with islet transplantation suggest that the dream of eliminating the major surgery of pancreas transplantation may soon be achieved. Meanwhile, pancreas transplantation continues to be done, and the lessons learned over many years at our own and other centers have contributed to the high success rate now achieved in both uremic and non-uremic diabetic recipients. Indeed, there is good evidence that a pancreas transplant prolongs survival of both nephropathic and neuropathic diabetic patients.

Our program differs from most in that we emphasized solitary pancreas transplants from the beginning and continued to do so. A few other programs have increased the proportion of solitary pancreas transplants in recent years, with results equivalent to our own.

Although there are very few aspects of our program that are original, much of what we have done has been adapted from the pioneering efforts of others. We were the first group to use an LD for a pancreas transplant, in 1979, with extension to identical twin donors in 1980. The LD option has been exercised only by a few other centers, but includes the use of an identical twin.

Our practice of splitting a CAD pancreas to give a segment to each of 2 recipients appears not to have been duplicated elsewhere. However, our use of LD kidney

E.A. Friedman and F.A. L'Esperance, Jr. (eds.), Diabetic Renal-Retinal Syndrome, 191–195.

transplants simultaneous with a CAD pancreas has been followed by many cases elsewhere. The introduction of an immediate retransplant for a primary technically failed pancreas graft (e.g. thrombosis) has been adopted by other groups. Even our use of enteric drained (ED) pancreas grafts to correct exocrine deficiency has been duplicated.

Regarding surgical techniques, segmental pancreas transplantation has largely disappeared except with living donors (LD), as has the use of duct management techniques other than bladder drained (BD) or ED. In the past decade, many groups have compared outcome with ED vs. BD and have concluded that results with the 2 techniques are equivalent, at least for SPK transplants. Although portal drainage of pancreas graft venous effluent was done in a few cases at several centers in the 1980s including our own, Rosenlof et al. reported its routine use for simultaneous pancreas-kidney (SPK) transplants in 1992, stimulating others to adopt the technique as well. Portal drainage has to be more physiologic than systemic drainage and the metabolic perturbations of systemic drainage includes pseudohyperinsulinemia, but the relevance is unknown.

Surgical complications of pancreas transplantation decreased at our center as we went from era to era. This decrease was paralleled at other centers. Chronic complications of BD, however, have persisted and our rate of conversion to ED was < 10% throughout the 1990s.

The most immediate and frequent posttransplant complication is pancreas graft thrombosis. Some groups have not found that heparinization reduces the risk, while in our experience it seems to have helped.

Infections after pancreas transplantation can be local or systemic. Our incidence of local infections has been reduced, but can necessitate graft removal. The most common systemic infection is due to cytomegalovirus (CMV) or Epstein Barr virus (EBV). We showed that drug prophylaxis is effective in reducing the incidence of CMV infections in pancreas recipients, as have others. The risk of posttransplant lymphoproliferative disorder (PTLD) following EBV infection is a risk in all organ allograft recipients. Our incidence of PTLD after pancreas transplantation has been < 2%, including only 0.6% in PTA recipients. Other groups report a similarly low rate of PTLD following pancreas transplantation.

Regarding immunosuppression, our center along with many others has evolved to use tacodimus (Tac) and/or mycophenolate mofetil (MMF) in all recipient categories. Rapamyacin is just beginning to be used for pancreas transplantation, but our experience is only with a few cases. We also use anti-T cell therapy routinely for induction, while its use is variable at other centers. We have employed adjunctive measures such as blood transfusion, as have others, but we have not adopted donor bone marrow administration. The immunosuppressive protocol changes initiated by us and others have been associated with a reduction in the previous high rejection (RE) rates seen even in SPK recipients. In pancreas transplant alone (PTA) recipients, however, the RE rate is still high, but reversal is more readily accomplished than in previous eras.

Early treatment of REs is important. SPK recipients of grafts from the same donor can be monitored by serum creatinine. For solitary pancreas transplant recipients, serum creatinine cannot be used as a surrogate marker for rejection. Thus, we still favor BD and use a decline in urine amylase as a marker for rejection. A decline in urine amylase is sometimes preceded or accompanied by a rise in serum pancreatic enzyme levels, but we have seen several REs where only the urine amylase declined; they would have been missed by relying on serum enzyme levels alone. Thus, it is our opinion that urine amylases is still superior for immunologic monitoring of solitary BD pancreas transplants and should be used, at least for PTA, until protocols that further lower the RE incidence have been developed.

Pancreas allograft biopsies and pathologic assessment are less important in SPK recipients than in solitary pancreas recipients for the reasons mentioned above. However, for solitary pancreas recipients, a pancreas graft biopsy has the same utility as a kidney graft biopsy in kidney transplant recipients. We have employed pancreas allograft biopsies for solitary pancreas recipients by one technique or another since the early 1980s. The introduction of the percutaneous needle technique by Allen et al. and Gaber et al. in the early 1990s made routine biopsies practical. Pathologic features and histologic grading of pancreatic allograft biopsies have been well described by many groups including our own. Our use of pancreas graft biopsies in the 1980s was critical in identifying disease recurrence (autoimmune isletitis). Although recurrence (selective beta cell loss) has occasionally been seen in human allografts, it is rare. One center that is very liberal in performing pancreatic allograft biopsies has never seen a case of recurrent disease.

In SPK recipients, documented discordant REs (i.e., involving only 1 organ) are rare, but they do occur and in our experience can lead to discordant graft loss as well. HLA matching reduced rejection failures in solitary pancreas transplants at a time when the overall results were not as good as now. Although not all agree, our own data reported here, as well as that of the Registry, indicates that HLA matching, particularly at the B locus, is still beneficial in the FK-MMF era.

If graft loss does occur for any reason, a pancreas retransplant is feasible. Although some have considered retransplantation a high-risk procedure, we have not been deterred. We have had a large number of candidates for retransplants because of the low success rate with primary transplants in the early eras. Pancreas graft survival rates were significantly lower for retransplants than for primary transplants in each era, but the current retransplant success rate is much higher than the primary pancreas transplant success rate in earlier eras. Thus, we have no hesitation in routinely offering a retransplant to recipients whose primary grafts fail.

Even apart from retransplantation, many risk factors influence outcome. Multivariate analyses have been done by others, looking at both recipient and donor risk factors, but with many fewer patients than in our analyses. Some groups have assessed individual risk factors, such as obesity or recipient race, the former has a moderate impact, but the latter does not seem to influence outcome.

The question as to whether early SPK transplants to preempt dialysis give an advantage in nephropathic diabetic recipeints is answered by our good results in non-diabetic patients. Other questions, such as the impact of vascular disease (VD), have not been well studied because most groups have excluded vasculopathic patients from pancreas transplantation in the first place. In our program, nearly all uremic diabetic patients undergo pretransplant coronary angiography followed by intervention (bypass or angioplasty), if indicated, before a pancreas transplant. Although patient survival rates are lower for those with vs. without preexisting VD, correcting diabetes is beneficial for both groups. A few other centers also do pancreas transplants in diabetic patients with coronary artery disease and believe it safe. Even uremic patients with Type II diabetes have routinely received SPK transplants in some programs. We, too, have found no difference in insulin independence rates in the few Type II diabetic patients we have transplanted.

As important as recipient risk are donor risk factors. Although pancreas grafts from both pediatric and older donors have been successfully transplanted, we and others are selective. Whatever the age range, one group has shown that the outcomes for paired grafts from the same donor are very similar in KTA and SPK recipients.

Restoration of normal metabolism is the immediate goal of pancreas transplantation. Although we have described delayed endocrine function, most recipients become insulin independent immediately posttransplant. Nearly all are euglycemic and have normal glycosylated hemoglobin levels as long as the graft functions. Several other groups have also performed sophisticated metabolic studies post-pancreas transplant. Even though metabolic perturbations are described, it is interesting that recipient lipid profiles usually improve after a successful pancreas transplant, whether the improvement in lipid profiles translates into a lower risk for VD in pancreas recipients has not been determined. It is clear from our and other studies, that micro-angiopathy can improve after a successful pancreas transplant including retinopathy if the intervention is early enough. Of course, advanced retinopathy is difficult to influence as we and others have found.

Every group that has reported on neuropathic studies has shown improvement after pancreas transplantation. Even autonomic dysfunction, including cardiopathy, vesicopathy and gastropathy, can improve.

As expected, diabetic nephropathy does not recur in kidney grafts of recipients with sustained insulin independence after an SPK transplant. The only real surprise in the area of kidney disease is our finding that advanced lesions of diabetic nephropathy in native kidneys can resolve over time after a successful PTA.

The improved metabolism after a pancreas transplant not only ameliorates secondary complications; concomitantly there is an improvement in quality of life (QOL). The independent studies done in our own patients have the same findings as those of other groups in their patients: recipients are more satisfied after than before the pancreas transplant.

Pancreas Transplantation is a highly effective therapy for diabetes mellitus. There are surgical complications, and immunosuppression is required, but QOL improves. At least in the short run, a pancreas transplant is more expensive than exogenous insulin treatment, but better treatment is worth the higher cost. The cost of pancreas transplantation in the short term has been studied by ourselves and others but the long-term overall economic impact of preventing or ameliorating secondary complications and thereby recouping initial start up costs have only been projected. Nevertheless, pancreas transplantation is so effective that a new American Diabetes Association (ADA) position statement says that SPK and PAK should be routine in diabetic kidney transplants recipients and that PTA is appropriate for nonuremic labile diabetic patients.

Our program is more liberal than the ADA. We have done pancreas transplants as prophylaxis for secondary complications or for adult patients who would rather manage immunosuppression and its risks than diabetes and its risks. We take it as a matter of informed consent as to which route patients want to take: the diabetes or immunosuppression. There are risks which each, but the benefits are greater with a transplant.

Of course if immunosuppression is required for other reasons in a diabetic patient, a pancreas transplant might as well be done, even in children. Certainly the current

ADA guidelines are appropriate for children. Whatever hesitation there may be to recommend endocrine transplantation as a treatment for diabetes, it will be less with free grafts of islets which eliminate surgical complications. If tolerance-inducing protocols become successful clinically, either pancreas or islet transplantation could be performed without the fear of immunosuppressive complications. If islets are as successful as the pancreas transplant technically, in the absence of immunosuppression (tolerance), virtually every diabetic patient would want to be treated. Because of the limited supply of human CAD donors, treating all diabetic patients would require the development of propagated beta cell lines that are suitable for transplantation or the application of xenografts. Ultimately, neither strategy may be needed if beta cell regeneration can be induced in the native pancreas and the autoimmune threat thwarted. When the last scenario will materialize is uncertain.

Certainly pancreas or islet transplantation will continue to be in the therapeutic armamentarium for diabetes in the immediate future. Our experience shows that the future is bright with current patient and graft survival rates of over 90% and 80% at 1 year posttransplant and thus large scale application is possible.

Anthony Cerami, Michael Brines, Cara Cerami, Pietro Ghezzi and Loretta Itri

18. Novel Applications for Recombinant Human Erythropoietin

Editor's Comment: Serendipity plus the prepared mind may lead to many intriguing though unexplored places. In this initial report, the Cerami team, previously known to those concerned with the complications of diabetes because of pioneering insights into the importance of advanced glycosylated end-products opens fresh vistas for the 'wonder drug' recombinant erythropoietin (r-HuEPO), a large glycosylated molecule. After detecting a specific neuronal receptor for r-HuEPO, rodent models of CNS disorders, were systemically studied. Administration of r-HuEPO 24 hours before or up to 6 hours after a focal ischemic stroke attenuated infarction size and also reduced the extent of concussive brain injury, autoimmune encephalomyelitis, and kainate-induced seizures. Although the relationship of these observations to the Diabetic Renal-Retinal Syndrome may appear remote, the investigator's mindset must be open no matter where the data lead.

Abstract: Erythropoietin (EPO) is mainly produced in the kidney and plays a key role in the physiologic response to hypoxia by promoting increased red blood cell production. In addition, astrocytes and neurons in the central nervous system (CNS) produce EPO in response to hypoxia/ischemia. Data from preclinical studies have demonstrated the ability of recombinant human erythropoietin (r-HuEPO) to protect neurons from hypoxic/ischemic stress when administered intracerebroventricularly, suggesting that EPO may also have a neuroprotective role. In animal models of CNS disorders, systemically administered r-HuEPO has not been investigated in depth because it was assumed that large glycosylated molecules were not able to cross the blood-brain barrier (BBB). A collaborative research program identified the expression of EPO receptors on human brain capillaries and detected a specific receptor-mediated transport of r-HuEPO across the BBB after a single intraperitoneal (IP) injection in rodents, with subsequent protection against several kinds of neuronal damage. For example, administration of r-HuEPO 24 hours before or up to 6 hours after focal ischemic stroke significantly attenuated the magnitude of the infarction. r-HuEPO also reduced the extent of concussive brain injury, autoimmune encephalomyelitis, and kainate-induced seizures. These preclinical data suggest that r-HuEPO may have therapeutic potential in clinical settings such as in the treatment of stroke, head trauma, and epilepsy. Further studies are required to confirm and expand on these promising observations in animal models.

INTRODUCTION

Endogenous erythropoietin (EPO) is a growth factor that stimulates red blood cell production (i.e., erythropoiesis) in response to tissue hypoxia (Jelkmann, 1992; Koury and Bondurant, 1992; Krantz, 1991). Although primarily a growth factor, EPO has been demonstrated to inhibit the apoptotic cell death of immature erythroblasts, thus allowing their progression to mature erythrocytes (Silva et al., 1999). This antiapoptotic activity

E.A. Friedman and F.A. L'Esperance, Jr. (eds.), Diabetic Renal-Retinal Syndrome, 197–206.

may arise from an interaction between EPO and its receptors, with subsequent upregulation of a member of the bcl-2 family of antiapoptotic genes (Gregory et al., 1999; Silva et al., 1999). In adults EPO production occurs almost entirely in the kidney, with limited EPO production occurring in the liver (Jelkmann, 1992). There is increasing preclinical evidence that EPO receptors are present in the central nervous system (CNS) and that astrocytes and neurons are able to produce EPO under hypoxic/ischemic conditions (Bernaudin et al., 2000; Digicaylioglu et al., 1995; Juul et al., 1998, 1999; Marti et al., 1996, 1997; Masuda et al., 1994; Morishita et al., 1997). However, the rate and extent of this physiologic EPO production probably are inadequate to counteract the acute and severe hypoxic/ischemic neuronal stress that would be associated with a blunt head trauma, cerebrovascular accident, or status epilepticus.

In the 1980s, recombinant human erythropoietin (r-HuEPO) was developed using the emerging recombinant DNA technology. This glycoprotein has an amino acid sequence identical to that of endogenous EPO, and identical biologic activity in terms of stimulating the growth, differentiation, and survival of erythroid progenitor cells (Egrie et al., 1985, 1986). In addition, animal models of CNS disorders involving ischemic insult and accelerated apoptosis have demonstrated that r-HuEPO appears to have some neuroprotective properties. Preclinical data have shown that r-HuEPO modulated the neuroexcitability of glutamate (Morishita et al., 1997; Sakanaka et al., 1998) and had neurotrophic activity typical of other cytokines and growth factors (Campana et al., 1998; Konishi et al., 1993; Tabira et al., 1995). In recent research undertaken in Europe and Japan using animal models of global or focal cerebral ischemia, pretreatment with r-HuEPO or recombinant mouse erythropoietin (r-MoEPO), administered as a direct intracerebroventricular (ICV) injection, has been shown to limit neuronal damage and functional disability (Bernaudin et al., 1999; Sadamato et al., 1998; Sakanaka et al., 1998). Conversely, in an animal model of mild ischemia (no neuronal damage), the administration of soluble EPO receptor (which neutralizes r-HuEPO) exacerbated ischemic stress and caused neuronal degeneration (Sakanaka et al., 1998). The discovery of EPO receptors in the CNS and the neuroprotective effects of r-HuEPO in preclinical studies of brain injury provided the rationale for further evaluation of r-HuEPO in animal models of hypoxic/ischemic CNS disorders.

Rationale for Investigating Systemic r-HuEPO in CNS Models

In the laboratory it is possible to deliver r-HuEPO directly into the brain. However, this is not a practical approach to managing CNS disorders in clinical practice. It had been assumed that the blood-brain barrier (BBB) would prevent large glycosylated proteins, such as r-HuEPO, from penetrating the CNS from the periphery. Consequently, systemic r-HuEPO was not extensively investigated in animal models of CNS hypoxia/ischemia. However, it is now apparent that certain large proteins may enter CNS after binding to their receptors on capillary endothelium (Broadwell et al., 1988; Duffy et al., 1988; Fishman et al., 1987; Golden et al., 1997; Pardridge, 1997).

This evidence prompted a collaborative research program designed specifically to demonstrate that r-HuEPO administered systemically can reach the CNS through an intact BBB, and subsequently alter neuronal function in animal models of concussive brain injury, ischemic focal stroke, and status epilepticus (Brines et al., 2000). In a preliminary investigation, EPO receptors were found within and around the capillaries of anatomically normal rodent human brain specimens (Brines et al., 2000). The most significant EPO receptor immunoreactivity was located within the astrocytic endfeet surrounding the capillaries and within or on the surface of capillary endothelial cells (Brines et al., 2000).

The widespread existence of EPO receptors within and around human brain capillaries suggests that specific receptor-mediated transport of EPO into the brain parenchyma occurs. To evaluate this transport process, biotinylated r-HuEPO 5,000 U/kg was administered to mice by intraperitoneal (IP) injection (Brines et al., 2000). Observation of brain sections 5 hours after the injection revealed a peroxidase reaction product surrounding the capillary lumens and extending into the brain parenchyma (Brines et al., 2000). When a 100-fold excess of unlabelled r-HuEPO was administered in conjunction with the biotinylated r-HuEPO, capillary staining was markedly reduced, suggesting that this apparent transport mechanism was saturable (Brines et al., 2000). Subsequent evaluation of the cerebral spinal fluid (CSF) provided further support that EPO is transported across the BBB. After a lag time of about 1 hour, there was a progressive increase of r-HuEPO in the CSF that reached statistical significance 1.5 hours after injection ($P < 0.01$) and reached a peak concentration approximately 3.5 hours postinjection ($P < 0.003$; Figure 1). Overall, these observations of immunocytochemical staining of human brain specimens, together with the studies of systemic r-HuEPO administration in mice, confirmed the hypotheses that EPO (a) undergoes specific receptor-mediated transport into the parenchyma of the brain, and (b) crosses

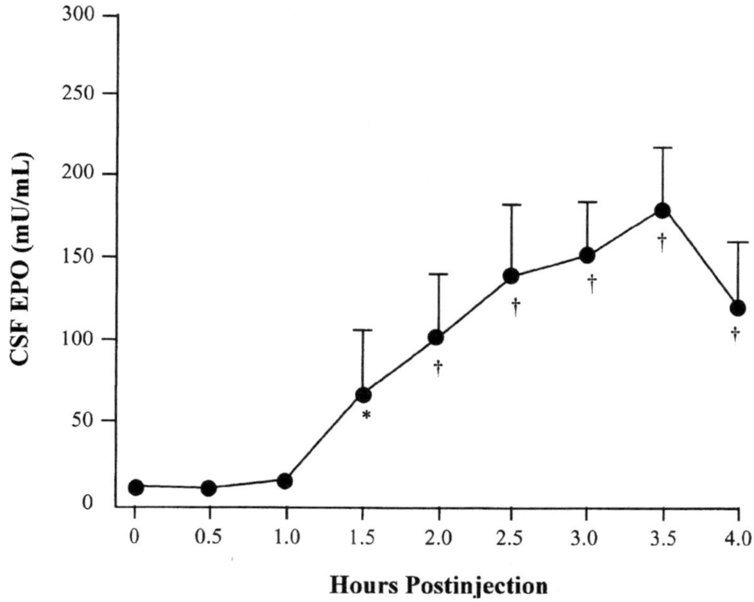

Figure 1. EPO concentrations in rodent cerebral spinal fluid following an intraperitoneal injection of r-HuEPO. * $P < 0.01$ versus baseline; † $P < 0.003$ versus baseline.

the BBB in the absence of injury/inflammation, where it quickly becomes measurable in the CSF.

Novel Uses for r-HuEPO

r-HuEPO increases hemoglobin levels in patients with anemia caused by cancer chemotherapy, surgery, chronic renal failure, and antiretroviral therapy for HIV infection (Abels, 1992; Eschbach et al., 1989, 1989; Demetri et al., 1998; Faris and Ritter, 1998; Glaspy et al., 1997; Henry, 1998). After several weeks of treatment with r-HuEPO, hemoglobin and hematocrit levels are increased, and transfusion requirements are reduced (Abels, 1992; Demetri et al., 1998; Glaspy et al., 1997; Eschbach et al., 1989; Faris and Ritter, 1998; Henry, 1998). In addition to the management of anemia, it is becoming apparent that r-HuEPO may have other therapeutic uses, as shown by the preclinical models described below of apoptosis, neurodegeneration, and/or neuronal excitability (Table 1). The degree to which these preclinical models translate into therapeutic applications remains to be determined.

Focal Ischemic Stroke

Intracerebroventricular injection of recombinant mouse EPO (r-MoEPO) in mice 24 hours before permanent focal cerebral ischemia has been shown to reduce the mean cortical infarct volume by 47% (mean volume 21 mm^3 in r-MoEPO group vs. 40 mm^3 in control group; $P < 0.0002$) (Bernaudin et al., 1999). Because the therapeutic potential of r-HuEPO for stroke would depend on its ability to confer neuroprotection from a transient occlusion, an animal model of human stroke was investigated that involved permanent right middle cerebral artery (MCA) occlusion in Sprague-Dawley male rats, followed by a reversible 1-hour occlusion of the left carotid artery (Brines et al., 2000). This sequence of events would be expected to produce a permanent central core of damage in the right frontal cortex, surrounded by an ischemic penumbra of dead cells

Table 1. Potential novel applications for r-HuEPO.

Neuroprotection
- Trauma (including surgery and radiotherapy)
- Epilepsy
- Stroke
- Cognitive dysfunction
- Neurodegenerative diseases
  - Parkinson's disease
  - Multiple sclerosis
  - Alzheimer's disease
  - HIV/AIDS
  - Amyotrophic lateral sclerosis

Cardioprotection
- Acute myocardial infarction

from the reversible occlusion. Animals were given r-HuEPO 250–5,000 U/kg or saline IP at one of the following time periods: 24 hours before MCA occlusion, at the time of MCA occlusion, or 3, 6, or 9 hours after MCA occlusion (Brines et al., 2000).

Results indicated that r-HuEPO had a protective effect against brain injury caused by focal ischemia. After 24 hours, animals in the control group had a large penumbra surrounding the central core of damage that was not evident in animals treated with r-HuEPO (Brines et al., 2000). r-HuEPO 5,000 U/kg administered 24 hours preocclusion dramatically reduced the size of the penumbra but not the ischemic core, indicating antiapoptotic activity in the reversible lesion (Brines et al., 2000). Furthermore, animals treated with r-HuEPO 24 hours before or up to 3 hours postocclusion had a significantly ($P < 0.01$) reduced volume of infarcted brain compared with controls (Figure 2) (Brines et al., 2000).

The apparent neuroprotective effect of r-HuEPO decreased to approximately 50% of the 24-hour pretreatment magnitude at 6 hours postocclusion, and was absent by 9 hours postocclusion (Brines et al., 2000). The minimum effective dose of r-HuEPO administered at the time of MCA occlusion was approximately 450 U/kg (Brines et al., 2000). The short duration of the study excluded any effect of erythropoiesis on cerebral ischemia, as improvement in hematopoietic parameters typically appear several weeks after administration of a single dose of r-HuEPO (Brines et al., 2000). Consequently, this model of focal ischemia suggested that r-HuEPO 450–5,000 U/kg IP had a neuroprotective effect against stroke, with an apparent therapeutic window of up to 6 hours after the onset of ischemia (Brines et al., 2000).

Further evidence of neuroprotection with r-HuEPO was provided by data from another study that used the same *in vivo* rat model for human stroke and *in vitro* measurement of apoptosis inhibition by r-HuEPO (Sirén et al., 2001). In this study, administration of r-HuEPO 5,000 U/kg to rats at the time of the occlusion conferred an approximately 75% reduction in the volume of the infarction after 24 h compared

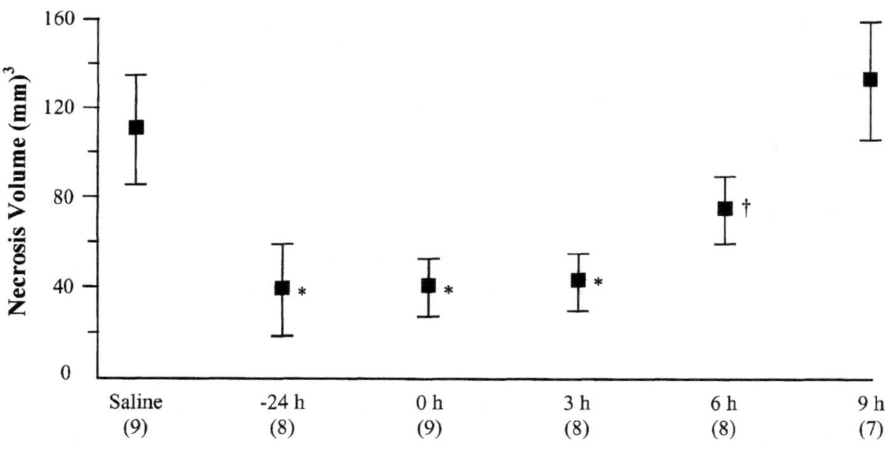

Figure 2. Effects of systemically administered r-HuEPO on infarct volume in an animal model of ischemic focal stroke. * $P < 0.01$ versus control group; † $P < 0.05$ versus control group. Reproduced with permission from Brines et al., Proc Nat Acad Sci USA 2000; 97: 10526–10531.

with controls. *In vitro* results demonstrated that r-HuEPO was a potent inhibitor of neuronal apoptosis induced by hypoxia, serum withdrawal and kainate exposure (Sirén et al., 2001).

Concussive Brain Injury

Blunt head trauma in humans causes a series of related events: neuronal ischemia, inflammation, and excitotoxicity. If severe, the trauma will lead to the development of a cavitary lesion after 7 to 10 days (Dixon et al., 1988). Because of its antiapoptotic effects in focal ischemic stroke, it was hypothesized that r-HuEPO would also have antiapoptotic effects in concussive brain injury. To test this hypothesis, a nonpenetrating concussive injury of moderate severity was made to the frontal cortex of 12 anesthetized BALB/c mice (Brines et al., 2000). The mice were injected with r-HuEPO 5,000 U/kg or saline IP once daily for 5 consecutive days. The first dose was injected at one of the following time points: 24 hours before impact, at the time of impact, 3 hours after impact, or 6 hours after impact (Brines et al., 2000).

Examination of brain sections after a 10-day recovery period revealed that r-HuEPO significantly reduced the size of the cavitary lesions resulting from the trauma compared with control animals, which had been treated with saline (Brines et al., 2000). Quantitative analysis of the brain sections confirmed that treatment with r-HuEPO 24 hours before trauma significantly ($P < 0.05$) reduced mean necrosis volume compared with saline-treated animals (Figure 3) (Brines et al., 2000).

Further qualitative examination of the brain sections indicated that administration of r-HuEPO 3 or 6 hours after trauma had a protective effect on frontal cortex injury similar to that achieved with 24-hour pretreatment (Brines et al., 2000). This result, analogous to that described above in the focal ischemic stroke model, suggests that there may be a therapeutic window for r-HuEPO administration in blunt trauma cases (Brines et al., 2000). Interestingly, animals treated with r-HuEPO also had a notable absence of prominent mononuclear cell infiltrate surrounding the area of necrosis (Brines et al., 2000). This apparent anti-inflammatory property of r-HuEPO in the CNS provided the basis for further investigation in an animal model of autoimmune inflammation (Brines et al., 2000).

Experimental Autoimmune Encephalomyelitis

To confirm the ability of r-HuEPO to reduce the infiltrative response seen in the concussive brain injury model, an animal model of neurodegeneration called experimental autoimmune encephalomyelitis (EAE) was used (Brines et al., 2000). This model has provided considerable biologic insight into the pathogenesis of multiple sclerosis. In EAE, the immune and inflammatory systems are activated, but apoptosis is not considered to have a major role in the clinical sequelae (Brines et al., 2000).

EAE was induced by the injection of myelin basic protein into the hind footpads of 36 Lewis rats, with the expectation that symptoms of disability would become apparent by Day 10 and a peak level of paralysis occur around Day 12.Brines et al 2000/10527/2/3, 10530/1/2 The animals received daily injections of r-HuEPO 5,000 U/kg or saline IP, starting 3 days after immunization with myelin and continuing for

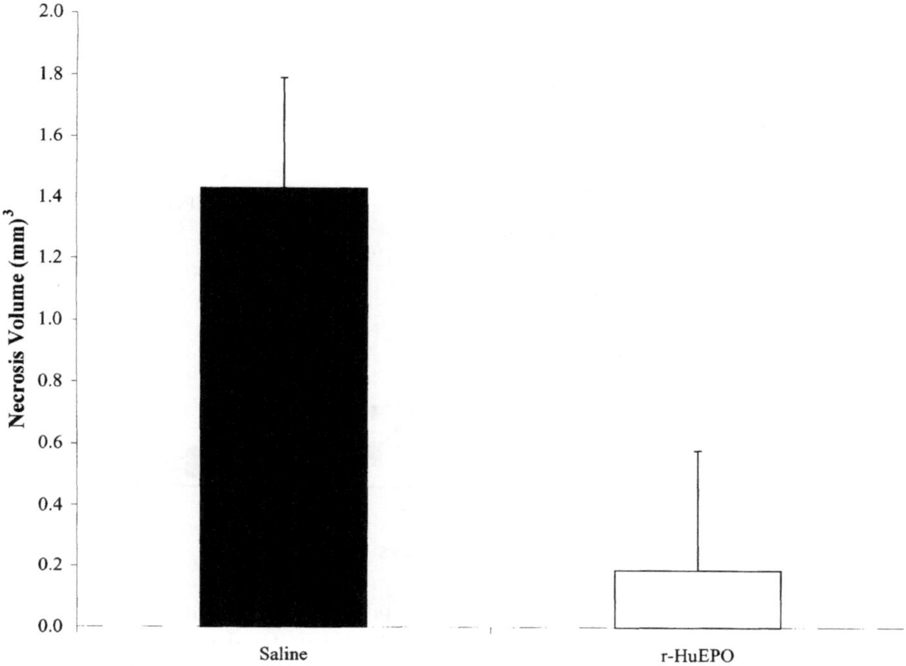

Figure 3. Effects of systemic administration of r-HuEPO on injury after blunt trauma. Mouse brains were evaluated 10 days after a nonpenetrating concussive injury to the brain. r-HuEPO injected 24 hours before impact significantly reduced necrosis volume compared with saline-treated controls ($P < 0.05$). Reproduced with permission from Brines et al., Proc Nat Acad Sci USA 2000; 97: 10526–10531.

15 consecutive days (Brines et al., 2000). In animals receiving r-HuEPO, the onset of EAE was significantly delayed, and the symptomatic severity reduced ($P < 0.01$; Figure 4) compared with controls (Brines et al., 2000).

Furthermore, there was no recurrence of symptoms up to 3 weeks after the last dose of r-HuEPO (Brines et al., 2000). This observation is in contrast with other EAE studies, in which a 'rebound' of clinical disease has been documented within weeks of discontinuing glucocorticoids or interferon-β (van der Meide et al., 1998). Overall, it appeared that r-HuEPO had marked and lasting antiinflammatory and immunomodulatory effects against this pathologic autoimmune process, properties that had not reported previously for this cytokine.

Kainate-Induced Seizures

An underlying component of many types of brain injury is neuronal excitability (Juurlink and Paterson, 1998; Martin et al., 1998). Preincubation with *in vitro* r-HuEPO at least

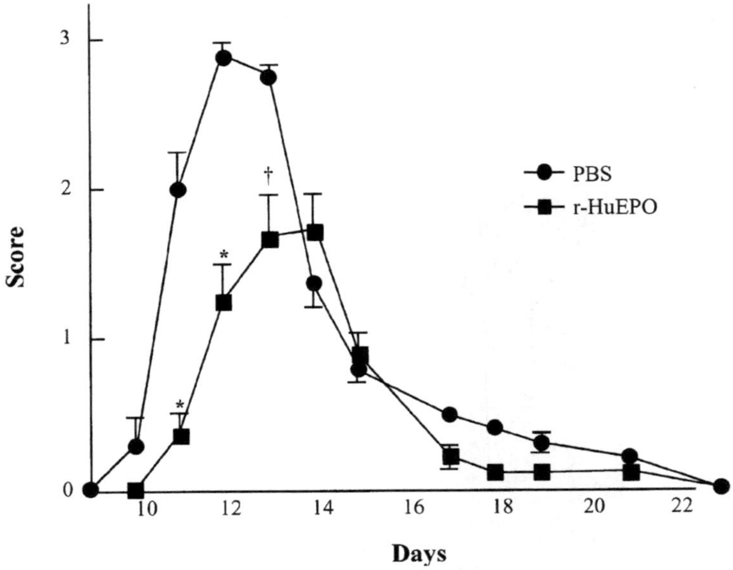

Figure 4. Effects of systemically administered r-HuEPO on clinical disability in experimental autoimmune encephalomyelitis. Lewis rats were evaluated daily for signs/symptoms of EAE, which were evaluated rated as: 0 = no limb symptoms; 1 = flaccid tail; 2 = ataxia; and 3 = complete hind limb paralysis with urinary incontinence. * $P < 0.01$ versus control group; † $P < 0.05$ versus control group. Reproduced with permission from Brines et al., Proc Nat Acad Sci USA 2000; 97: 10526–10531.

8 hours prior to exposure to glutamate (an excitatory neurotransmitter), has been shown to protect hippocampal and cortical neurons from glutamate-induced excitability and cell death (Morishita et al., 1997). Consequently, a series of experiments was undertaken to investigate whether r-HuEPO has the ability to reduce neuronal excitability associated with kainate, a glutamate analog that rapidly induces status epilepticus (Brines et al., 2000).

r-HuEPO 5,000 U/kg IP or saline was administered to 100 female BALB/c mice 24 hours before administration of kainate 20 mg/kg IP (Brines et al., 2000). The results showed that pretreatment with r-HuEPO had a beneficial effect on survival. The overall survival rate for mice pretreated with r-HuEPO was 65%, versus 40% for the control mice, providing a reduction in overall mortality of approximately 45% ($P < 0.01$) (Brines et al., 2000). Mean durations of survival were 25.8 min and 18.2 min in the r-HuEPO-pretreated and saline-pretreated groups, respectively ($P < 0.0002$) (Brines et al., 2000). When a single dose of r-HuEPO 5,000 U/kg IP was administered to mice 30 min or 1, 2, 3 or 7 days before kainate injection, protection against kainate-induced seizures was provided by r-HuEPO administered 1, 2, and 3 days ($P < 0.05$ vs. controls) with a trend at 7 days before kainate injection (Brines et al., 2000). However, this was not the case for mice who received r-HuEPO 30 min before kainate, or at any point after seizure onset (Brines et al., 2000). The inability of r-HuEPO to protect against seizures immediately before or after their onset suggests that r-HuEPO stimulates the induction of genes that modulate neuronal excitability (Brines et al., 2000). Thus, the mechanism by which r-HuEPO exerts its anticonvulsant activity may differ from that of standard antiepileptic drugs, which are able to terminate seizures but require therapeutic levels to attain and sustain their clinical efficacy (Brines et al., 2000).

SUMMARY

Studies reported here and elsewhere have demonstrated that the biologic activity of r-HuEPO is more extensive than its effects on erythroid progenitor cells. Preclinical studies have shown that r-HuEPO is able to penetrate the BBB and reach the CSF and brain parenchyma via a specific receptor-mediated transport process. Consequently, it is reasonable to assume that endogenous EPO is transported from the periphery to the CNS, as well as being produced by astrocytes and other brain cells. In the preclinical studies described here, r-HuEPO administered systemically reduced the severity of experimentally induced stroke, blunt trauma, and autoimmune encephalitis; and prolonged survival in an animal model of status epilepticus. Further investigations are required to determine the mechanism by which r-HuEPO exerts its protective activity in these models of CNS disorders, and to assess whether r-HuEPO has therapeutic potential beyond erythropoiesis, in human CNS-related disorders.

REFERENCES

Abels RI. Recombinant human erythropoietin in the treatment of the anaemia of cancer. Acta Haematol 1992; 87 (suppl 1): 4–11.

Bernaudin M, Marti HH, Roussel S, et al. A potential role for erythropoietin in focal permanent cerebral ischemia in mice. J Cereb Blood Flow Metab 1999; 19: 643–651.

Bernaudin M, Bellail A, Marti HH, et al. Neurons and astrocytes express EPO mRNA: oxygen-sensing mechanisms that involve the redox-state of the brain. Glia 2000; 30: 271–278.

Brines ML, Ghezzi P, Keenan S, et al. Erythropoietin crosses the blood-brain barrier to protect against experimental brain injury. Proc Natl Acad Sci USA 2000; 97: 10526–10531.

Broadwell RD, Balin BJ, Salcman M. Transcytotic pathway for blood-borne protein through the blood-brain barrier. Proc Natl Acad Sci USA 1988; 85: 632–636.

Campana WM, Misasi R, O'Brien JS. Identification of a neurotrophic sequence in erythropoietin. Int J Mol Med 1998; 1: 235–241.

Demetri GD, Kris M, Wade J, et al. Quality-of-life benefit in chemotherapy patients treated with epoetin alfa is independent of disease response or tumor type: results from a prospective community oncology study. J Clin Oncol 1998; 16: 3412–3425.

Digicaylioglu M, Bichet S, Marti HH, et al. Localization of specific erythropoietin binding sites in defined areas of the mouse brain. Proc Natl Acad Sci USA 1995; 92: 3717–3720.

Dixon CD, Lighthall JW, Anderson TE. Physiologic, histopathologic, and cineradiographic characterization of a new fluid-percussion model of experimental brain injury in the rat. J Neurotrauma 1988; 5: 91–104.

Duffy KR, Pardridge WM, Rosenfeld RG. Human blood-brain barrier insulin-like growth factor receptor. Metabolism 1988; 37: 136–140.

Egrie JC, Browne J, Lai P, et al. Characterization of recombinant monkey and human erythropoietin. Prog Clin Biol Res 1985; 191: 339–350.

Egrie JC, Strickland TW, Lane J, et al. Characterization and biological effects of recombinant human erythropoietin. Immunobiology 1986; 172: 213–224.

Eschbach JW, Abdulhadi MH, Browne JK, et al. Recombinant human erythropoietin in anemic patients with end-stage renal disease. Ann Intern Med 1989; 111: 992–1000.
Eschbach JW, Kelly MR, Haley NR, et al. Treatment of the anemia of progressive renal failure with recombinant human erythropoietin. N Engl J Med 1989; 321: 158–163.
Faris PM, Ritter MA. Epoetin alfa: a bloodless approach for the treatment of perioperative anemia. Clin Orthop 1998; 357: 60–67.
Fishman JB, Rubin JB, Handrahan JV, et al. Receptor-mediated transcytosis of transferrin across the blood-brain barrier. J Neurosci Res 1987; 18: 299–304.
Glaspy J, Bukowski R, Steinberg D, et al. Impact of therapy with epoetin alfa on clinical outcomes in patients with nonmyeloid malignancies during cancer chemotherapy in community oncology practice. J Clin Oncol 1997; 15: 1218–1234.
Gregory T, Yu C, Ma A, et al. GATA-1 and erythropoietin cooperate to promote erythroid cell survival by regulating bcl-x_L expression. Blood 1999; 94: 87–96.
Golden PL, Maccagnan TJ, Pardridge WM. Human blood brain barrier leptin receptor: binding and endocytosis in isolated human brain microvessels. J Clin Invest 1997; 99: 14–18.
Henry DH. Experience with epoetin alfa and acquired immunodeficiency syndrome anemia. Semin Oncol 1998; 25 (suppl 7): 64–68.
Jelkmann W. Erythropoietin: structure, control of production, and function. Physiol Rev 1992; 72: 449–489.
Juul SE, Anderson DK, Li Y, et al. Erythropoietin and erythropoietin receptor in the developing human central nervous system. Pediatr Res 1998; 43: 40–49.
Juul SE, Stallings SA, Christensen RD. Erythropoietin in the cerebrospinal fluid of neonates who sustained CNS injury. Pediatr Res 1999; 46: 543–547.
Juurlink BHJ, Paterson PG. Review of oxidative stress in brain and spinal cord injury: suggestions for pharmacological and nutritional management strategies. J Spinal Cord Med 1998; 21: 309–334.
Konishi Y, Chui D-H, Hirose H, et al. Trophic effect of erythropoietin and other hematopoietic factors on central cholinergic neurons *in vitro* and *in vivo*. Brain Res 1993; 609: 29–35.
Koury MJ, Bondurant MC. The molecular mechanism of erythropoietin action. Eur J Biochem 1992; 210: 649–663.
Krantz SB: Erythropoietin. Blood 1991; 77: 419–434.
Marti HH, Gassmann M, Wenger RH, et al. Detection of erythropoietin in human liquor: Intrinsic erythropoietin production in the brain. Kidney Int 1997; 51: 416–418.
Marti HH, Wenger RH, Rivas LA, et al. Erythropoietin gene expression in human, monkey and murine brain. Eur J Neurosci 1996; 8: 666–676.
Martin LJ, Al-Abdulla NA, Brambrink AM, et al. Neurodegeneration in excitotoxicity, global cerebral ischemia, and target deprivation: a perspective on the contributions of apoptosis and necrosis. Brain Res Bull 1998; 46: 281–309.
Masuda S, Okano M, Yamagishi K, et al. A novel site of EPO production: oxygen-dependent production in cultured rat astrocytes. J Biol Chem 1994; 269: 19488–19493.
Morishita E, Masuda S, Nagao M, et al. Erythropoietin receptor is expressed in rat hippocampal and cerebral cortical neurons, and erythropoietin prevents *in vitro* glutamate-induced neuronal death. Neuroscience 1997; 76: 105–116.
Pardridge WM. Drug delivery to the brain. J Cereb Blood Flow Metab 1997; 17: 713–731.
Sadamoto Y, Igase K, Sakanaka M, et al. Erythropoietin prevents place navigation disability and cortical infarction in rats with permanent occlusion of the middle cerebral artery. Biochem Biophys Res Commun 1998; 253: 26–32.
Sakanaka M, Wen T-C, Matsuda S, et al. *In vivo* evidence that erythropoietin protects neurons from ischemic damage. Proc Natl Acad Sci USA 1998; 95: 4635–4640.
Silva M, Benito A, Sanz C, et al. Erythropoietin can induce the expression of Bcl-x_L through Stat5 in erythropoietin-dependent progenitor cell lines. J Biol Chem 1999; 274: 22165–22169.
Sirén A-L, Fratelli M, Brines M, et al. Erythropoietin prevents neuronal apoptosis after cerebral ischemia and metabolic stress. Proc Natl Acad Sci 2001; 98: 4044–4049.
Tabira T, Konishi Y, Gallyas F Jr. Neurotrophic effect of hematopoietic cytokines on cholinergic and other neurons *in vitro*. Int J Dev Neurosci 1995; 13: 241–252.
van der Meide PH, de Labie MCDC, Ruuls SR, et al. Discontinuation of treatment with IFN-β leads to exacerbation of experimental autoimmune encephalomyelitis in Lewis rats. Rapid reversal of the antiproliferative activity of IFN-β and excessive expansion of autoreactive T cells as disease promoting mechanisms. J Neuroimmunol 1998; 84: 14–23.

ELI A. FRIEDMAN

19. Lessons Learned Since the Last Renal-Retinal Conference

Editor's Comment: Diabetes is the disorder most often linked with development of end-stage renal disease (ESRD) in the USA, Europe, South America, Japan, India, and Africa. Kidney disease is as likely to develop in long-duration non-insulin dependent diabetes (type 2) as in insulin-dependent diabetes mellitus (type 1). Nephropathy in diabetes – as usually managed – follows a predictable course starting with microalbuminuria through proteinuria, azotemia and culminating in ESRD. Renal functional decline in diabetic nephropathy is slowed by normalization of hypertensive blood pressure, establishment of euglycemia, and a reduced dietary protein intake. Compared with other causes of ESRD, the diabetic patient sustains greater mortality and morbidity due to co-morbid systemic disorders especially coronary artery and cerebrovascular disease. A functioning kidney transplant provides the uremic diabetic patient better survival with superior rehabilitation than does either CAPD or maintenance hemodialysis. For the minority (< 5%) of diabetic ESRD patients who have type 1 diabetes, a combined pancreas and kidney transplant may cure diabetes and permit full rehabilitation. No matter which ESRD therapy has been elected, optimal rehabilitation in diabetic ESRD patients requires that effort be devoted to recognition and management of co-morbid conditions.

Survival in treating ESRD in diabetes by dialytic therapy and renal transplantation is continuously improving. This inexorable progress in therapy reflects multiple small advances in understanding of the pathogenesis of extrarenal micro- and macrovasculopathy in an inexorable disease, coupled with safer immunosuppression. Recognizing the perturbed biochemical reactions underlying the pathogenesis of diabetic vasculopathy – especially the adverse impact of accumulated advanced glycosylated end-products (AGEs) – raises the possibility of blocking end-organ damage without necessarily correcting hyperglycemia.

Diabetes mellitus the most prevalent cause of end-stage renal disease (ESRD) in the industrialized world. As tabulated in the United States Renal Data System (USRDS) 2001 Report, of 88,091 patients begun on therapy for ESRD during 1999, 38,160 (43%) had diabetes, an *incidence* rate of 83 per million population [1]. Reflecting their relatively higher death rate compared to other causes of ESRD, the *prevalence* of US diabetic ESRD patients on December 31, 1999, was 34% (114,478 of 340,261 patients). Both glomerulonephritis and hypertensive renal disease rank below diabetes in frequency of diagnosis among new ESRD patients, substantiating Mauer and Chavers contention that 'Diabetes is the most important cause of ESRD in the Western world' [2].

According to the National Diabetes Fact Sheet of the Centers for Disease Control [3], more than 16 million people in the United States have diabetes – one third of whom are unaware of theird disorder. During 2001 in the United States 798,000 people will have newly diagnosed diabetes while 187,000 people will die from diabetes. Depending

E.A. Friedman and F.A. L'Esperance, Jr. (eds.), Diabetic Renal-Retinal Syndrome, 207–242.

on age, race, and gender, diabetes in 1996 ranked from 8th (White men 45 to 65 years) to 4th (Black women 45 years and over) leading cause of death [4]. Health care expenditures for diabetes in the United States amount to a minimum of $98 billion and may be as high as $150 annually. The full impact of diabetic complications is unmeasured but includes in addition to 34,874 new cases of ESRD, 56,000 lower limb amputations, and 24,000 cases of blindness.

Projections by the World Health Organization call attention to a growing pandemic of type 2 diabetes that is anticipated to crest in 2025 when a striking 300 million people will have diabetes. Some explain the remarkable expanding base of diabetes, by the concept of a 'thrifty gene' now speculatively identified as peroxisome proliferator-activated receptor-gamma (PPARgamma), a novel nuclear receptor, that enhances insulin-mediated glucose uptake. During food deprivation it is theorized PPARgamma enhances efficient utilization of limited calorie intake. When food is abundant, as in industrialized nations, the thrifty gene promotes obesity and type 2 diabetes. In the rat, PPARgamma activation may directly attenuate diabetic glomerular disease, possibly by inhibiting mesangial growth, which occurs early in the process of diabetic nephropathy, or by inhibiting PAI-1 expression inhibiting activation of plasmin and matrix metalloproteinase, which degrade extracellular matrix in the glomerulus [5].

Whereas earlier impressions of the prevalence of diabetes suggested that only a small minority of Asians contracted the disease. But, according to the World Health Organization, the coming pandemic will hit hardest in China and India where over 100 million people will be afflicted by 2025.

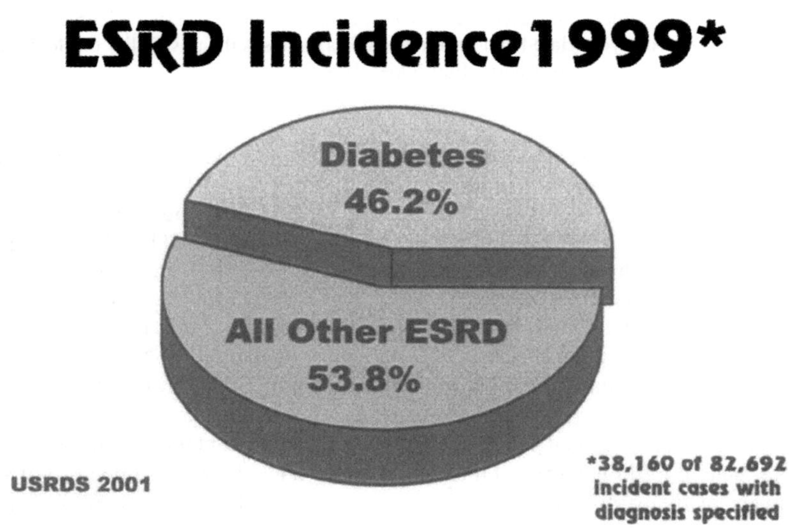

Figure 1.

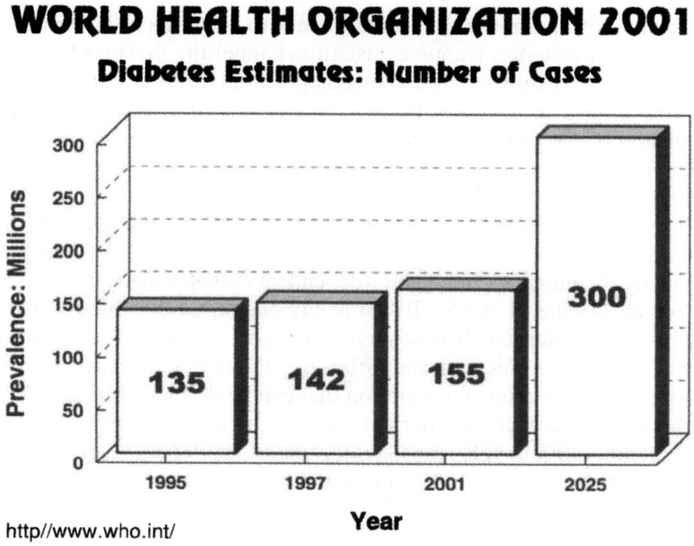

Figure 2.

OPTIONS FOR ESRD TREATMENT IN DIABETES

As listed in Table 1, diabetic ESRD patients are managed similarly to nondiabetic ESRD patients with two exceptions: 1) simultaneous pancreas and kidney transplantation is a diabetes-specific therapy and 2) no treatment, meaning passive suicide, is the choice more often selected for and by diabetic than by nondiabetic individuals. While the goal

Table 1. Options in uremia therapy for diabetic ESRD patients.

1. No Specific Uremia Intervention = Passive Suicide
2. Peritoneal Dialysis
   - Intermittent Peritoneal Dialysis (IPD)
   - Continuous Ambulatory Peritoneal Dialysis (CAPD)
   - Continuous Cyclic Peritoneal Dialysis (CCPD)
3. Hemodialysis
   - Facility Hemodialysis
   - Home Hemodialysis
4. Renal Transplantation
   - Cadaver Donor Kidney
   - Living Donor Kidney
5. Pancreas plus Kidney Transplantation
   - Type 1
   - ?Type 2

of uremia therapy is to permit an informed patient to select from a menu of available regimens, realities of program resources usually channel the diabetic ESRD patient to that treatment preferred by the supervising nephrologist. As a consequence, CAPD may be the first choice in Toronto, home hemodialysis in Seattle, and a renal transplant in Minneapolis. No prospective, controlled trials of dialytic therapy – of any type – versus kidney transplantation have been reported. Therefore, what follows reflects an acknowledged bias in interpreting the bias of others.

Confusion over diabetes type is frequent when evaluating diabetic ESRD patients. Confounding diabetes type distinction is the realization that in Sweden, as many as 14% of cases originally diagnosed as noninsulin-dependent diabetes mellitus (type 2 diabetes) progressed to type 1 diabetes, while 10% of newly diagnosed diabetic individuals could not be classified [6]. Islet β-cell dysfunction in type 2 diabetes – including 56,059 with diabetes (27.2%) – varies with the different genetic defects associated with characteristic patterns of altered insulin secretion that can be defined clinically [7]. Subjects with mild glucose intolerance and normal fasting glucose concentrations and normal glycosylated hemoglobin levels consistently manifest defective β-cell function, a component of type 2 diabetes that is present before onset of overt hyperglycemia. Hyperglycemia assessed by the level of hemoglobin A_{1c} (HbA_{1c}) is the best predictor of microvascular and macrovascular complications of diabetes [8].

At the other extreme, it is well established that some patients with type 1 diabetes maintain a measurable level of pancreatic β-cell activity for many years after onset of the disease [9] thwarting the utility of C-peptide measurements to distinguish type 1 diabetes from type 2 diabetes [10].

Diabetes worldwide is overwhelmingly type 2, fewer than seven percent of diabetic Americans are insulinopenic, C-peptide negative persons who have type 1 diabetes. ESRD in diabetic persons reflects the demographics of diabetes *per se* [11] in that: 1. The incidence [12] is higher in women, blacks [13], Hispanics [14], and native

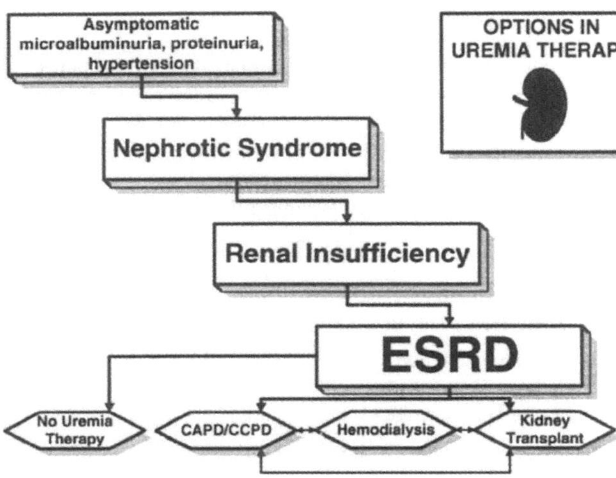

Figure 3.

Americans [15]. 2. The peak incidence of ESRD occurs from the 5th to the 7th decade. Inferred from these relative attack rates, is the reality that blacks over the age of 65 face a seven times greater risk of diabetes-related renal failure than do whites. In the United States, it is not surprising, therefore, that ESRD associated with diabetes is mainly a disease of poor, elderly blacks [16]. By telephone survey of hemodialysis units in New York, Chicago, Oklahoma City, San Antonio and Detroit, I determined that one-third to more than one-half of newly diagnosed *inner city* ESRD patients starting maintenance hemodialysis are diabetic black or Hispanic persons – predominantly women – over the age of 50.

Vasculopathic complications of diabetes including the onset and severity of hypertension are at least as severe intype 2 diabetes as in type 1 diabetes [17]. In fact, recognition of the high prevalence of proteinuria and azotemia in carefully followed individuals with type 2 diabetes contradicts the view that type 2 diabetes only infrequently induces nephropathy [18]. While there are differences between type 1 diabetes and type 2 diabetes in genetic predisposition [19] and racial expression, other aspects of the two disorders – particularly manifestations of nephropathy – are remarkably similar.

Careful observation of the course of nephropathy in type 1 and type 2 diabetes indicates strong similarities in rate of renal functional deterioration [20] and onset of comorbid complications. Early nephromegaly, as well as both glomerular hyperfiltration and microalbuminuria, previously thought limited to type 1 diabetes are now recognized as equally prevalent in type 2 diabetes [21]. Not yet included in USRDS reports is any distinction between type 1 and type 2 diabetes in terms of dialysis morbidity and mortality or posttransplant patient and allograft survival.

Lack of precision in diabetes classification provokes confusing terms like 'insulin requiring' to explain treatment with insulin in persons thought to have *resistant* type 2 diabetes. In fact, present criteria are unable to classify as many as one-half of diabetic

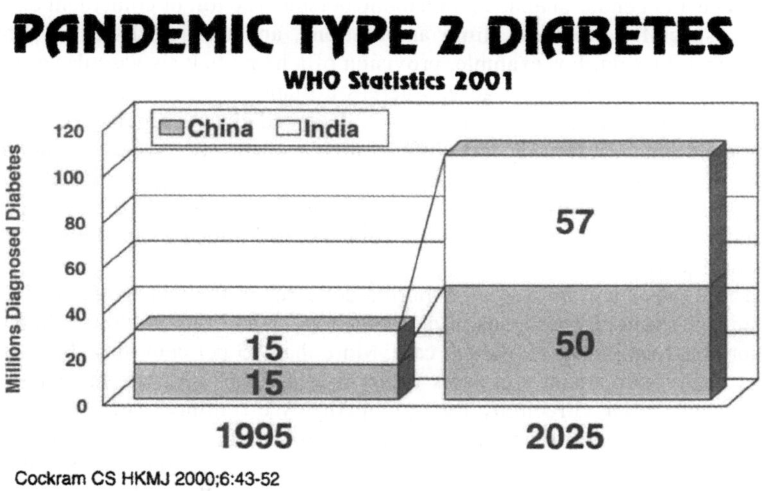

Figure 4.

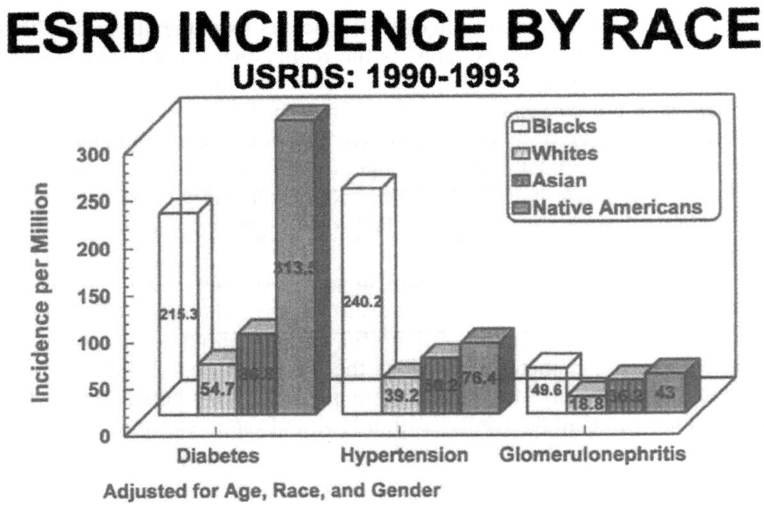

Figure 5.

persons as specifically type 1 or type 2 diabetes [22, 23]. Consequently, literature reports of the outcome of ESRD therapy by diabetes type are few and imprecise.

CO-MORBID RISK FACTORS

Management of a diabetic person with progressive renal insufficiency is more difficult than in an age and gender matched nondiabetic person. The toll of coincident extrarenal disease – especially blindness, limb amputations, and cardiac disease – limits or preempts rehabilitation. For example, provision of a hemodialysis vascular access in a nondiabetic patient is minor surgery, whereas a diabetic patient after even minimal surgery risks major morbidity from infection or deranged glucose regulation. As a group, diabetic patients manifesting ESRD suffer a higher death rate due to cardiac decompensation, stroke, sepsis and pulmonary disease than do nondiabetic ESRD patients. Sadly, depression caused by multiple complications during dialytic therapy prompts a substantially higher rate of withdrawal from therapy (suicide) in diabetic than in nondiabetic ESRD patients.

Listed in Table 2 are the major co-morbid concerns in the management of diabetic ESRD patients. Diabetic retinopathy ranks at the top – with heart and lower limb disease – as major concerns in overall patient care. More than 95 per cent of diabetic individuals who begin maintenance dialysis or receive a renal allograft have undergone laser treatment and/or vitrectomy for retinopathy. We routinely collaborate with an ophthalmologist skilled in retinal disorders. By this tact, laser and/or vitreous surgery can be integrated as a component of comprehensive management [24]. Similarly, we consult – even in asymptomatic patients – a cardiologist familiar with uremia in diabetic patients. Coronary angiography (if indicated), is performed as a valuable maneuver to

Table 2. Diabetic complications which persist and/or progress during ESRD.

1. Retinopathy, glaucoma, cataracts.
2. Coronary artery disease. Cardiomyopathy.
3. Cerebrovascular disease.
4. Peripheral vascular disease: limb amputation.
5. Motor neuropathy. Sensory neuropathy.
6. Autonomic dysfunction: diarrhea, dysfunction, hypotension.
7. Myopathy.
8. Depression.

detect those for whom prophylactic coronary artery angioplasty or bypass surgery is likely to extend life. Included on our renal team is a podiatrist who delivers routine foot care. The podiatrist has by regular surveillance of patients at risk of major lower extremity disease sharply reduced the chance of amputations, a complication noted in about 20% who do not receive podiatric care. Autonomic neuropathy – expressed as gastropathy, cystopathy, and orthostatic hypotension – is a frequently overlooked, highly prevalent disorder impeding life quality in the diabetic with ESRD. Diabetic cystopathy, though common, is frequently unrecognized and confused with worsening diabetic nephropathy and is sometimes interpreted as allograft rejection in diabetic kidney transplant recipients. In 22 diabetic patients who developed renal failure – 14 men and 8 women of mean age 38 years – an air cystogram detected cystopathy in 8 (36%) manifested as detrusor paralysis in 1 patient; severe malfunction in 5 patients (24%); and mild impairment in 1 patient. Gastroparesis afflicts one-quarter to one-half of azotemic diabetic persons when initially evaluated for renal disease [25]. Other expressions of autonomic neuropathy – obstipation and explosive nighttime diarrhea – often coexists with gastroparesis [26]. Obstipation responds to daily doses of cascara, while diarrhea is treated with psyllium seed dietary supplements one to three times daily plus loperamide [27] in repetitive 2 mg. doses to a total dose of 18 mg daily.

Pregnancy in a diabetics with proteinuria or azotemia, previously regarded as an unavoidable prelude to disaster – in terms of fetal loss and/or maternal risk of death – is now managed with a high probability of successful outcome. Miodovnik et al. followed 182 pregnant women withtype 1 diabetes, 46 of whom had overt nephropathy for a minimum of 3 years after delivery and concluded that pregnancy neither increases the risk of subsequent nephropathy nor accelerates progression of preexisting renal disease [28]. In an equally encouraging series from Finland, Kaaja et al. followed 28 diabetic women for 7 years after delivery compared with 17 nulliparous controls matched for age, duration of diabetes, and severity of vasculopathy and concluded that: 'pregnancy does not seem to affect development or progression of diabetic nephropathy [29].'

SELECTING THERAPY

Depending on age, severity of co-morbid disorders, available local resources, and patient preference, the uremic diabetic patient may be managed according to different protocols. Diabetic ESRD patients select the no further treatment option, equivalent to passive suicide, more frequently than do nondiabetic patients [30]. Such a decision is understandable for blind, hemiparetic, bed-restricted limb amputees for whom life quality has been reduced to what is interpreted as unsatisfactory. On the other hand, attention to the total patient may restore a high quality of life that was unforeseen at the time of ESRD evaluation [31].

Unfortunately, in both Europe and the US, so called 'preterminal care in diabetic patients with ESRD' is deficient in amount and quality [32] with inadequate attention to control of hypertension, hyperlipidemia or ophthalmologic intervention [33]. For the large majority – over 80% of diabetic persons who develop ESRD in the United States – maintenance hemodialysis is the only renal replacement regimen that will be employed. Approximately 12% of diabetic persons with ESRD will be treated by peritoneal dialysis while the remaining 8% will receive a kidney transplant. To perform maintenance hemodialysis requires establishment of a vascular access to the circulation. Creation of what has become the *standard access* – an internal arteriovenous fistula in the wrist – is often more difficult in a diabetic than in a nondiabetic person because of advanced systemic atherosclerosis. For many diabetic patients with peripheral vascular calcification and/or atherosclerosis, creation of an access for hemodialysis necessitates resort to synthetic (Dacron) prosthetic vascular grafts. The typical hemodialysis regimen requires three weekly treatments lasting 4 to 5 hours each, during which extracorporeal blood flow must be maintained at 300 to 500 ml/min. Motivated patients trained to perform self-hemodialysis at home gain the longest survival and best rehabilitation afforded by any dialytic therapy for diabetic ESRD. When given hemodialysis at a facility, however, diabetic patients fare less well, receiving significantly less dialysis than nondiabetic patients due, in part, to hypotension and reduced access

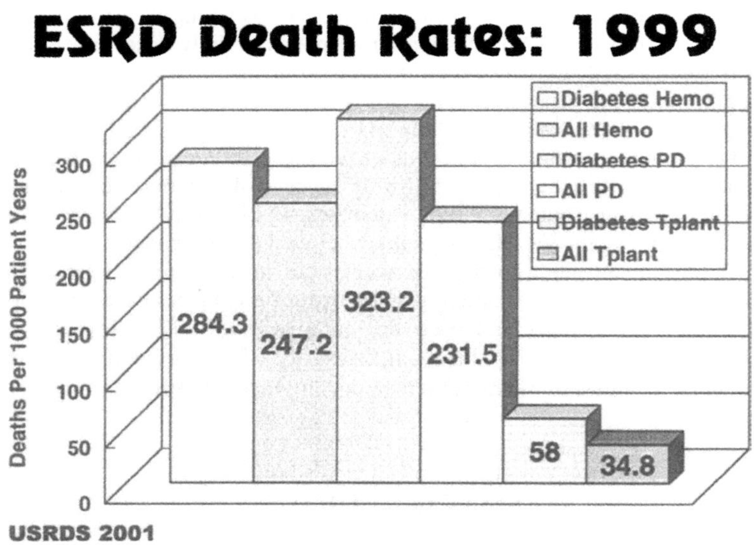

Figure 6.

blood flow [34]. Maintenance hemodialysis does not restore vigor to diabetic patients as documented by Lowder et al., in 1986, who reported that of 232 diabetics on maintenance hemodialysis only seven were employed while 64.9 per cent were unable to conduct routine daily activities without assistance [35]. Approximately 50% of diabetic patients begun on maintenance hemodialysis die within two years of their first dialysis.

PERITONEAL DIALYSIS

In the US, peritoneal dialysis sustains the life of about 12% of diabetic ESRD patients. Continuous ambulatory peritoneal dialysis (CAPD) holds the advantages of freedom from a machine, performance at home, rapid training, minimal cardiovascular stress and avoidance of heparin [36]. To permit CAPD, an intraperitoneal catheter is implanted one or more days before CAPD is begun. Even blind diabetic patients learn to perform CAPD at home within 10 to 30 days. Typically, CAPD requires exchange of 2 to 3 liters of sterile dialysate, containing insulin, antibiotics, and other drugs, 3 to 5 times daily. Mechanical cycling of dialysate, termed continuous cyclic peritoneal dialysis (CCPD) can be performed during sleep. CAPD and CCPD pose the constant risk of peritonitis as well as a gradual decrease in peritoneal surface area. Some clinicians characterize CAPD as 'a first choice treatment' for diabetic ESRD patients [37]. A less enthusiastic judgment of the worth of CAPD in diabetic patients was made by Rubin et al. in a largely black diabetic population treated with CAPD in Jackson, Mississippi [38]. Only 34% of patients remained on CAPD after two years, and at three years, only 18% continued on CAPD. According to the USRDS, survival of diabetic ESRD patients treated by CAPD is significantly less than on hemodialysis. A decision to select CAPD, therefore, must be individual-specific after weighing its benefits including freedom from a machine and electrical outlets, and ease of travel against the disadvantages of unremitting attention to fluid exchange, constant risk of peritonitis, and disappearing exchange surface. As concluded in a Lancet editorial: 'Until the frequency of peritonitis is greatly reduced, most patients can expect to spend only a few years on CAPD before requiring a different form of treatment, usually haemodialysis [39].'

KIDNEY TRANSPLANTATION

As predicted by Sutherland et al. [40], following a renal transplant, patient survival at one and two years is now equivalent in diabetic and nondiabetic recipients [41], though graft survival, in some large series, remains marginally lower in diabetic persons. At its best, as illustrated by a single center retrospective review of all kidney transplants performed between 1987 and 1993, there is no significant difference in actuarial 5-year patient or kidney graft survival between diabetic and nondiabetic recipients overall or when analyzed by donor source. Furthermore, no difference in mean serum creatinine levels at 5 years was discerned between diabetic and nondiabetic recipients [42]. Statistical superiority in survival after a renal transplant, compared with dialytic therapy, does not tell the whole story as rehabilitation is incomparably better. The enhanced life quality effected prompts selection of a kidney transplant as the distinctly favored treatment presented to newly evaluated diabetic persons with ESRD under the age of 60. More than half of diabetic kidney transplant recipients in most series live for at least three years: many survivors return to occupational, school and home responsibilities.

Positioning the option of a combined pancreas and kidney transplant for the diabetic ESRD patient is presently difficult. Though still regarded as investigational by some [43] and, even when successful, applicable to no more than the 6% subset of uremic diabetic patients who have type 1 diabetes, pancreatic transplantation has attained acceptability and technical success [44] In one remarkable series, survival one year post-renal transplant, in 995 diabetic kidney recipients who also received a pancreas transplant renal allograft survival was a remarkable 84% [45]. World-wide results in simultaneous kidney-pancreas transplants show that more than 90% of recipients were alive at 1 year, more than 80% had functioning kidney grafts, and more than 70% no longer required insulin [46].

Combining pancreas and kidney transplants does not raise perioperative mortality; but perioperative morbidity is greatly increased, mainly due to mechanical and inflammatory problems diverting pancreatic exocrine secretions into the urinary system. When restricted to type 1 recipients younger than 45 years old, as reported by the University of Texas, a simultaneous pancreas and kidney transplant permits a five year patient survival of 78% with kidney and pancreas function at five years of 69% and 62% respectively [47]. By sharp contrast, however, the pioneer series of pancreas transplants at the University of Minnesota had a three year patient survival in 54 patients given a simultaneous kidney and pancreas transplant of only 68% versus a 90% survival in 46 patients given a cadaver kidney alone [48].

The exceptional benefit of the euglycemia restored by a pancreas transplant was reported by Fioretto et al. who found disappearance of diabetic glomerulopathy 10 years after a pancreas transplant in type 1 diabetic individuals. There is expectation that pancreatic transplantationwill interdict diabetic micro- and macrovascular extrarenal complications. Several reports of kidney biopsies in patients who have received sequential kidney and later pancreas allografts indicate that a functioning pancreas does impede progression or recurrence of diabetic nephropathy. Likewise, the course of diabetic neuropathy following combined pancreas and kidney transplantation underscores stabilization and, in some patients, improvement in diabetic motor neuropathy [49]. Unfortunately, pancreas transplantation in patients with extensive extrarenal disease, has neither arrested nor reversed diabetic retinopathy, diabetic cardiomyopathy, or extensive peripheral vascular disease [50, 51]. Nevertheless, a functioning pancreas transplant does free patients with type 1 diabetes from the dreaded daily sentence of balancing diet, exercise, and insulin dosage. In 1997, consensus of clinical nephrologists is that the ESRD patient with type 1 diabetes should consider a simultaneous kidney and pancreas transplant as at least a temporary cure of inexorable disease [52].

PATIENT SURVIVAL DURING TREATMENT OF ESRD

All reported comparisons (retrospective and prospective) of the fate of diabetic patients treated for ESRD by different modalities lack balanced treatment groups in terms of equalities in age, race, diabetes type, severity of complications, and degree of metabolic control. Prospective studies of renal transplantation compared with peritoneal or hemodialysis do not overcome limitations imposed by patient and physician refusal to permit random assignment to one treatment over another. As a generalization, younger patients with fewer complications are assigned to renal transplantation while residual older, sicker patients are treated by dialysis. Combined kidney/pancreas transplants are restricted to those with type 1 diabetes who are younger than age 50.

Reports from the European Dialysis and Transplant Association (EDTA) Registry, summarized by Brunner et al., demonstrate the singular and understandable effect of age on survival during treatment for ESRD 'irrespective of treatment modality and of primary renal disease [53].' At 10 and 15 years after starting treatment, 58% and 52% respectively of patients who were 10 to 14 years old when begun on ESRD therapy were alive, compared to 28% and 16% who were alive at 10 and 15 years of those who were 45 to 54 years old when starting ESRD therapy. A similar effect of increasing age is noted in recipients of living related donor kidney transplants. In the early 1980s, kidney recipient survival was 92% at 5 years for patients younger than 15, 87% for the 15 to 44 year old cohort and 72% for those aged 45 or older.

Diabetes adds a severe restriction on life anticipation, imparting a threefold rise in risk of dying compared with either chronic glomerulonephritis or polycystic kidney disease. In England, diabetic and nondiabetic patients starting CAPD or hemodialysis in seven large renal units between 1983–1985 were monitored prospectively over four years. Of 610 new patients (median age 52 years, range 3–80 years) beginning CAPD and 329 patients (median age 48 years, range 5–77 years) starting hemodialysis, patient survival estimates at 4 years were 74% for hemodialysis and 62% for CAPD [54]. Survival on CAPD and maintenance hemodialysis is lower in the US than in Europe. Diabetic dialysis patients previously were thought to have better survival in Europe than in the U.D. though recent comparisons of carefully matched cohorts indicate equivalent survival. In Tassin, France, a program renowned for superior survival attained by

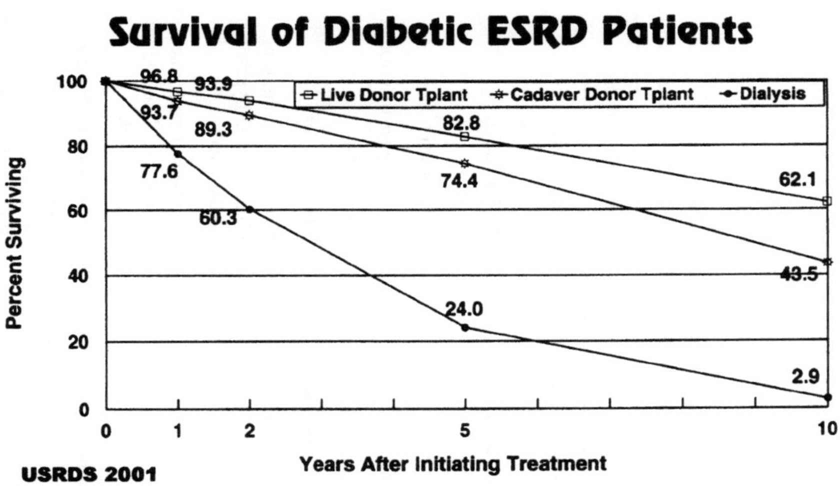

Figure 7.

Table 3. Comparison of ESRD options for diabetic patients.

Factor	Peritoneal dialysis	Hemodialysis	Kidney transplant
Extensive extrarenal disease	No limitation	No limitation except for hypotension	Excluded in cardiovascular insufficiency
Geriatric patients	No limitation	No limitation	Arbitrary exclusion as determined by program
Complete rehabilitation	Rare, if ever	Very few individuals	Common so long as graft functions
Death rate	Much higher than for nondiabetics	Much higher than for nondiabetics	About the same as nondiabetics
First year survival	About 75%	About 75%	> 90%
Survival to second decade	Almost never	Fewer than 5%	About 1 in 5
Progression of complications	Usual and unremitting. Hyperglycemia and hyperlipidemia accentuated.	Usual and unremitting. May benefit from metabolic control.	Interdicted by functioning pancreas + kidney. Partially ameliorated by correction of azotemia.
Special advantage	Can be self-performed. Avoids swings in solute and intravascular volume level.	Can be self-performed. Efficient extraction of solute and water in hours.	Cures uremia. Freedom to travel.
Disadvantage	Peritonitis. Hyperinsulenemia, hyperglycemia, hyperlipidemia. Long hours of treatment. More days hospitalized than either hemodialysis or transplant.	Blood access a hazard for clotting, hemorrhage and infection. Cyclical hypotension, weakness. Aluminum toxicity, amyloidosis.	Cosmetic disfigurement, hypertension, personal expense for cytotoxic drugs. Induced malignancy. HIV transmission.

Patient acceptance	Variable, usual compliance with passive tolerance for regimen.	Variable, often noncompliant with dietary, metabolic, or anti-hypertensive component of regimen.	Enthusiastic during periods of good renal allograft function. Exalted when pancreas proffers euglycemia.
Bias in comparison	Delivered as first choice by enthusiasts though emerging evidence indicates substantially higher mortality than for hemodialysis.	Treatment by default. Often complicated by inattention to progressive cardiac and peripheral vascular disease.	All kidney transplant programs preselect those patients with fewest complications. Exclusion of those older than 45 for pancreas + kidney simultaneous grafting obviously favorably predjudices outcome.
Relative cost	Most expensive over long run	Less expensive than kidney transplant in first year, subsequent years more expensive.	Pancreas + kidney engraftment most expensive uremia therapy for diabetic. After first year, kidney transplant – alone – lowest cost option.

its hemodialysis program, mean half-life of type 2 diabetic dialysis patients was 2.7 years [55].

The case for or against CAPD as a preferred therapy is still open. On the positive side, for example, is the report of Maiorca et al. who detailed an 8 year experience at a single center in Italy which offered 'all treatments' for ESRD [56]. Survival at 5 years was equivalent for CAPD and hemodialysis patients but 98% of those started on hemodialysis continued hemodialysis while only 71% of CAPD treated patients remained on CAPD ($P < 0.01$). Contending that survival on hemodialysis or CAPD is now equivalent, Burton and Walls determined life-expectancy using the Cox Proportional Hazards statistical methodology for unequal group analysis in 389 patients accepted for renal replacement therapy in Leicester between 1974 and 1985 [57]. There were no statistically significant differences between the relative risk of death for patients on CAPD (1.0), those on hemodialysis (1.30), and those who received a kidney transplant (1.09). CAPD, the authors concluded 'is at least as effective as haemodialysis or transplantation in preserving life.' Based on a large Canadian experience treating diabetic ESRD patients by dialysis, it was concluded that: 'Diabetic patients undergoing PD and HD probably have similar survival, and those undergoing CAPD have lower survival and technique success rates than nondiabetic patients of comparable age.' Residual renal function judged by urea clearance after commencing CAPD deteriorates to negligible more rapidly in diabetic (35.3 months) than in nondiabetic (50.5 months) patients [58].

Substantiation of the superiority of one ESRD treatment over another is lacking whether for the total population of ESRD patients or for the subset with diabetic nephropathy (Table 3). Overall, survival of diabetic patients with ESRD has been improving annually over the past decade whether treated by peritoneal dialysis, hemodialysis, or a kidney transplant. Illustrating this point is the increase in five year allograft function of diabetic cadaver kidney transplant recipients that reached 57% by 1997 versus a five year allograft function of 58% of all recipients reported to the USRDS [59].

REHABILITATION

Inferences extracted from the study of rehabilitation in the diabetic ESRD patient are that: 1) Patients fare best when participating in their treatment regimen. 2) A functioning renal transplant permits markedly superior rehabilitation than that attained by either peritoneal dialysis or hemodialysis. Unfortunately, bias in assignment to a specific treatment may have prejudiced the favorable view of kidney transplants to the extent that statistical corrections (Cox Proportional Hazards technique) cannot compensate for group differences. Studies in which the mean age of transplant patients is a decade younger than the CAPD or hemodialysis groups are likely to discern better functional status in the younger group. Another variable affecting the magnitude of rehabilitation attained in diabetic and nondiabetic ESRD patients is the progressive increase in age of newly treated patients. In the United States, for example, patients over the age of 69 years who comprised 27% of all dialysis patients in 1979 increased by 450% between 1974 and 1981, will make up 60% of all dialysis patients by the year 2010. An ageing ESRD population has a declining rate of employment and increasingly prevalent comorbid complications. An extremely optimistic picture of rehabilitation during maintenance hemodialysis was projected by a state-wide longitudinal prospective study of 979 ESRD patients in Minnesota in which the Karnofsky scoring system [60] was employed to assess patient well being [61]. Initial Karnofsky scores showed that 50% of all patients were able to care for themselves when starting treatment. After two years of maintenance hemodialysis, a remarkable 78% of patients maintained or improved their functional status. Kidney transplant recipients, however, had higher initial Karnofsky scores than did those relegated to long-term dialysis. Selection for a kidney transplant gleaned the most functional patients leaving a residual population of less functional patients. Thereafter, comparisons of relative rehabilitation in transplant and dialysis groups are flawed by selection bias favoring kidney transplant recipients.

The Minnesota description of well being on maintenance hemodialysis is highly atypical. Sustaining this point, for example, is the nationwide survey of maintenance hemodialysis patients, in which Gutman, Stead and Robinson measured functional assessment in 2,481 dialysis patients irrespective of location or type of dialysis [62]. Diabetic patients achieved very poor rehabilitation; only 23% of diabetic patients (versus 60% of nondiabetic patients) were capable of physical activity beyond caring for themselves. Lowder et al. discerned the same very low level of rehabilitation [23]. More recent confirmation of this point was afforded by Ifudu et al. who documented pervasive failed rehabilitation in a multicenter studies of diabetic and nondiabetic [63], and elderly inner-city [64] hemodialysis patients. The inescapable conclusion of studies to date is that maintenance hemodialysis – in many instances – does not permit return to life's responsibilities for diabetic individuals.

CO-MORBID INDEX FOR DIABETIC PATIENTS

To aid in grading the course of diabetic patients over the course of ESRD treatment we inventory the type and severity of common co-morbid problems. Numerical ranking of this inventory constitutes a co-morbid index (Table 4). As remarked above, comparison between treatments (hemodialysis versus CAPD versus renal transplantation versus combined kidney and pancreas transplantation) demands that patient subsets be equivalent in severity of illness before application of the treatment modality under study.

PRE-ESRD INTERVENTION

Screening for microalbuminuria should be performed annually and a spot urine albumin:creatinine ratio should be calculated to identify those diabetics at risk for nephropathy, retinopathy, and cardiovascular disease according to 1995 recommendations of the National Kidney Foundation [65]. Once such high-risk individuals are noted, treatment with an angiotensin-converting enzyme (ACE) inhibitor should be started. Ravid et al. in a double-blinded randomized control study of 94 normotensive type 2 diabetes patients with microalbuminuria treated with enalapril for 7 years reported that the drug stabilizes renal function with an 'absolute risk reduction of 42% for nephropathy [66].' Other signs have been proposed for risk factors of renal deterioration. For example, a prolonged QT interval in a standard 12-lead electrocardiogram was

Table 4. Variables in morbidity in diabetic kidney transplant recipients the co-morbidity index.

1) Persistent angina or myocardial infarction.
2) Other cardiovascular problems, hypertension, congestive heart failure, cardiomyopathy.
3) Respiratory disease.
4) Autonomic neuropathy (gastroparesis, obstipation, diarrhea, cystopathy, orthostatic hypotension.
5) Neurologic problems, cerebrovascular accident or stroke residual.
6) Musculoskeletal disorders, including all varieties of renal bone disease.
7) Infections including AIDS but excluding vascular access-site or peritonitis.
8) Hepatitis, hepatic insufficiency, enzymatic pancreatic insufficiency.
9) Hematologic problems other than anemia.
10) Spinal abnormalities, lower back problems or arthritis.
11) Vision impairment (minor to severe – decreased acuity to blindness) loss.
12) Limb amputation (minor to severe – finger to lower extremity).

Mental or emotional illness (neurosis, depression, psychosis). To obtain a numerical Co-Morbidity Index for an individual patient, rate each variable from 0 to 3 (0 = absent, 1 = mild – of minor import to patient's life, 2 = moderate, 3 = severe). By proportional hazard analysis, relative significance of each variable isolated from the other 12.

a predictor of an increased risk of death in 85 proteinuric type 1 diabetes patients followed for 5–13 years [67]. For the present, however, microalbuminuria and hypertension are the most reliable indicators of impending renal failure. As reviewed by Nathan [68], both nephropathy and retinopathy are delayed in onset by a regimen termed *intensive therapy of diabetes* which consists of striving for euglycemia (tight control), dietary protein restriction, and blood pressure reduction. Sustained euglycemia reduces enlarged kidney size typical of early hyperfiltration [69]. Prevention of the synthesis of sorbitol and other alcohols by inhibition of aldose reductase interdicts nephropathy, neuropathy and retinopathy in rats, an approach that may be applicable to human diabetes [70]. Intensive attention to striving for euglycemia reduces the cumulative incidence and overall risk for development of microalbuminuria as well as clinical albuminuria (defined as $\geq$ 208 μg/min), a derivative finding in the Diabetes Control and Complications Trial (DCCT) [71]. Another strategy retarding diabetic microvasculopathy – in trial in Spain and Russia – is a reduction in erythrocyte stiffness, a hemorrheological alteration universally noted in diabetes, by administration of pentoxifylline [72, 73] in type 1 and type 2 diabetes.

Hypertension is a major confounding factor in the genesis and progression of nephropathy. In hypertensive subjects with type 2 diabetes > 10 years, 36% had impaired renal function defined as a glomerular filtration rate < 80 ml/min/1.73 m^2 or a serum creatinine concentration > 1.4 mg/dl and 75% had microalbuminuria or clinical proteinuria [74]. Control of hypertension [75] – increasingly by angiotensin-converting enzyme inhibition [76] – and hyperglycemia [77] are the main components of contemporary treatment. It is now clear that use of an ACE inhibitor proffers unique benefit to halting progression of both microalbuminuria and proteinuria or in diabetic patients [78]. Studies with beta-blockers, calcium antagonists, diuretics, and AGE inhibitors in hypertensive diabetics with microalbuminuria have all shown significant reduction in urinary albumin excretion rates. When applied as monotherapy for 12 weeks in 31

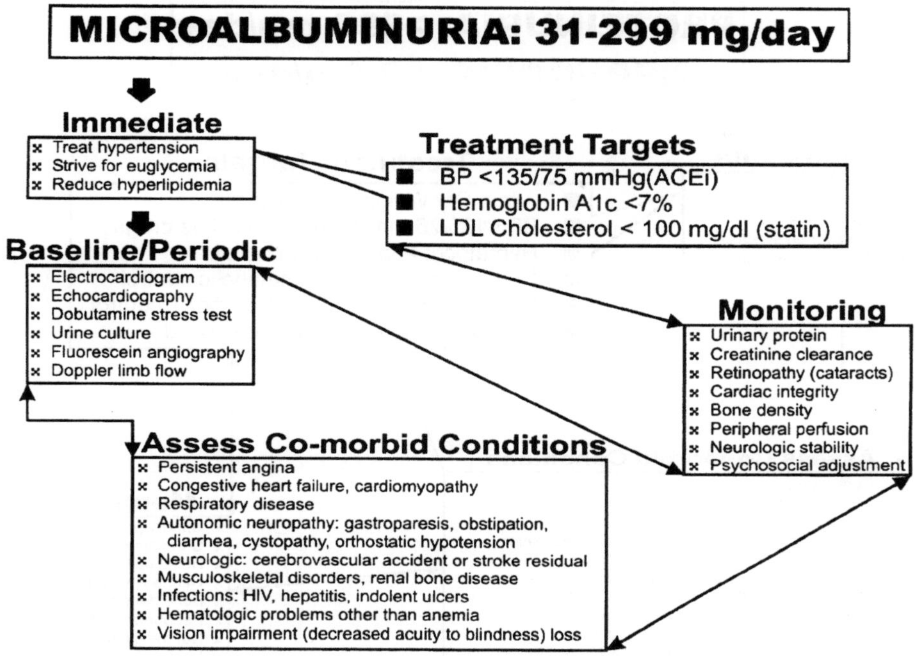

Figure 8.

diabetic patients with established microalbuminuria, captopril and indapamide were equivalent in blood pressure reduction and decrease of proteinuria [79]. Depending on the choice of calcium antagonist, sodium intake may modulate reduction of albumin excretion; diltiazem which decreased proteinuria in patients fed a diet containing 50 mEq/day of sodium was ineffective in reducing proteinuria when sodium intake was increased to 250 mEq/day while nifedipine decreased proteinuria independent of dietary sodium intake [80]. Treatment with captopril, an ACE inhibitor administered for 18 months to 24 type 2 diabetes patients with proteinuria > 500 mg/day reduced proteinuria and prevented decrease in GFR compared with 18 type 2 diabetes treated with 'conventional' antihypertensive drugs [81]. Recently, the renal protective effects of ACE inhibitors have been extended to reducing proteinuria, limiting GFR decline, and preventing ESRD in proteinuric, nondiabetic renal disorders except for polycystic kidney disease [82]. Consensus thinking advocates treatment with an ACE inhibitor when microalbuminuria is the only manifestation of incipient diabetic nephropathy [83]. Despite the now established benefit of early detection and treatment of incipient diabetic nephropathy, however, in one survey of 4,623 Americans with diabetes without evidence of nephropathy, only 16.5% had an annual urinalysis while 2.1% of those without proteinuria were screened for microalbuminuria [84]. Cigarettes increases systolic blood pressure and proteinuria in both micro- and macroalbuminuric type 1 diabetes patients and should be counted as a risk for faster progression of diabetic nephropathy [85].

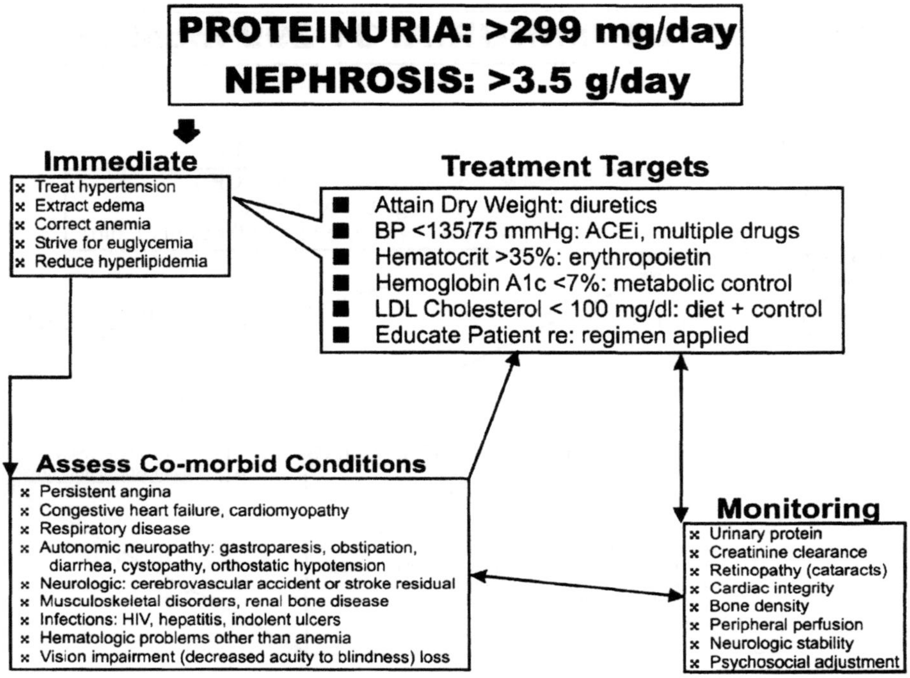

Figure 9.

Wang, Lau, and Chalmers conducted a meta-analysis of the effects of intensive blood glucose control on the development of late complications of type 1 diabetes concluding that 'Long-term intensive blood glucose control significantly reduces the risk of diabetic retinopathy and nephropathy progression [86].' Dietary protein restriction, previously thought to be beneficial in retarding loss of renal function [87] was inefficacious in a prospective multicenter trial in nondiabetic patients in Italy. Furthermore, in type 2 diabetes, dietary protein intake does not correlate with the degree of proteinuria [88]. By contrast, in a meta-analysis of five randomized, controlled or time-controlled with nonrandomized crossover design studies comprising 108 patients with type 1 diabetes with a mean follow-up of 9 to 35 months, 'a low-protein diet significantly slowed the increase in urinary albumin level or the decline in glomerular filtration rate or creatinine clearance [89].' By consensus, most nephrologists now advise limitation in dietary protein in the belief that the rate of deterioration of renal function will be slowed.There are no data supporting an advantage other than well being for strict metabolic control once uremia has developed. On the other hand, it is reasonable to anticipate that all of the benefits to native kidneys of blood pressure and blood glucose control should be conferred on a renal transplant, retarding the recurrence of diabetic nephropathy in the kidney allograft. In a comparison of renal transplant biopsies taken ≥ 2.5 years post-transplant, 92% of recipients of a combined pancreas and renal transplant but only

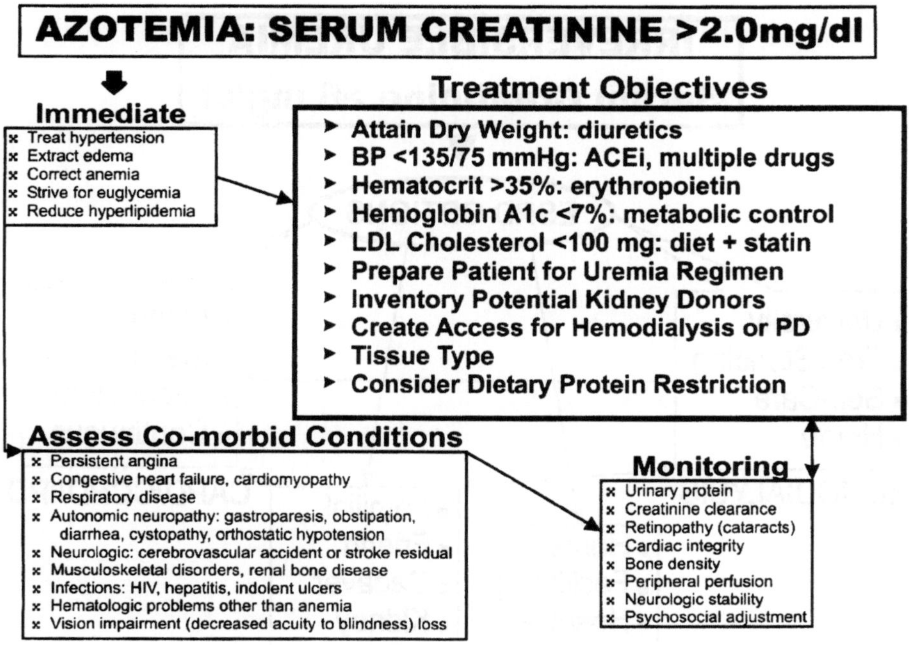

Figure 10.

35% of recipients with renal transplant alone had normal glomerular basement membrane thickness [90]. Glomerular mesangial volume expansion in the renal transplant, another early sign of recurrent diabetic nephropathy, is also retarded by the presence of a functioning pancreatic transplant. Anemia in azotemic diabetic patients adds to comorbidity and is responsive to treatment with recombinant erythropoietin. Concern over an increase in severity of hypertension as red cell mass increases is based in the finding that ambulatory maintenance hemodialysis patients evince such a change [91]. To expedite management of the myriad micro- and macrovascular complications that are manifested as azotemia increases, an orderly approach is advised. Subsequent selection of ESRD therapy for a diabetic individual whose kidneys are failing requires appreciation of the patient's family, social, and economic circumstances. Home hemodialysis, for example, is unworkable for a blind diabetic who lives alone. Deciding upon a kidney transplant requires knowledge of the patient's family structure, including its willingness to participate by donating a kidney. Without premeditation, the diabetic ESRD patient is subjected to repetitive, inconclusive studies instead of implementation of urgently required treatment (such as panretinal photocoagulation or arterial bypass surgery).

A *Life Plan* may elect 'no treatment' when life extension is unacceptable. Illustrating this point, a blind, hemiparetic diabetic patient experiencing daily angina and nocturnal diarrhea, who is scheduled for bilateral lower limb amputation may chose death

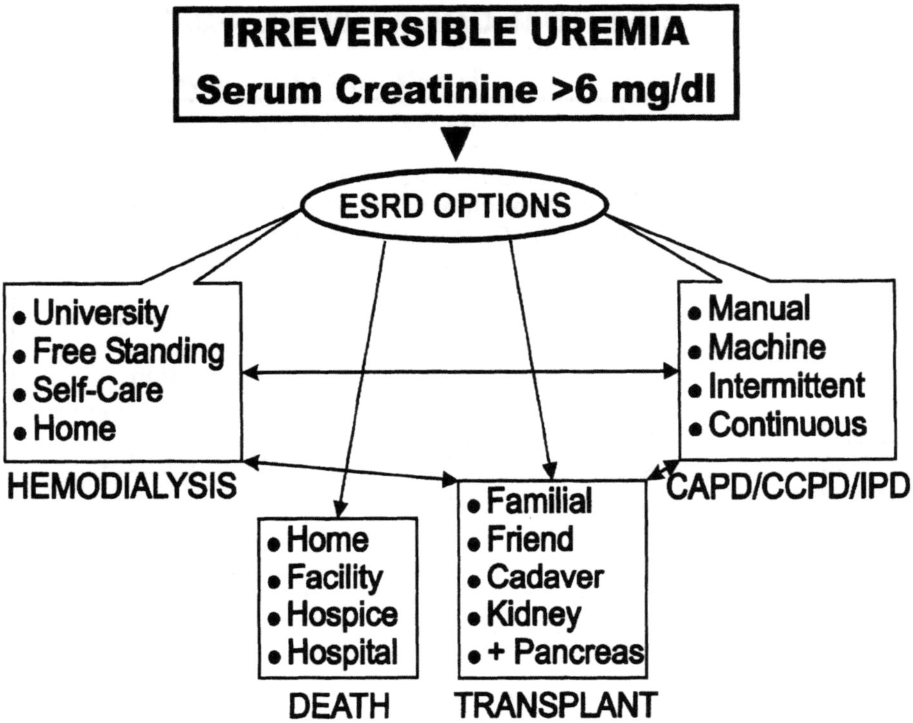

Figure 11.

despite his family's plea that he start maintenance dialysis. Because azotemic diabetic patients typically are depressed, however, a rational decision to die must be distinguished from temporary despair over a current setback. Despondent diabetics on occasion respond to visits by rehabilitated dialysis patients or transplant recipients by reversing their decision to die. It is unwise to coerce acceptance of dialysis or a kidney transplant, when life has minimal (or even negative) value. Diabetic patients forced into uremia therapy by family or the health care team are often noncompliant to dietary and drug regimens, thereby expressing behavior culminating in passive suicide.

Future Directions of Therapy

Perturbed micro- and macrovascular function is strongly implicated in the cellular and molecular abnormalities of vascular endothelium in diabetes [92]. Hyperglycemia is increasingly linked to the pathogenesis of nephropathy, retinopathy, and atherosclerosis in individuals with long-duration diabetes [93]. The metabolic pathway between a high ambient glucose concentration and end organ damage in diabetes is under intensive investigation. Three candidate mechanisms are [94]: 1. activation of the aldose-reductase pathway leading to toxic accumulation of sorbitol in nerves; 2. accelerated nonenzymatic glycosylation with deposition of advanced glycosylated endproducts [95]; 3. activation of isoform(s) of protein kinase C in vascular tissue initiating a cascade of events culminating in diabetic complications [96]. PKC activity is increased in renal glomeruli, retina, aorta, and heart of diabetic animals, probably because of increased synthesis de novo of diacylglycerol (DAG), a major endogenous activator of PKC [97].

PKC in Diabetic Complications

Vascular damage in diabetes develops slowly over years to decades in the presence of continuous hyperglycemia [98]. Sustained alterations in PKC-regulated gene expression in vascular cells may contribute to the onset and progression of vascular abnormalities in diabetes. Activation of PKC begins a complex network of intracellular signaling that could change gene expression [99]. A result of altered gene expression might be changed transcription factor binding to promoter regions on responsive genes as for example, the collagenase gene promoter that responds to the PKC agonist, phorbol ester, by binding both AP-1 and PEA3 transcription factor proteins [100].

A role for PKC in the pathogenesis of diabetic nephropathy is inferred from experiments in induced-diabetic rats. Evidence for this linkage includes findings that: 1. PKC is activated in glomeruli isolated from diabetic rats [101]. 2. Activation of PKC by either intravitreal injection of the PKC agonist, phorbol 12,13-dibutyrate [102] or by exposure of granulation tissue to the PKC agonist, 12-O-tetradecanolylphorbol3-acetate in normal rats reproduces the vascular abnormalities induced by diabetes and high glucose levels [103]. 3. Mesangial cells cultured in high (27.8 mM) concentration of glucose for 5 days increase PKC and mitogen-activated protein kinase activity in their membrane fraction supporting the hypothesis that hyperglycemia induces abnormalities in the glomerular mesangium [104]. Furthermore, PKC promotes elevated levels of mRNA encoding matrix components – matrix synthesis is accelerated in early diabetic glomerulopathy – in glomeruli isolated from streptozotocin-induced diabetic rats [105]. The metalion vanadate, advocated in human diabetes for its insulin-like effects, undesirably also stimulates PKC activity in human mesangial cells in vitro thereby limiting its application in clinical trials [106].

Inhibitors of PKC

A small synthetic organic molecule belonging to the class of napthopyrans, 2-amino-4-(3-nitrophenyl)-4H-naphtho(1,2-B)-pyran-3-carbonitrile (LY290181)when applied topically (20–50 mol/l) to newly grown vessels in granulation tissue, blocked glucose-induced increases in both blood flow and permeability. Additionally, when included as 0.1% of the diet for 8 weeks in male Sprague-Dawley streptozotocin-induced diabetic rats, LY290181 prevented diabetes-induced increases in albumin permeation in the retina, nerve, and aorta, but did not effect albumin permeation in muscle or brain. Because LY290181 inhibited phorbol ester-stimulated activation of the porcine urokinase plasminogen activator (uPA) promoter (–4600/+398 linked to the chloramphenicol acetyltransferase (CAT) reporter gene (p4660CAT), it was inferred that this unique chemical may block diabetes-induced vascular dysfunction by inhibiting transcription factor binding to specific PKC-regulated genes involved in vascular functio [107].

Based on the observation that it is the PKC β_2 isoenzyme that is preferentially activated in the retina, heart, and aorta of diabetic rats [6] an orally effective inhibitor of PKC β_2 was synthesized and found to ameliorate vascular dysfunction in streptozotocin-induced diabetic rats [108]. The macrocyclic bis(indolyl)maleimidestructure (LY333531) selectively inhibits PKC β. LY333531 corrected renal functional perturbations in diabetic rats but had no effect on renal function in normal rats. GFR was 3.0 ± 0.2 ml/min in nondiabetic rats, increased to 4.6 ± 0.4 ml/min in diabetic rats, and returned to normal after oral treatment with LY333531 (1.0 and 10 mg/dk) for 8 weeks. Concomitantly, urinary albumin excretion rate was decreased in diabetic rats treated with LY333531 from 11.7 ± 0.5 mg/day to 4.9 ± 1.6 mg/day (normal 1.6 ± 0.5 mg/day). No change in glycosylated hemoglobin or DAG content in the retina or glomeruli of diabetic rats resulted from treatment with LY333531. These early trials of PKC inhibition underscore potential treatment of diabetes with pharmaceutical agents that block organ damage due to hyperglycemia without the requirement for euglycemia. The fact that no adverse reactions followed 8 weeks of oral administration in rats is an encouraging first step along the long road that may culminate in clinical trials.

ALDOSE REDUCTASE AND SORBITOL

The role of the sorbitol pathway in diabetic complications has been termed 'the longest running controversy among researchers and clinicians studying this disease [109].' The *polyol/sorbitol hypothesis* is an attractive explanation for the mechanism by which hyperglycemia induces diabetic complications because it raises the therapeutic possibility of preventing injury by blocking a biochemical pathway without the necessity for establishing often difficult to attain euglycemia. The finding of excess sorbitol in cataracts in diabetic rats [110], stimulated the hypothesis that high ambient glucose levels damage cells by increasing intracellular osmolality, decreasing myoinositol levels thereby altering Na+/K+ ATPase activity (nerves) or shifting redox potential in cells. Aldose reductase, an enzyme present in most tissues, converts glucose to sorbitol, which is further processed to fructose. The enzyme has a low affinity for glucose, and under physiologic conditions little substrate is processed. In experimental diabetes, sorbitol production is markedly enhanced by hyperglycemia leading to its accumulation and injury to cells [111]. type 2 diabetes patients with either proliferative or nonproliferative retinopathy have significantly higher levels of erythrocyte aldose reductase than do type 2 diabetes patients without retinopathy despite equivalent mean HBA1c and blood pressure levels. On the other hand, erythrocyte aldose reductase activity in type 2 diabetes patients is not correlated with age, duration of diabetes, fasting blood glucose or HbA1 [112].

Drugs which block aldose reductase activity include spirohydantoins (sorbinil), carboxylic acid derivatives (tolrestat and ponalrestat), and flavonoids. Clinical trials over the past decade have assessed the efficacy of sorbinil, tolrestat, and ponalrestat in the treatment of diabetic retinopathy, neuropathy, and nephropathy. Overall, though some positive results have been reported, the benefits have been minimal and of no clinical moment. Due to its severe and frequent toxicity, sorbinil has been withdrawn from further use. Ponalrestat has not been effective in clinical trials.

Disappointment over the unfulfilled promise of aldose reductase inhibitors may reflect selection of too low a drug study dose in anticipation of toxic reactions at higher doses and/or insufficient duration of study. From the DCCT it is evident that the complications of diabetes develop slowly, often taking a decade or longer to become evident clinically. Therefore, drug evaluations in the early stages of nephropathy, neuropathy, and retinopathy may not be detect statistically different differences between treatment groups for years. Large scale, long-term drug trials in a slowly evolving disorder such as diabetic vasculopathy are difficult to design and expensive to conduct. Presently, tolrestat is the only aldose reductase inhibitor being tested in the US, in a multicenter trial for diabetic nephropathy, neuropathy, and retinopathy.

ADVANCED GLYCOSYLATED ENDPRODUCTS

In health, protein alteration resulting from a nonenzymatic reaction between ambient glucose and primary amino groups on proteins to form glycated residues called Amadori products is termed the Maillard reaction. After a series of dehydration and fragmentation reactions, Amadori products are transformed to stable covalent adducts called advanced glycosylation endproducts (AGEs). In diabetes, accelerated synthesis and tissue deposition of AGEs is proposed as a contributing mechanism in the pathogenesis of clinical complications [113]. Accumulation of AGEs in the human body is implicated in aging and in complications of renal failure [114] and diabetes [115]. AGEs are bound to a cell surface receptor (RAGE) inducing expression of vascular cell adhesion molecule-1 (VCAM-1), an endothelial [116] cell surface cell-cell recognition protein that can prime diabetic vasculature for enhanced interaction with circulating monocytes thereby initiating vascular injury [117]. Specific AGEs are now being identified within tissues injured in diabetes. N-(carboxymethyl)valine (CMV), for example, induces higher levels of CMV-hemoglobin in diabetic individuals than in healthy persons suggesting that it may be a valuable marker for progression of diabetic nephropathy [118].

Glomerular hyperfiltration, characteristic of the clinically silent early phase of diabetic nephropathy may be induced by Amadori protein products – in rats, infusion of glycated serum proteins induces glomerular hyperfiltration [119]. Nitric oxide, produced by endothelial cells, the most powerful vasodilator influencing glomerular hemodynamics [120], has enhanced activity in early experimental diabetes [121]. Subsequently, AGEs, by quenching nitric oxide synthase activity, limit vasodilation and reduce glomerular filtration rate [122].

AGEs in atherosclerosis

An additional avenue of AGE research concerns their potential to promote rapidly progressive atherosclerosis in patients with diabetes and renal insufficiency. The azotemic diabetic manifests elevation in the plasma level of apoprotein B (ApoB), very low density lipoprotein (VLDL) and low density lipoprotein (LDL). Under this circumstance, circulating high levels of AGEs react directly with plasma lipoproteins preventing their recognition by tissue LDL receptors significantly increasing the level of AGE-modified LDL in the plasma of diabetic or nondiabetic uremic patients compared with normal controls, possibly contributing to the accelerated atherosclerosis that is typical of diabetes and uremia [123]. LDL modified in vitro by AGE-peptides to the level present in azotemic diabetic patients markedly impaired LDL clearance kinetics when injected into transgenic mice expressing the human LDL receptor indicating that AGE modification of LDL receptors promotes elevated LDL levels in azotemic diabetic patients. Immunohistochemical analysis of coronary arteries obtained from type 2 diabetes patients stained with anti-AGE antibodies showed high levels of AGE reactivity within atherosclerotic plaques [124]. From these observations, a linkage between hyperglycemia, hyperlipidemia, nitric oxide activity, and atherosclerosis in diabetes becomes obvious. Central to all this pathologic processes is the primary event of hyperglycemia.

AGE formation probably also contributes to development of diabetic complications by changing the structure and function of extracellular matrix in the glomerular mesangium and elsewhere. When segregated by type, it has been noted that on the surface of type IV collagen from basement membrane, AGE formation decreases binding of the noncollagenous NC1 domain to the helix-rich domain, interfering with the lateral association of these molecules into a normal lattice structure [125]. By contrast, denaturation of type I collagen, a substance also found in glomerular mesangium, by AGEs expands molecular packing [126]. Changing the integrity of collagen adversely affects biological functions important to normal vascular tissue integrity such as reaction to endothelium-derived relaxing factor (nitric oxide) and antiproliferative factor [17]. In this context, as cited above, studies in rodents suggest that AGEs exert their toxicity by impairing nitric oxide-mediated vital processes including neurotransmission [127], wound healing [128] and blood flow in small vessels [129]. Thus, AGEs by blocking the synthesis of nitric oxide, almost certainly interfere with maintenance of normal physiologic processes such as autoregulation of blood flow [130]. The myriad actions of nitric oxide pertinent to nephrologists have been recently reviewed journal [131].

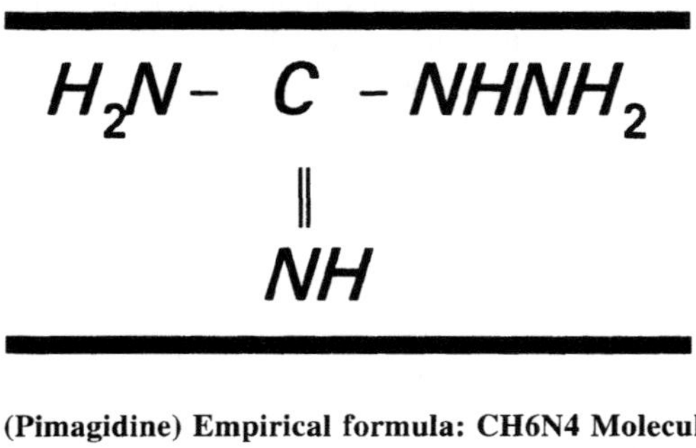

Aminoguanidine (Pimagidine) Empirical formula: CH6N4 Molecular weight: 110.5

Figure 12.

Pimagidine

Pharmacologic prevention of AGE formation is an attractive means of preempting diabetic microvascular complications because it bypasses the necessity of having to attain euglycemia, an often unattainable goal. Pimagidine (aminoguanidine), interferes with non-enzymatic glycosylation [132] and reduces measured AGE levels leading to its investigation as a potential treatment. Pimagidine was selected because its structure is similar to α-hydrazinohistidine, a compound known to reduce diabetes-induced vascular leakage, while having opposite effects on histamine levels [133].

Pimagidine treatment in rats made diabetic with streptozotocin preempts complications viewed as surrogates for human diabetic complications. Representative examples from the recent literature include: 1) Preventing development of cataracts in rats 90 days after being made 'moderately diabetic' (< 350 mg/dl plasma glucose); lens soluble and insoluble AGE fractions were inhibited by 56% and 75% by treatment with aminoguanidine 25 mg/kg body weight starting from the day of streptozotocin injection [134]. 2) Blocking AGE accumulation (measured by tissue fluorescence) in glomeruli and renal tubules in rats 32 weeks after induction of diabetes 32 weeks earlier; ponalrestat, an aldose reductase inhibitor, did not block AGE accumulation [135]. Treatment of streptozotocin-induced diabetic rats with pimagidine prevents glomerular basement membrane thickening typical of renal morphologic changes noted in this model of diabetic nephropathy [136]. 3) Reducing severity of experimental diabetic retinopathy as judged by a decrease in the number of acellular capillaries by 50% and complete prevention of arteriolar deposition of PAS-positive material and microthrombus formation after 26 weeks of induced diabetes in spontaneous hypertensive rats [137]. 4) Ameliorating slowing of sciatic nerve conduction velocity dose dependently after treatment at three doses of 10, 25, and 50 mg/kg for 16 weeks [138]. Autonomic neuropathy (neuroaxonal dystrophy), however, was not prevented by treatment with pimagidine [139]. 5) Preventing development of the 'stiff myocardium' that is a main component of diabetic cardiomyopathy; in a dose of 7.35 mmol/kg/dl for 4 months, aminoguanidine prevented both increased myocardial collagen fluorescence indicative of AGE accumulation and decreased left ventricular end-diastolic compliance characteristic of early cardiomyopathy [140]. 6) Preventing the diabetes-induced 24% impairment in maximal endothelium-dependent relaxation to acetylcholine for phenylephrine precontracted aortas by treatment for 2 months in a dose of 1 g/kg/day [141]. The strategic potential of blocking AGE formation to impede development of diabetic complications has been reviewed [142, 143]. An attractive aspect of this approach to impeding diabetic complications is the elimination of the necessity for euglycemia [144]. Neither animal models of pancreatic beta cell toxicity-induced diabetes [145–148] nor inbred spontaneous rodent strains with sustained hyperglycemia emulate the sequential intraglomerular hyperfiltration, glomerulopathy and renal insufficiency of human diabetic nephropathy [149]. In normal rats, however, daily injection of AGE-modified rat albumin 25 mg per kg per day i.v. induced both albuminuria and glomerulosclerosis – perturbations characteristic of diabetic nephropathy [150]. The nephropathy was AGE-specific as treatment with pimagedine ameliorated its severity.

Mechanism of pimagidine action

Pimagidine treatment significantly prevents NO activation and limits tissue accumulation of AGEs. Corbett et al. speculate that pimagidine inhibits interleukin-1 beta-induced nitrite formation (an oxidation product of NO) [151]. In a derivative study, pimagidine but not methylguanidine, inhibited AGE formation from L-lysine and G6P while both guanidine compounds were equally effective in normalizing albumin permeation in induced-diabetic rats [152]. A role for a relative or absolute increase in NO production in the pathogenesis of early diabetic vascular dysfunction was also Inferred as was the possibility that inhibition of diabetic vascular functional changes by pimagidine may reflect inhibition of NO synthase activity rather than, or in addition to, prevention of AGE formation. An alternative role assigned to pimagidine is that of a glucose competitor for the same protein-to-protein bond [153] that becomes the link for the formation and accumulation of irreversible and highly reactive advanced glycation end-products (AGE) over long-lived fundamental molecules such as the constituents of arterial wall collagen, GBM, nerve myelin, DNA and others. The mechanism by which pimagidine prevents renal, eye, nerve, and other microvascular complications in animal models of diabetes is under investigation [154]. Separate multicenter clinical trials of pimagidine intype 1 diabetes and type 2 diabetes whose proteinuria is attributable to diabetic nephropathy are in progress. Adults with documented diabetes, fixed proteinuria ≥ 500 mg/day and a serum creatinine concentration ≥ 1.2 mg/dl for women and ≥ 1.5 mg/dl for men, will be randomly assigned to treatment with pimagidine or a placebo for four years. The effect of treatment on the amount of proteinuria, progression of renal insufficiency, and the course of retinopathy will be monitored.

AGEs in renal failure

Uremia in diabetes is associated with both a high serum level of AGEs and accelerated macro and microvasculopathy. The renal clearance of AGE-peptides is 0.72 ± 0.23 ml/min for normal subjects and 0.61 ± 0.2 ml for diabetics with normal glomerular filtration (*P* value NS) [155]. Diabetic uremic patients accumulate advanced glycosylated end-products in 'toxic' amounts that are not decreased to normal by hemodialysis or peritoneal dialysis [156] but fall sharply, to within the normal range, within 8 hours of restoration of half-normal glomerular filtration by renal transplantation [157]. It follows that the higher mortality of hemodialysis treated diabetic patients compared with those given a renal transplant may relate – in part – to persistent AGE toxicity. Trials of pimagidine in diabetic hemodialysis patients begun in 1996 are designed to test this hypothesis. Due to a high incidence of adverse effects including elevated liver enzymes, the trial of pimagidine in type 2 diabetes was discontinued. is pending.

Epilogue: Ten 'Take Home' Points

There is hazard in simplifying complex matters into unadorned extractable truths. Diabetes is a disorder that generates approximately one-quarter of hospitalizations in large medical services while sustaining legions of physicians, nurse-educators, physical therapists, nutritionists, and even focused malpractice attorneys. No fewer than 18 journal focus on diabetes, in English. It may be an exercise in arrogance to attempt to distill a huge information data base into compact digestible parcels. Nevertheless, there are identifiable themes that can be summarized at any one time:

1. *Definition.* Diabetes mellitus comprises a family of diseases, not a single or pair of disorders. At the least, tissue and organ complications may vary with diabetes type (variety) and environmental exposures (obesity, smoking as examples). Ketoacidosis occurs in type 2 diabetes as do low C-peptide levels blurring distinction between types. Renal failure due to diabetes is in the large majority (> 90%) caused by type 2 diabetes. Terms that can be confusing in the classification of diabetes include: *insulin-dependent*, meaning that the patient will die without insulin rather than that the physician opted for an insulin-based regimen and *insulin-requiring*, again signaling no more than that a choice to include insulin was made by the health care team. Throughout its early years, the USRDS accepted insulin treatment as sufficient evidence to define type 1 diabetes, adding to the overestimate of type 1 patients in the renal failure population.
2. *Risk Factors.* Nephropathy in diabetes correlates with genetic predisposition, type of diabetes, and control of compounding risk factors, especially hypertension and hyperglycemia.
3. *Reversibility of Nephropathy.* Early perturbations of diabetic nephropathy including glomerular hyperfiltration, microalbuminuria, and glomerular mesangial expansion are reversible with phramacologic therapy (angiotensin converting enzyme inhibitors) and/or a functioning pancreas transplant (type 1 diabetes).
4. *Effect of Age.* There is minimal support for the view that the course of nephropathy in diabetes is slower in older individuals. Former belief that type 2 diabetes is inherently a mild affliction has yielded to present understanding that loss of glomerular filtration rate is at least as rapid in type 1 as in type 2 diabetes. A prospective study of the course of nephropathy in large cohorts of Japanese diabetic individuals found more rapid renal functional deterioration in type 2 than in type 1 subjects.
5. *Impact of Comorbidity.* Managing an individual with progressive kidney disease due to diabetes involves identifying and treating extrarenal manifestations of diabetes (complications) especially cardiovascular, cerebrovascular, and retinopathy. Rehabilitation and survival are contingent on recognition and treatment of these comorbid complications. Supervision of individuals with progressive diabetic nephropathy necessitates timely and repetitive collaboration with a podiatrist, an ophthalmologist specializing in laser surgery and a cardiologist skilled in assessing coronary artery disease.
6. *Effective Interventions.* Solid evidence justifies pursuit of normotension and euglycemia as prime objectives in management of diabetes. Suggestive inferences support correcting hyperlipidemia while limiting dietary protein though neither case is closed. Whether antihypertensive drug regimens in diabetes must be based on angiotensin converting enzyme inhibitors (ACEi) and/or an AT1 receptor blocker derives more from consensus than evidence based controlled trials. It must be recalled that the most impressive preservation of declining renal function by blood pressure control was achieved by Parving et al. prior to introduction of ACEi or calcium blockers.

7. *Options in Uremia Therapy.* There are four main treatment choices open to diabetic patients who lapse into end-stage renal disease (ESRD): 1. No further renal therapy (death). 2. Peritoneal dialysis. Hemodialysis. Kidney (kidney plus pancreas) transplantation. Matching an individual patient to optimized treatment requires consideration of personal preference as well as physical reality. As examples: Solid organ transplants are unwise in a patient with unstable cardiovascular disease. Peritoneal dialysis is preempted in the presence of an ileostomy. Home hemodialysis is impractical in the absence of a willing partner. Recipients of a kidney transplant attain the best survival and most complete rehabilitation of any diabetic ESRD patients. Whether this superior outcome results from selection bias (cherry picking) in choosing the most fit for a kidney transplant leaving those with excessive comorbidity behind in the cohort treated by peritoneal dialysis or hemodialysis. An alternative explanation may lie in the improved metabolic milieu produced by a kidney transplant that extracts nitrogenous molecules such as advanced glycosylated end-products that are retained in 'toxic' amounts by dialysis patients.
8. *ESRD Treatment Outcome.* Throughout the past decade, survival of diabetic ESRD patients whether treated by dialysis or a kidney transplant has improved each year. But survival after five years falls off precipitously in those treated by dialysis: fewer than 1 in 20 live ten years whether treated by peritoneal or hemodialysis. Benefits of increased dose of dialysis and higher hematocrits afforded by erythropoietin and intravenous iron replenishment may be evident in current registry reports.
9. *Pancreas Transplantation.* As this is written, a fresh approach to islet transplantation has gained attention because of the avoidance of steroids and the sequence of ten successes in a row in inducing freedom from insulin in type 1 diabetes. But, each recipient has required islets harvested from more than one donor (as many as four pancreases per recipient) immediately raising the problem of sufficient donor pancreases for broad application. Furthermore, type 1 diabetes is responsible for no more than 5% of ESRD attributed to diabetes. Whether the type 2 diabetic individual will benefit from pancreas or islet transplantation is unknown. Not far down the road, however, growth of immortalized Beta cells and humanized porcine pancreata and islets may solve the supply dilemma just as insertion of the insulin gene into yeast cells ended worry over an insufficient supply of insulin.
10. *Brightening Prognosis.* Whether due to incremental advances in metabolic and hypertensive normalizing of diabetes' main perturbations or bold new therapies based on molecular manipulation, there is pervasive hope and optimism palpably present among those treating diabetes. Particularly appealing are regimens that do not enslave the diabetic patient with unending daily blood tests and unattainable dietary regulations. In this regard, stimulation of the insulin receptor by oral medications and interdiction of oxidative stress by blockade of kinin release caused by advanced glycosylated end-products are promises now under test. An entirely fresh approach to attaining euglycemia is now being pursued. After elucidation of the function of the insulin receptor, a search was initiated through 50,000 compounds to unearth any that might stimulate the receptor's phosphorylating action activating it's intrinsic tyrosine kinase. A nonpeptidyl five-ringed quinone (L-783,281) extracted from the Kenyan fungus Pseudomassaria was discovered to have the capacity to mimic insulin action in biochemical and cellular assays acting

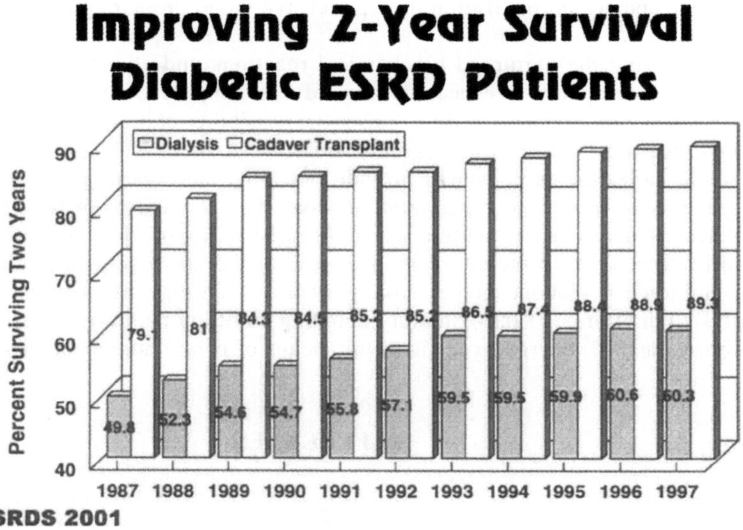

Figure 13.

as an 'insulin pill'clviii. In two mutant mouse models of diabetes, L-783,281 administered orally lowered blood glucose levels by up to 50% – equivalent to the reduction attained with current oral anti-diabetic therapies.

The novel concept of cell receptor modification to alter downstream signaling molecules is also being extended to bone marrow [159], and solid organ transplant recipients in what has been termed 'the new immunology' [160]. For individuals with diabetes, there is reasonable promise of long-term successful 'tolerance' of kidney and pancreas allografts in those whose vasculopathy is best managed by organ substitution.

When I entered medical school, wise old professors advocated intensified study of first syphilis and later tuberculosis as systemic diseases that encompassed all aspects of organ injury that might be encountered by a practitioner. Both diseases have receded limiting exposure of 21st Century urban physicians to old enemies that comprised so much of medicine's heritage. It is not unreasonable to suggest that diabetes will likely follow the same path to relative obscurity. In a year when the entire human genome has been sequenced it is far from fantasy to think that diabetes may well be erased as a threat to longevity before this decade ends.

While the specter of a potentially overwhelming 21st century burden of type 2 diabetes is frightening, the progress of the past decade in interdicting diabetic complications provides reason to remain hopeful that medical progress will prove equal to the challenge. Gradually, almost unperceptively, the formerly stark prognosis for combined diabetic nephropathy and retinopathy has been transformed from a tragedy signaling near-term death to a clarion call for effective interventive measures. Managing diabetes mandates normalization of hypertensive blood pressure plus attaining normoglycemia.

Recognizing the perturbed biochemical reactions underlying the pathogenesis of diabetic vasculopathy – especially the adverse impact of accumulated advanced glycosylated end-products (AGEs) – raises the possibility of blocking end-organ damage without necessarily correcting hyperglycemia.

While molecular biology will provide blocking and diverting techniques to mute the panoply of tissue and organ complications the elusive goal of total prevention of diabetes as a disease is within reach. Promising approaches include: (1) insulin sensitizers including protein tyrosine phosphatase-1B and glycogen synthase kinase 3, (2) inhibitors of gluconeogenesis like pyruvate dehydrogenase kinase inhibitors, (3) lipolysis inhibitors, (4) fat oxidation including carnitine palmitoyltransferase I and II inhibitors, and (5) energy expenditure by means of beta 3-adrenoceptor agonists.

REFERENCES

1. United States Renal Data System, USRDS 2001 Annual Data Report, The National Institutes of Health, National Institute of Diabetes and Digestive and Kidney Diseases. Bethesda, MD, June, 2001.
2. Mauer SM, Chavers BM. A comparison of kidney disease in type I and type II diabetes. Adv Exp Med Biol 1985; 189: 299–303.
3. Centers for Disease Control and Prevention. National Diabetes Fact Sheet: National estimates and general information on diabetes in the United States, Revised edition. US Department of Health and Human Services, Centers for Disease Control and Prevention, Atlanta, GA, 1998.
4. National Center for Health Statistics. Health, United States, 1998. Public Health Service, Hyattsville, Maryland, 1999.
5. Nicholas SB, Kawano Y, Wakino S, Collins AR, Hsueh WA. Expression and function of peroxisome proliferator-activated receptor-gamma in mesangial cells. Hypertension 2001; 37 (2 Part 2): 722–727.
6. Blohme G, Nyström L, Arnqvist HG, et al. Male predominance of type 1 (insulin-dependent) diabetes in young adults: results from a 5-year prospective nationwide study of the 15–34 age group in Sweden. Diabetologia 1993; 35: 56–62.
7. Polonsky KS. The β-cell in diabetes: from molecular genetics to clinical research. Diabetes 1995; 44: 705–717.
8. Harris MI, Eastmen RC. Early detection of undiagnosed non-insulin-dependent diabetes mellitus. JAMA 1996; 276: 1261–1262.
9. Madsbad S., McNair P, Faber OK, Binder C, Christiansen Transbol J. β-cell function and metabolic control in insulin-dependent treated diabetics. Acta Med Endocrinol 1980; 1980: 196–200.
10. Wahren J, Johansson B-L, Wallberg-Henriksson H, Linde B, Fernqvist-Forbes E, Zierath JR. C-peptide revisited – new physiological effects and therapeutic implications. J Intern Med 1996; 240: 115–124.
11. Zimmet PZ. Kelly West Lecture 1991. Challenges in diabetes epidemiology – from West to the rest. Diabetes Care 1992; 15: 232–252.
12. Harris M, Hadden WC, Knowles WC, et al. Prevalence of diabetes and impaired glucose tolerance and plasma glucose levels in US population aged 20–74 yr. Diabetes 1987; 36: 523–534.
13. Stephens GW, Gillaspy JA, Clyne D, Mejia A, Pollak VE. Racial differences in the incidence of end-stage renal disease in Types I and II diabetes mellitus. Am J Kidney Dis 1990; 15: 562–567.
14. Haffner SM, Hazuda HP, Stern MP, Patterson JK, Van-Heuven WA, Fong D. Effects of socio-economic status on hyperglycemia and retinopathy levels in Mexican Americans with NIDDM. Diabetes Care 1989; 12: 128–134.
15. National Diabetes Data Group. Diabetes in America. NIH Publication No. 85-1468, August 1985.
16. Council on Ethical and Judicial Affairs. Black-white disparities in health care. JAMA 1990; 163: 2344–2346.
17. Melton LJ, Palumbo PJ, Chu CP. Incidence of diabetes mellitus by clinical type. Diabetes Care 1983; 6: 75–86.

18. Ritz E, Stefanski A. Diabetic nephropathy in Type II diabetes. Am J Kidney Dis 1996; 2: 167–194.
19. Sheehy MJ. HLA and insulin-dependent diabetes. A protective perspective. Diabetes 1992; 41: 123–129.
20. Biesenback G, Janko O, Zazgornik J. Similar rate of progression in the predialysis phase in type I and type II diabetes mellitus. Nephrol Dial Transplant 1994; 9: 1097–1102.
21. Wirta O, Pasternack A, Laippala P, Turjanmaa V. Glomerular filtration rate and kidney size after six years disease duration in non-insulin-dependent diabetic subjects. Clinical Nephrology 1996; 45: 10–17.
22. Abourizk NN, Dunn JC. Types of diabetes according to National Diabetes Data Group Classification. Limited applicability and need to revisit. Diabetes Care 1990; 13: 1120–1122.
23. Sims EAH, Calles-Escandon J. Classification of diabetes. A fresh look for the 1990s? Diabetes Care 1990; 13: 1123–1127
24. Berman DH, Friedman EA, Lundin AP. Aggressive ophthalmological management in diabetic ESRD: a study of 31 consecutively referred patients. Amer J Nephrol 1992; 12: 344–350.
25. Clark DW, Nowak TV. Diabetic gastroparesis. What to do when gastric emptying is delayed. Postgrad Med 1994; 95: 195–198, 201–204.
26. Battle WM, Cohen JD, Snape WJ Jr. Disorders of colonic motility in patients with diabetes mellitus. Yale J Biol Med 1983; 56: 277–283.
27. Lux G. Disorders of gastrointestinal motility – diabetes mellitus. Leber Magen Darm 1989; 19: 84–93.
28. Miodovnik M, Rosenn BM, Khoury JC, Grigsby JL, Siddiqi TA. Does pregnancy increase the risk for development and progression of diabetic nephropathy? Am J Obstet Bynecol 1996; 174: 1180–1191.
29. Kaaja R, Sjoberg L, hellsted T, Immonenen I, Sane T, Teramo K. Long-term effects of pregnancy on diabetic complications. Diabet Med 1996; 13: 165–169.
30. Meisel A. The Right to Die. John Wiley and Sons, New York, 1989, p 122.
31. Kjellstrand CM. Practical aspects of stopping dialysis and cultural differences. In: Kjellstrand CM, Dossetor JB (eds), Ethical Problems in Dialysis and Transplantation. Kluwer Academic Publishers, Dordrecht, 1992.
32. Pommer W, Bressel F, Chen F, Molzahn. There is room for improvement of preterminal care in diabetic patients with end-stage renal failure – The epidemiological evidence in Germany. Nephrol Dial Transplant 1997; 12: 1318–1230.
33. Passa P. Diabetic nephropathy in the NIDDM patient on the interface between diabetology and nephrology. What do we have to improve? Nephrol Dial Transplant 1997; 12: 1316–1317.
34. Cheigh J, Raghavan J, Sullivan J, Tapia L, Rubin A, Stenzel KH. Is insufficient dialysis a cause for high morbidity in diabetic patients? J Amer Soc Nephrol 1991; 317 (abstract).
35. Lowder GM, Perri NA, Friedman EA. Demographics, diabetes type, and degree of rehabilitation in diabetic patients on maintenance hemodialysis in Brooklyn. J Diabetic Complications 1988; 2: 218–226.
36. Lindblad AS, Nolph KD, Novak JW, Friedman EA. A survey of the NIH CAPD Registry population with end-stage renal disease attributed to diabetic nephropathy. J Diabetic Complications 1988; 2: 227–232.
37. Legrain M, Rottembourg J, Bentchikou A, et al. Dialysis treatment of insulin dependent diabetic patients. Ten years experience. Clin Nephrol 1984; 21: 72–81.
38. Rubin J, Hsu H. Continuous ambulatory peritoneal dialysis: ten years at one facility. Amer J Kidney Dis 1991; 17: 165–169.
39. Prevention of peritonitis in CAPD. Lancet 1991; 337: 22–23.
40. Sutherland DER, Morrow CE, Fryd DS, Ferguson R, Simmons RL, Najarian JS. Improved patient and primary renal allograft survival in uremic diabetic recipients. Transplantation 1982; 34: 319–325.
41. Sollinger HW, Ploeg RJ, Eckhoff DE, Stegall MD, Isaacs R, Pirsch JD, D'Alessandro AM, Knechtle SJ, Kalayoglu M Belzer FO. Two hundred consecutive simultaneous pancreas-kidney transplants with bladder drainage. Surgery 1993; 114: 736–743.
42. Shaffer D, Simpson MA, Madras PN, Sahyoun AI, Conway PA, Davis CP, Monaco AP. Kidney transplantation in diabetic patients using cyclosporine. Five-year follow-up. Arch Surg 1995; 130: 287–288.
43. Robertson RP. Pancreas transplantation in humans with diabetes mellitus. Diabetes 1991; 40: 1085–1089.
44. Diagnostic and therapeutic technology assessment (DATTA). Pancreatic transplantations. JAMA 1991; 265: 510–514.
45. Cecka JM, Terasaki PI. 1991. The UNOS Scientific Renal Transplant Registry – 1991. Clinical Transplants 1991, (Terasaki P.I. ed.) UCLA Tissue Typing Lab, Los Angeles CA, pp 1–11, 1992.
46. Sutherland D, Gruessner A, Moudry-Munns K. International Pancreas Transplant Registry report. Transplant Proc 1994; 26: 407–411.
47. Douzdjian V, Ricd JC, Gugliuzza KK, Fish JC, Carson RW. Renal allograft and patient outcome after

transplantation: Pancres-kidney versus kidney-alone transplant in Type 1 diabetic patients versus kidney-alone transplants in nondiabetic patients. Amer J Kidney Dis 1996; 27: 106–116,
48. Manske CL, Wang Y, Thomas W. Mortality of cadaveric kidney transplantation versus combined kidney-pancreas transplantation in diabetic patients. Lancet 1995; 346: 1658–1662.
49. Kennedy WR, Navarro X, Goetz FC, et al. Effects of pancreatic transplantation on diabetic neuropathy. N Engl J Med 1990; 322: 1031–1037.
50. Ramsay RC, Goetz FC, Sutherland DE, Mauer SM, Robison LL, Cantril HL, Knobloch WH, Najarian JS. Progression of diabetic retinopathy after pancreas transplantation for insulin-dependent diabetes mellitus. N Engl J Med 1988; 318: 208–214.
51. Remuzzi G, Ruggenenti P, Mauer SM. Pancreas and kidney/pancreas transplants: experimental medicine or real improvement? Lancet 1994; 353: 27–31.
52. Taylor RJ, Bynon JS, Stratta RJ. Kidney/pancreas transplantation: a review of the current status. Urol Clin North Am 1994; 21: 343–354.
53. Brunner FP, Fassbinder W, Broyer M, Oules R, Brynger H, Rizzoni G, Challah S, Selwood NH, Dykes SR, Wing AJ. Survival on renal replacement therapy: data from the EDTA Registry. Nephrol Dial Transplant 1988; 3: 109–122.
54. Gokal R, Jakubowski C, King J, Hunt L, Bogle S, Baillod R, Marsh F, Ogg C, Oliver D, Ward M, et al. Outcome in patients on continuous ambulatory peritoneal dialysis and haemodialysis: 4-year analysis of a prospective multicentre study. Lancet 1988; 2: 1105–1109.
55. Charra B, VoVan C, Marcelli D, Ruffet M, Jean G, Hurot JM, Terrat JC, Vanel T, Chazot C. Diabetes mellitus in Tassin, France: remarkable transformation in incidence and outcome of ESRD in diabetes. Adv Ren Replace Ther 2001; 8: 42–56.
56. Maiorca R, Cancarini G, Manili L, Brunori G, Camerini C, Strada A, Feller P. CAPD is a first class treatment: results of an eight-year experience with a comparison of patient and method survival in CAPD and hemodialysis. Clin Nephrol 1988; 30 (Supp 1): S3–S7.
57. Burton PR, Walls J. Selection-adjusted comparison of life-expectancy of patients on continuous ambulatory peritoneal dialysis, haemodialysis, and renal transplantation. Lancet 1987; 1: 1115–1119.
58. Tzamaloukas AH, Murata GH, Malhotra D. Renal clearances in continuous ambulatory peritoneal dialysis: differences between diabetic and non-diabetic subjects. Int J Artif Organsr 2001; 24: 203–207.
59. US Renal Data System, USRDS 1997 Annual Data Report, The National Institutes of Health, National Institute of Diabetes and Digestive and Kidney Diseases, Bethesda, MD, April 1997.
60. Karnofsky DA, Burchenal JH. The clinical evaluation of chemotherapeutic agents in cancer. In: MacLeod CM (ed), Evaluation of Chemotherapeutic Agents. Columbia University Press, New York, 1949, pp 191–205.
61. Carlson DM, Johnson WJ, Kjellstrand CM. Functional status of patients with end-stage renal disease. Mayo Clin Proc 1987; 62: 338–344.
62. Gutman RA, Stead WW, Robinson RR. Physical activity and employment status of patients on maintenance dialysis. N Engl J Med 1981; 304: 309–313.
63. Ifudu O, Paul H, Mayers JD, Cohen LS, Brezsynyak WF, Herman AI, Avram MM, Friedman EA. Pervasive failed rehabilitation in center-based maintenance hemodialysis patients. Am J Kidney Die 1994; 23: 394–400.
64. Ifudu O, Mayers J, Matthew J, Tan CC, Cambridge A, Friedman EA. Dismal rehabilitation in geriatric inner-city hemodialysis patients. JAMA 1994; 271: 29–33.
65. Bennett PH, Haffner S, Kasiske BL, Keane WF, Mogensen CE, Parving HH, Steffes MW, Striker GE. Screening and management of microalbuminuria in patients with diabetes mellitus: recommendations to the Scientific Advisory Board of the National Kidney Foundation from an ad hoc committee of the Council on Diabetes Mellitus of the National Kidney Foundation. Amer J Kidney Dis 1995; 25: 107–112.
66. Ravid M, Lang R, Rachmani R, Lishner M. Long-term renoprotective effect of angiotensin-converting enzyme inhibition in non-insulin-dependent diabetes mellitus. A 7-year follow-up study. Ann Intern Med 1996; 156: 286–289.
67. Sawicki PT, Dahne R, Bender R, Berger M. Prolonged QT interval as a predictor of mortality in diabetic nephropathy. Diabetologia 1996; 39: 77–81.
68. Nathan DM. Long-term complications of diabetes mellitus. N Engl J Med, 1993, 328: 1676–1682.
69. Tuttle KR, Bruton JL, Perusek MC, Lancaster JL, Kopp DT, DeFronzo RA. Effect of strict glycemic control on renal hemodynamic response to amino acids and renal enlargement in insulin-dependent diabetes mellitus. N Engl J Med 1991; 324: 1626–1632.

70. Robison WG Jr, Tillis TN, Laver N, Kinoshita JH. Diabetes-related of the rat retina prevented with an aldose inhibitor. Exp Eye Res 1990; 50: 355–366.
71. The Diabetes Control and Complications (DCCT) Research Group. Effect of intensive therapy on the development and progression of diabetic nephropathy in the Diabetes Control and Complications Trial. Kidney Internat 1995; 47: 1703–1720.
72. Solerte SB, Fioravanti M, Patti AL, Schifino N, Zanoletti MG, Inglese V, Ferrari E. Pentoxifylline, total urinary protein excretion rate and arterial blood pressure in long-term insulin-dependent diabetic patients with overt nephropathy. Acta Diabetol Lat 1987; 24:229-239.
73. Solerte SB, Ferrari E. Diabetic retinal vascular complications and erythrocyte filterability; results of a 2-year follow-up study with pentoxifylline. 1985; 4: 341–350.
74. Chaiken RL, Palmisano J, Norton ME, Banerji MA, Bard M, Sachimechi I, Lebovitz HE. Interaction of hypertension and diabetes on renal function in black NIDDM subjects. Kidney Internat 1995; 47: 1697–1702.
75. Christlieb AR. Treatment selection considerations for the hypertensive diabetic patient. Arch Intern Med 1990; 150: 1167–1174.
76. Keane WF, Shapiro BE. Renal protective effects of angiotensin-converting enzyme inhibition. Am J Cardiol 1990; 65: 49I–59I.
77. Siegel EG. Normoglycemia as a therapy goal in diabetes treatment – concept and realization. Klin Wochenschr 1990; 68: 306–312.
78. Euclid study group. Randomised pacebo-controlled trial of lisinopril in normotensive pateints with insulin-dependent diabetes and normoalbuminuria or microalbuminuria. Lancet 1997; 349: 1787–1792.
79. Donnelly R, Molyneaux LM, Willey KA, Yue DK. Comparative effects of indapamide and captopril on blood pressure and albumin excretion rate in diabetic microalbuminuria. Am J Cardiol 1996; 77: 26B–30B.
80. Bakris GL, Smith A. Effects of sodium intake on albumin excretion in patients with diabetic nephropathy treated with long-acting calcium antagonists. Ann Intern Med 1996; 125: 201–204.
81. Liou HH, Huang TP, Campese VM. Effect of long-term therapy with on proteinuria and renal function in with non-insulin-dependent diabetes and with non-diabetic renal diseases. Nephron 1995; 69: 41–48.
82. Gissen Group. Randomised placebo-controlled trial of effect of ramipril on decline in glomerular filtration rate and risk of terminal renal failure in proteinuric, non-diabetic nephropathy. Lancet 1997; 349: 1857–1863.
83. Should all patients with type 1 diabetes mellitus and microalbuminuria receive angiotensin-converting enzyme inhibitors? A meta-analysis of individual patient data. The ACE Inhibitors in Diabetic Nephropathy Trialist Group. Ann Intern Med 2001;134: 370–379.
84. Mainous AG 3rd, Gill JM. The lack of screening for diabetic nephropathy: evidence from a privately insured population. Fam Med 2001; 33: 115–119.
85. Sawicki PT, Muhlhauser I, Bender R, Pethke W, Heinemann L, Berger M. Effects of smoking on blood pressure and proteinuria in patients with diabetic nephropathy. J Intern Med 1996; 239: 345–352.
86. Wang PH, Lau J, Chalmers TC. Meta-analysis intensive blood glucose control on late complications of type I diabetes. Lancet 1993; 341: 1306–1309.
87. Evanoff G, Thompson C, Brown J, Weinman. Prolonged dietary protein restriction in diabetic nephropathy. Arch Intern Med 1989; 149: 1129–1133.
88. Jameel N, Pugh JA, Mitchell BD, Stern MP. Dietary protein intake is not correlated with clinical proteinuria in NIDDM. Diabetes Care 1992; 15: 178–183.
89. Pedrinini MT, Levey AS, Lau J, Chalmers TC, Wang PH. The effect of dietary protein restriction on the progression of diabetic and nondiabetic renal diseases: a meta-analysis. Ann Inatern Med 1996; 124: 627–632.
90. Wilczek HE, Jaremko G, Tyden G, Groth CG. Evolution of diabetic nephropathy in kidney grafts. Evidence that a simultaneously transplanted pancreas exerts a protective effect. Transplantation 1995; 59: 51–57.
91. Lebel M, Kingma I, Grose JH, Langlois S. Effect of recombinant human erythropoietin therapy on ambulatory blood pressure in normotensive and in untreated borderline hypertensive hemodialysis patients. Amer J Hypertension 1995; 8: 545–551.
92. King GL, Shiba T, Oliver J, Inoguchi T, Bursell SE. Cellular and molecular abnormalities in the vascular endothelium of diabetes mellitus. Annu Rev Med 1994; 45: 179–188.
93. Wang PH, Korc M. Searching for the Holy Grail: the cause of diabetes. Lancet 1995; 346: s4.

94. Porte Jr D, Schwartz MW. Diabetes complications: Why is glucose potentially toxic? Science 1996; 272: 699–670.
95. Soulis-Liparota T, Cooper ME, Dunlop M, Jerums G. The relative roles of advanced glycation, oxidation and aldose reductase inhibition in the development of experimental diabetic nephropathy in the Sprague-Dawley rat. Diabetologia 1995; 38: 387–394.
96. Craven PA, Studer RK, Negrete H, DeRubertis FR. Protein kinase C in diabetic nephropathy. J Diabetes Complications 1995; 9: 241–245.
97. Inoguchi T, Battan R, Handler E, Sportsman JR, Heath W, King GL. Preferential elevation of protein kinase C isoform bII and diacylglycerol levels in the aorta and heart of diabetic rats: differential reversibility to glycemic control by islet cell transplantation. Proc Natl Acad Sci USA 1992; 89: 11059–11063.
98. The Diabetes Control and Complications Trial Research Group: the effect of intensive treatment of diabetes on the development and progression of long-term complications in insulin-dependent diabetes mellitus. N Engl J Med 1993; 329: 977–986.
99. Karin M. The AP-1 complex and its role in transcriptional control by protein kinase C. In: Cohen P, Foulkes JG (eds), Molecular Aspects of Cellular Regulation. Elsevier, New York, 1991, pp 235–253.
100. Gutman A, Wasylyk B. The collagenase gene promoter contains a TPA and oncogene-responsive unit encompassing the PEA3 and AP-1 binding sites. EMBO J 1990; 9: 2241–2246.
101. Craven PA, DeRubertis FR. Protein kinase C is activated in glomeruli from streptozotocin diabetic rats: possible mediation by glucose. J Clin Invest 1989; 87: 31–38.
102. Shiba T, Inoguchi T, Sportsman JR, Heath WF, Bursell S, King GL. Correlation of diacylglycerol level and protein kinase C activity in rat retina to retinal circulation. Am J Physiol 1993; 265: E783–E793.
103. Wolf BA, Williamson JR, Eamon RA, Chang K, Sherman WR, Turk J. Diacylglycerol accumulation and microvascular abnormalities induced by elevated glucose levels. J Clin Invest 1991; 87: 31–38.
104. Kikkawa HM, Sugimoto T, Koya D, Araki S, Togawa M, Shigeta Y. Abnormalities in protein kinase C and MAP kinase cascade in mesangial cells cultured under high glucose conditions. J Diabetes Complications 1995; 9: 246–248.
105. Ziyadeh FN, Fumo P, Rodenberger CH, Kuncio GS, Neilson EG. Role of protein kinase C and cyclic AMP/protein kinase A in high glucose-stimulated transcriptional activation of collagen alpha 1(IV) in glomerular mesangial cells. J Diabetes Complications 1995; 4: 255–261.
106. Wenzel UP, Fouqueray B, Biswas P, Grandaliano G, Choudhury GG, Abboud HE. Activation of mesangial cells by the phosphatase inhibitor vanadate. Potential implications for diabetic nephropathy. J Clin Invest 1995; 95: 1244–1252.
107. Birch KA, Heath WF, Hermeling RN, Johnston CM, Stramm L, Dell C, Smith C, Williamson JR, Reifel-Miller A. LY290181, an inhibitor of diabetes-induced vascular dysfunction, blocks protein kinase C-stimulated transcriptional activation through inhibition of transcription factor binding to a phorbol response element. Diabetes 1996; 45: 642–650.
108. Ishii H, Jirouske MR, Koya D, Takagi C, Xia P, Clermont A, Bursell S-E, Kern TS, Ballas LM, heath WF, Stramm LE, Feener EP, King GL. Amelioration of vascular dysfunction in diabeti rats by an oral PKC β_2 inhibitor. Science 1996; 272: 728–731.
109. Frank RN. The aldose reductase controversy. Diabetes 1994; 43: 169–172.
110. Van Heyningen R. Formation of polyols by the lens of the rat with 'sugar' cataract. Nature 1959; 184: 194–195.
111. Tomlinson DR, Stevens EJ, Diemel LT. Aldose reductase inhibitors and their potential for the treatment of diabetic complications. Trends Pharmacol Sci 1994; 15: 293–297.
112. Nishimura C, Saito T, Ito T, Omori Y, Tanimoto T. High levels of erythrocyte aldose reductase and diabetic retinopathy in NIDDM patients. Diabetologia 1994; 37: 328–330.
113. Brownlee M, Cerami A, Vlassara H. Advanced glycosylation end products in tissue and the biochemical basis of diabetic complications. N Engl J Med 1988; 318: 1315–1321.
114. Sell DR, Monnier VM. End stage renal disease and diabetes catalyze the formation of a pentose-derived crosslink from aging human collagen. J Clin Invest 1990; 85: 380–384.
115. Vlassara H, Bucala R, Striker L. Pathogenic effects of advanced glycosylation: biochemical, biological, and clinical implications for diabetes and aging. J Lab Invest 1994; 70: 138–151.
116. chimura T, Nakano K, Hashiguchi T, Iwamoto H, Miura K, Yoshimura Y, Hanyu N, Hirata K, Imakuma M, Motomiya Y, Maruyama I. Elevation of N-(carboxymethyl)valine residue in hemoglobin of diabetic patients. Its role in the development of diabetic nephropathy. Diabetes Care 2001; 24: 891–896.

117. Schmidt AM, Hori O, Chen JX, Li JF, Crandall J, Zhang J, Cao R, Yan SD, Brett J, Stern D. Advanced glycation endproducts interacting with their endothelial receptor induce expression of vascular cell adhesion molecule-1 (VCAM-1) in cultured human endothelial cells and in mice. J Clin Invest 1995; 96: 1395–1403.
118. Uchimura T, Nakano K, Hashiguchi T, Iwamoto H, Miura K, Yoshimura Y, Hanyu N, Hirata K, Imakuma M, Motomiya Y, Maruyama I. Elevation of N-(carboxymethyl)valine residue in hemoglobin of diabetic patients. Its role in the development of diabetic nephropathy. Diabetes Care 2001; 24: 891–896.
119. Sabbatini M. Sansone G, Uccello F, Giliberti A, Conte G, Andreucci VE. Early glycosylation products induce glomerular hyperfiltration in normal rats. Kidney Int 1992; 42: 875–881.
120. Moncado S, Palmer RMJ, Higgs EA. Nitric oxide: physiology, pathophysiology, and pharmacology. Pharmacological Reviews 1991; 43: 109–142.
121. Bank N, Aynedjian HS. Tole of EDRF (nitric oxide) in diabetic renal hyperfiltration. Kidney Int 1993; 43: 1306–1312.
122. Bucala R, Tracey KJ, Cerami A. Advanced glycosylation products quench nitric oxide and mediate defective endothelium-dependent vasodilation in experimental diabetes. J Clin Invest 1991; 87: 432–438.
123. Bucala R, Makita Z, Vega G, Grundy S, Koschinsky T, Cerami A, Vlassara H. Modification of low density lipoprotein by advanced glycation end products contributes to the dyslipidemia of diabetes and renal insufficiency. Proc Natl Acad Sci USA 1994; 91: 9441–9445.
124. Nakamura Y, Horil Y, Nishino T, Shiiki H, Sakaguchi Y, Kagoshima T, Dohi K, Makita Z, Vlassara H, Bucala R. Immunohistochemical localization of advanced glycosylation end products in coronary atheroma and cardiac tissue in diabetes mellitus. Amer J Path 1993; 143: 1649–1656.
125. Tsilbary EC, Charonis AS, Reger LA, Wohlhueter RM, Furcht LT. The effect of nonenzymatic glycosylation on the binding of the main noncollagenous NC1 domain to type IV collagen. J Biol Chem 1988; 263: 4302–4308.
126. Tanaka S, Avigad G, Brodsky B, Eikenberry EF. Glycation induces expansion of the molecular packing of collagen. J Mol Biol 1988; 203: 495–505.
127. Way KJ, Reid JJ. Effect of aminoguanidine on the impaired nitric oxide-mediated neurotransmission in anococcygeus muscle from diabetic rats. Neuropharmacology 1994; 33: 1315–1322.
128. Knowx LK, Stewart AG, Hayward PG, Morrison WA. Nitric oxide synthase inhibitors improve skin flap survival in the rat. Microsurgery 1994; 15: 708–711.
129. Tilton RG, Chang K, Corbett JA, Misko TP, Currie MG, Bora NS, Kaplan HJ, Williamson JR. Endotoxin-induced uveitis in the rat is attenuated by inhibition of nitric oxide production. Invest Ophthalmol Vis Sci 1994; 35: 3278–3288.
130. Hogan M, Cerami A, Bucala R. Advanced glycosylation endproducts block the antiproliferative effect of nitric oxide. Role in the vascular and renal complications of diabetes mellitus. J Clin Invest 1992; 90: 1110–1105.
131. Wong GKT, Marsden PA. Nitric oxide synthases: regulation in disease. Nephrol Dial Transplant 1996; 11: 215–220.
132. Edelstein D, Brownlee M. Mechanistic studies of advanced glycosylation end product inhibition by aminoguanidine. Diabetes 1992; 41: 26–29.
133. Brownlee M, Vlassara H, Kooney T, Ulrich P, Cerami A. Aminoguanidine prevents diabetes-induced arterial wall protein cross-linking. Science 1986; 232: 1629–1632.
134. Swamy-Mruthinti S, Green K, Abraham EC. Inhibition of cataracts in moderately diabetic rats by aminoguanidine. Experimental Eye Research 1996; 62: 505–510.
135. Soulis-Liparota T, Cooper ME, Dunlop M, Jerums G. The relative roles of advanced glycation, oxidation and aldose reductase inhibition in the development of experimental diabetic nephropathy in the Sprague-Dawley rat. Diabetologia 1995; 38: 387–394.
136. Ellis EN, Good BH. Prevention of glomerular basement membrane thickening by aminoguanidine in experimental diabetes mellitus. Metabolism 1991; 40: 1016–1019.
137. Hammcs HP, Brownlee M, Edelstein D, Saleck M, Martin S, Federlin K. Aminoguanidine inhibits the development of accelerated diabetic retinopathy in the spontaneous hypertensive rat. Diabetologia 1994; 37: 32–35.
138. Miyauchi Y, Shikama H, Takasu T, Okamiya H, Umeda M, Hirasaki E, Ohhata I, Nakayama H, Hakagawa S. Slowing of peripheral motor nerve conduction was ameliorated by aminoguanidine in streptozotocin-induced diabetic rats. European J Endocrinology 1996; 134: 467–473.
139. Schmidt RE, Dorsey DA, Beaudet LN, Reiser KM, Williamson JR, Tilton RG. Effect of aminoguani-

dine on the frequency of neuronal dystrophy in the superior mesenteric sympathetic autonomic ganglia of rats with streptozotocin-induced diabetes. Diabetes 1996; 45: 284–290.
140. Norton GR, Candy G, Woodiwiss AJ. Aminoguanidine prevents the decreased myocardial compliance produced by streptozotocin-induced diabetes mellitus in rats. Circulation 1996; 93: 1905–1912.
141. Archibald V, Cotter MA, Keegan A, Cameron NE. Contraction and relaxation of aortas from diabetic rats: effects of chronic anti-oxidant and aminoguanidine treatments. Naunyn-Schmiedbergs Arch Pharm 1996; 353: 584–591.
142. Brownlee M. Pharmacological modulation of the advanced glycosylation reaction. Prog Clin Biol Res 1989; 304: 235–248.
143. Nicholls K, Mandel TE. Advanced glycosylation end-products in experimental murine diabetic nephropathy: effect of islet isografting and of aminoguanidine. Lab Invest 1989; 60: 486–491.
144. Lyons TJ, Dailie KE, Dyer DG, Dunn JA, Baynes JW. Decrease in skin collagen glycation with improved glycemic control in patients with insulin-dependent diabetes mellitus. J Clin Invest 1991; 87: 1910–1915.
145. Wald H, Markowitz H, Zevin S, Popovtzer MM. Opposite effects of diabetes on nephrotoxic and ischemic acute tubular necrosis. Proc Soc Exp Biol Med 1990; 195: 51–56.
146. Soulis-Liparota T, Cooper M, Papazoglou D, Clarke B. Jerums G. Retardation by aminoguanidine of development of albuminuria, mesangial expansion, and tissue fluorescence in streptozotocin-induced diabetic rat. Diabetes 1991; 40: 1328–1334.
147. Makino H, Yamasaki Y, Hironaka K, Ota Z. Glomerular extracellular matrices in rat diabetic glomerulopathy by scanning electron microscopy. Virchows Arch B Cell Pathol 1992; 62: 19–24.
148. Petersen J, Ross J, Rabkin R. Effect of insulin therapy on established diabetic nephropathy in rats. Diabetes 1988; 37: 1346–1350.
149. Stockand JD, Sansom SC. Regulation of filtration rate by glomerular mesangial cells in health and diabetic renal disease. Amer j Kidney Dis 1997; 29: 971–981.
150. Vlassara H, Striker LJ, Teichberg S, Fuh H, Li YM, Steffes M. Advanced glycation end products induce glomerular sclerosis and albuminuria in normal rats. Proc Natl Acad Sci USA 1994; 91: 11704–11708.
151. Corbett JA, Tilton RG, Chang K, Hasan KS, Ido Y, Wang JL, Sweetland MA, Lancaster JR Jr, Williamson JR, McDaniel ML. Aminoguanidine, a novel inhibitor of nitric oxide formation, prevents diabetic vascular dysfunction. Diabetes 1992; 4: 552–556.
152. Tilton RG, Chang K, Hasan KS, Smith SR, Petrash JM, Misko TP, Moore WM, Currie MG, Corbett JA, McDaniel ML, et al. Prevention of diabetic vascular dysfunction by guanidines. Inhibition of nitric oxide synthase versus advanced glycation end-product formation. Diabetes 1993; 42: 221–232.
153. Sensi M, Pricci F, Andreani D, DiMario U. Advanced nonenzymatic glycation endproducts (AGE): their relevance to aging and the pathogenesis of late diabetic complications. Diabetes Res 1991; 16: 1–9.
154. Eika B, Levin RM, Longhurst PA. Collagen and bladder function in streptozotocin-diabetic rats: effects of insulin and aminoguanidine. J Urol 1992; 148: 167–172.
155. Vlassara H. Serum advanced glycosylation end products: a new class of uremic toxins? Blood Purif 1994; 12: 54–59.
156. Papanastasiou P, Grass L, Rodela H, Patrikarea A, Oreopoulos D, Diamandis EP. Immunological quantification of advanced glycosylation end-products in the serum of patients on hemodialysis or CAPD. Kidney Internat 1994; 46: 216–222.
157. Makita Z, Radoff S, Rayfield EJ, Yang Z, Skolnik E, Delaney V, Friedman EA, Cerami A, Vlassara H. Advanced glycosylation end products in patients with diabetic nephropathy. New Engl J Med 1991; 325: 836–842.
158. Zhang B, Salituro G, Szalkowski D, Li Z, Zhang Y, Royo I, Vitella D, Diez MT, Pelaez F, Ruby C, Kendall RL, Mao X, Griffin P, Calaycay J, Zierath JR, Heck JV, Smith RG, Moller DE. Discovery of a small molecule insulin mimetic with antidiabetic activity in mice. Science 1999; 284: 974–977.
159. Guinan EC, Boussiotis VA, Neuberg D, Brennan LL, Hirano N, Nadler LM, Gribben JG, Harris MI, Hadden WC, Knowles WC, Bennett PH. Transplantation of anergic histoincompatible bone marrow allografts. N Engl J Med 1999; 340: 1704–1714.
160. Schwartz RS. The new immunology – the end of immunosuppressive drug therapy? N Engl J Med 1999; 340: 754–755.

Index

MIX
Papier aus verantwortungsvollen Quellen
Paper from responsible sources
FSC® C105338

If you have any concerns about our products,
you can contact us on
ProductSafety@springernature.com

In case Publisher is established outside the EU,
the EU authorized representative is:
Springer Nature Customer Service Center GmbH
Europaplatz 3, 69115 Heidelberg, Germany

Printed by Libri Plureos GmbH
in Hamburg, Germany